OECD DOCUMENTS

Wider Application and Diffusion of Bioremediation Technologies

The Amsterdam '95 Workshop

Th ... cost.

O ... MENT

ORGANISATION FOR ECONOMIC CO-OPERATION AND DEVELOPMENT

Pursuant to Article 1 of the Convention signed in Paris on 14th December 1960, and which came into force on 30th September 1961, the Organisation for Economic Co-operation and Development (OECD) shall promote policies designed:

- to achieve the highest sustainable economic growth and employment and a rising standard of living in Member countries, while maintaining financial stability, and thus to contribute to the development of the world economy;
- to contribute to sound economic expansion in Member as well as non-member countries in the process of economic development; and
- to contribute to the expansion of world trade on a multilateral, non-discriminatory basis in accordance with international obligations.

The original Member countries of the OECD are Austria, Belgium, Canada, Denmark, France, Germany, Greece, Iceland, Ireland, Italy, Luxembourg, the Netherlands, Norway, Portugal, Spain, Sweden, Switzerland, Turkey, the United Kingdom and the United States. The following countries became Members subsequently through accession at the dates indicated hereafter: Japan (28th April 1964), Finland (28th January 1969), Australia (7th June 1971), New Zealand (29th May 1973), Mexico (18th May 1994), the Czech Republic (21st December 1995) and Hungary (7th May 1996). The Commission of the European Communities takes part in the work of the OECD (Article 13 of the OECD Convention).

FOREWORD

The OECD Workshop Amsterdam '95 on Wider Application and Diffusion of Bioremediation Technologies took place on 19-21 November 1995. It was sponsored by the Government of the Netherlands (Ministry of Economic Affairs), chaired by Dr. Wim Harder (TNO, Delft), and attended by more than 100 participants from government, industry and academia from many countries. It was the second in a series of OECD conferences, approved by the OECD Working Party on Biotechnology, on the scientific, technological and policy implications of bioremediation/bioprevention.

While the first of these, the OECD Workshop Tokyo '94 on Bioremediation (see *Bioremediation: The Tokyo '94 Workshop*, OECD 1995), presented a wide-ranging review of both remediation and prevention technologies in all environmental media, the Amsterdam workshop was more closely focused: the presentations covered bioremediation for air/off-gas and soil, and accorded particular attention to issues of industrial application and diffusion.

This reflected an important evolution: in a relatively short time, bioremediation evolved from a collection of technologies into an expanding business. Not all problems have been solved; on the contrary, the future progress of bioremediation will continue to depend critically on improving the scientific underpinnings and on an appropriate public policy framework. The OECD already has broad experience in dealing with these issues, and Member countries continue to encourage its efforts to help resolve the problems raised by bioremediation issues.

The present report contains the Summary and Conclusions of the Amsterdam workshop, drafted by Mr. M. Griffiths and Ms. D. Poole, rapporteurs, the speakers' papers, and summaries of the question-and-answer sessions. The Summary and Conclusions, revised and approved by the Chairman and the four co-Chairs (Mr. A. Bull, Mr. T. Egli, Mr. J. Grimes, Mr. D. Janssen), do not necessarily represent government policies. The papers present the views of the authors.

In the OECD, the responsibility for preparing the workshop and this publication lay with Mr. S. Wald and Mr. T. Hirakawa.

The Summary and Conclusions were submitted by written procedure to all participants and to the Working Party on Biotechnology, which requested the Committee on Scientific and Technological Policy to recommend their publication, together with the speakers' papers and the summaries of the question-and-answer sessions, on the responsibility of the Secretary General. The Committee agreed to derestriction. It is published on the responsibility of the Secretary-General of the OECD.

Particular thanks are due to the Government of the Netherlands (Ministry of Economic Affairs) for its generous contribution to the funding of the Amsterdam workshop, and to Dr. Harder, his collaborators and Dr. Mosmuller for their intellectual leadership. Thanks are also due to the European Commission (DG XII) and Japan for extra-budgetary financial contributions that helped the OECD to organise this workshop.

AVANT-PROPOS

La réunion de travail de l'OCDE "Amsterdam 95" sur l'application et la diffusion à plus grande échelle des technologies de biodépollution s'est tenue du 19 au 21 novembre 1995 sous le parrainage du Gouvernement des Pays-Bas (ministère des Affaires économiques). Elle était présidée par le Dr. Wim Harder (TNO, Delft) et plus de cent personnes représentant les pouvoirs publics, l'industrie et les milieux universitaires de nombreux pays y ont assisté. C'était le deuxième volet d'une série de conférences de l'OCDE, approuvées par le Groupe de travail de l'OCDE sur la biotechnologie, relatives aux implications scientifiques et technologiques et aux répercussions sur l'action des pouvoirs publics de la biodépollution et de la bioprévention.

La première de ces conférences, la réunion de travail de l'OCDE "Tokyo 94" sur la biodépollution (voir : *Bioremediation: The Tokyo 94 Workshop*, OCDE, 1995) donnait un aperçu général des techniques de dépollution et de prévention dans tous les milieux de l'environnement. La réunion de travail d'Amsterdam avait pour sa part une orientation plus précise : les exposés ont porté sur la biodépollution de l'air, des effluents gazeux et du sol, et ont accordé une attention particulière aux problèmes d'application industrielle et de diffusion.

Ce choix reflète une importante évolution : en peu de temps, la biodépollution, de simple assortiment de techniques, s'est transformée en une branche d'activité en expansion. Tous les problèmes n'ont certes pas été résolus. Bien au contraire, les progrès futurs de la biodépollution continueront à dépendre de façon critique de l'amélioration des fondements scientifiques et de l'existence d'un cadre approprié d'action des pouvoirs publics. L'OCDE possède déjà une large expérience de ces questions et les pays Membres continuent à encourager ses efforts pour contribuer à résoudre les problèmes liés à la biodépollution.

Ce rapport présente le Résumé et les Conclusions de la réunion de travail d'Amsterdam, qu'ont établis les rapporteurs, M. M. Griffiths et Mme D. Poole, ainsi que les exposés des orateurs et les synthèses des séances de questions et réponses. Le Résumé et les Conclusions, révisés et approuvés par le Président et les quatre co-présidents (MM. A. Bull, T. Egli, J. Grimes et D. Janssen) ne reflètent pas nécessairement les politiques des gouvernements. Les exposés présentent les points de vue de leurs auteurs.

Au sein du Secrétariat de l'OCDE, MM. S. Wald et T. Hirakawa ont assuré la préparation de la réunion et de cette publication.

Le Résumé et les Conclusions ont été soumis par la procédure écrite à tous les participants et au Groupe de travail sur la biotechnologie, qui a demandé au Comité de la politique scientifique et technologique de recommander leur publication, accompagnée des exposés des orateurs et des synthèses des séances de questions et réponses, sous la responsabilité du Secrétaire général. Le Comité a donné son accord pour la mise en diffusion générale. Le document est publié sous la responsabilité du Secrétaire général de l'OCDE.

Des remerciements particuliers sont adressés au Gouvernement des Pays-Bas (ministère des Affaires économiques) pour sa généreuse contribution au financement de la réunion de travail d'Amsterdam, ainsi qu'au Dr. Harder, à ses collaborateurs et au Dr. Mosmuller pour leur contribution intellectuelle. Des remerciements sont dus aussi à la Commission européenne (DG XII) et au Japon pour leurs contributions financières extrabudgétaires qui ont aidé l'OCDE à organiser cette réunion.

TABLE OF CONTENTS

SUMMARY AND RECOMMENDATIONS 13

RÉSUME ET RECOMMANDATIONS 29

OPENING REMARKS

"Opening remarks"
Mr. W. Harder, Chairman, Director of TNO-MW, the Netherlands 49

"Welcoming speech"
Mr. P.G. Winters, Deputy Director-General for Industry, Ministry of Economic Affairs, the Netherlands 51

"OECD's bioremediation work"
Mr. R. Nezu, Director, Directorate for Science, Technology and Industry, OECD 55

SETTING THE SCENE

"The microbe's contribution to sustainable development"
Mr. W. Harder, Chairman, Director of TNO-MW, the Netherlands 61

"Environmental biotechnology: Business opportunities and bottlenecks"
Mr. J. Pâques, Director of Pâques B.V., the Netherlands 69

Question-and-Answer Session

"Long-term opportunities and sustainable technology in bioremediation and bioprevention"
Mr. K. Nakajima, Deputy Director-General of MITI, Japan 75

"R&D in the field of bioremediation technologies"
Mr. R. Unterman, Vice-President, Envirogen Inc., United States 81

Question-and-Answer Session

"Information transfer and the public context"
Mr. B. Dixon, Contributing Editor Bio/Technology, United Kingdom 93

Question-and-Answer Session

SUMMARY AND RECOMMENDATIONS

Introduction

In 1994, the OECD published a report entitled "Biotechnology for a Clean Environment -- Prevention, Detection, Remediation". This report raised a number of issues, scientific, industrial and economic, relating to the bottlenecks which inhibit the wider application of biotreatment and bioremediation.

A major OECD workshop on bioremediation, hosted by the Japanese government, was held in Tokyo in November 1994, to address some of the scientific issues, specifically those related to the efficacious and safe reduction of pollutant hazards. The proceedings of this workshop have been published as "Bioremediation: the Tokyo '94 Workshop".

OECD Member countries agreed to the holding of a second workshop, this time to be hosted by the government of the Netherlands, which would approach the subject of bioremediation from the perspective of wider industrial diffusion and application. This workshop, to be held in Amsterdam, would take as its subject matter the bioremediation of soil, air and off-gases, leaving the subject of water treatment, use and conservation to be reviewed by the planned workshop to be held in Mexico in 1996. The Netherlands workshop would examine the business opportunities and bottlenecks; the efficacy, reliability and predictability, the focus and trends in R&D; standardization and best practice; and information transfer and dissemination. The outcomes and recommendations of this workshop are presented in this report.

More than 100 participants, industrialists, scientists and administrators from many countries attended the workshop which was held over two and a half days, 19-21 November 1995.

Setting the scene

The Workshop Chairman, Professor Wim Harder (TNO, the Netherlands), set the scene at the beginning of the plenary session.

Technologies for the more distant future will be based on renewable resources, the recycling of products and the wastes generated by their manufacture, and the use of renewable energy to drive production processes. In the shorter term, however, the focus of environmental biotechnology will be on bioprevention, biotreatment and bioremediation.

Bioremediation relies upon the ability of natural processes to degrade organic molecules. Microbes play a pivotal role in these processes and did so even before photosynthesis emerged on the planet. Over time, microbes have evolved and have become remarkably effective to the point where they can mineralise most organic substances and by doing so close, or partly close, many material cycles.

Industrial applications require the harnessing of microbiological activities and their incorporation into the engineered applications and bioreactors that together make up viable commercial systems. Wider application and diffusion of these technologies requires a better understanding of both the underlying science and the economics of application. Government policy and frameworks do and will play a major role in the diffusion of the technology, and the OECD is an effective platform for discussing policies and information transfer to and between governments.

In order to provide direction to the workshop, the Chairman posed five questions:

1. What are the key environmental problems?

2. What is the potential of bioremediation with respect to the state-of-the-art?

3. What are the bottlenecks to wider application -- technical, economic, government policy and public perception?

4. How can we facilitate the introduction of bioremediation in industry?

5. What actions can be recommended to governments and industry?

The speakers in the plenary session re-emphasized the chairman's questions while addressing a wide range of issues from industrial and governmental perspectives. In particular, the business opportunities and the value to solution providers of regulations actually enforced, were stressed; links were built from bioremediation to the longer term future represented by bioprevention; and focus was placed on cost-effectiveness as the driving force for the wider use of biotreatment systems. The need was stressed to disseminate information about the potential of the technology, its wider implications and potential hazards, together with the need to involve the community at large.

There are many niche markets for biotreatment and bioremediation -- taking care of the environment can be a profitable business. The problem is that the current competitive technology can be as simple as digging out and transporting contaminated soil to dump in a landfill site. Scale-up to a commercial process is one of the main technical barriers to be addressed. Data are accumulating on the capital expenditure and operating costs of bioremediation. In a number of cases these show a pay-back advantage over alternative treatments of 1-2 years.

While the initial investment that has to be made for bioprocesses is higher, the running costs are usually lower. However, while biotechnology provides good solutions to many problems, performance in practice is still often unpredictable. The treatment market is driven by government regulation, but only if those regulations are actually enforced. The positive factors, as perceived by the solution-providing community, include environmental legislation and financial regulations, stricter controls, new technologies and improved public awareness. The negative factors include unpredictable policies, economic recession and a low return on investment. Government support of R&D is important, and R&D should always be undertaken in close co-operation with the end users. There is also a need for co-operation in process scale-up, integrated applications within the total production process, global marketing and business diversifications strategies.

In situ degradation should always be the first choice, for both economic and other reasons, such as reduced site disturbance. Better understanding is emerging of the factors affecting the survival time of organisms used for augmentation and the induction of natural populations. The engineering solutions associated with co-metabolism are being developed and considerable advances are being made in combining metabolic activities by genetic manipulation. However, the demand for remediation is always changing. For example, MTBE (methyl tertiary butyl ether) is now the third largest product by quantity of any chemical in the United States today. While this is a brand new target for remediation, already a biological solution to MTBE pollution has appeared.

Governments are becoming more pragmatic as they realise that the costs of meeting the high remediation standards initially set will exceed any reasonably available budget. Guidelines are required for intrinsic remediation (the default situation) in order to define socially acceptable endpoints. International collaboration is required to assist those countries whose resources are insufficient to cope with their environmental damage. More synergistic interaction is required since joint action can tackle problems more effectively.

Environmental treatments need three main characteristics: they should be ecologically acceptable, not radically new, and such as can be commended to the public. Bioremediation clearly meets all three criteria and, in addition, is particularly appropriate to developing countries, with diverse cultural and religious traditions. Public acceptance is, nonetheless, not inevitable. It is unwise to conclude that a better informed public is necessarily more supportive. Public information, demands and acceptance do underlie new regulations and there is a close link between what the press says and how politicians respond. Anticipating public attitudes is vitally important.

Session A1.1: Business opportunities and bottlenecks (air/off-gas)

The 1970s and 1980s saw the scientific underpinning of industrial biotreatment of gaseous pollutants. The 1990s have seen the full commercialisation of these processes. While biofilters remain the most widely used system, there has been a progression from biofilters to biotrickling filters (where water is circulated over the biofilm -- see also Session A2.2) as the technology currently being studied for possible application. Biological air and off-gas cleaning processes are of outstanding ecological quality in that they will not transfer compounds of environmental concern from air to other compartments, but rather degrade them to carbon dioxide and water.

A wide range of gaseous wastes have been identified as treatable by biotechnological means, and for most of these commercial processes are already available. Biofiltration equipment can treat waste gas from many industrial sources under conditions of fluctuating load. However, biofilters have often been designed and built by the user companies themselves, and hence the technology of biofilters is difficult to evaluate. This gives rise to a lack of product development. Application know-how is the real prerequisite for success -- a solution provider may have to spend many man-years to gather the know-how to be able to operate in the market successfully. Economics for this biological process are very favourable compared with most competitive non-biological treatment processes.

There is a large difference between the state-of-the-art in research in the laboratory and in the field, and scale-up is complex. For example, the first stage of commercialisation may be a demonstration of a process using a pilot plant but a full-scale plant may require significant design changes and engineering input, leaving only the biotreatment stage intact.

Session A1.2: Business opportunities and bottlenecks (soil)

In situ bioremediation should be more cost-effective than other remediation methods as it avoids excavation, requires less energy, and does not generate waste streams. However, to-date, much "remediation" has been by removal to landfill. Cost and landfill availability has made this an attractive option, particularly in the United Kingdom, but increasing environmental pressures mean that this route is becoming increasingly unacceptable. Standards set by government are essential to generate actual markets for soil remediation.

Analysis of contaminated soil treatable by *in situ* methods in Europe has identified the following major potential markets (in billion $): Germany (16), Italy (10), Russian Federation (9), Romania (6), France (6), the Netherlands (5), Great Britain (2), Spain (2), Poland (2), the Czech Republic (1), and Belgium (1). Population density is a good indicator of market potential.

Bioremediation may become more advantageous as naturally occurring organisms are found with a high capacity to degrade such highly chlorinated compounds. During the past 20 years, PCB (polychlorinated biphenyls) levels in most environmental compartments have been declining, mediated by four types of processes: physical transport, aerobic microbial biodegradation, photolytic (solar) dechlorination/degradation and anaerobic microbial dechlorination. The contributions of various remediation processes in any given case may be assessed and such assessments are useful in evaluating alternative remediation strategies and for addressing risk.

A number of lessons can be learnt from full-scale treatments. In particular, there is an absolute need for independent validation of treatments and for critical inspection of vendor claims and performance. For the technology as a whole to survive and prosper it is crucial that treatment claims be realistic, and that problem owners develop confidence in the reality of such claims.

Biological techniques may be more advantageous than physico-chemical alternatives from an ecological point of view because they make use of natural processes. Nevertheless, new strategies are required to approach the problems of (non)-availability of pollutants in soil in order to extend the field of applications. These should be based on research into the interactions between micro-organisms, contaminants and soil, and the development of agents such as enzymes or emulsifiers to accelerate the degradation process, in order to extend the field of application. Use of combined treatment methods, i.e. physico-chemical and biological should be extended; new monitoring and sampling techniques will be required. Variability of data is a major problem. Statistics and sampling methods are important as are new analytical techniques.

Session A2.1: Efficacy, reliability and predictability (soil)

It is essential to know the form of the pollutant and to ask basic questions such as what is the lowest achievable concentration of a pollutant in soil. Economic feasibility depends on accurate predictions. Tests have been developed which allow the estimation of both the final concentration of a pollutant, following bioremediation, and the time necessary to reach it. A mobile system has been developed and used for determining heterogeneity on a large scale. The test system can be used to predict the behaviour of pollutant plumes, provided that adequate models are used. A key factor that determines predictability of the treatment process is the differences between various sites, and site biogeochemistry as well as the nature of the pollutants and their availability for biological treatment should be carefully considered when choosing a treatment method.

Distribution of contamination in soil is very diverse, ranging from dumped drums of pollutants, right through to dispersed fine particulates. Contaminants may be persistent in the environment for various reasons, including their non-natural molecular structure, their strong affinity for soil, and the absence of a suitable population of degrading micro-organisms. Each contaminated site has its own hydrogeological conditions which must be evaluated in each case, along with investigation of soil characteristics, indigenous micro-organisms and site history.

Soil may contain on average two tonnes of living organisms per hectare. Many species play a vital role in bioremediation but little is known about how to influence their performance. Interactions with higher organisms may influence community structure and the question is whether one can manipulate the composition of the microbial populations. For the future, it will be necessary to construct reliable and functional microbial communities and improve microbial associations via horizontal gene flow. An additional approach may be actively to decrease the overall level of bio-availability and hence eco-toxicity.

Inoculation with micro-organisms has had some success in a limited number of cases to-date but the technique may eventually increase the potential for degradation. It may be better to introduce catabolic and promoter genes into indigenous organisms, while at the same time researching those factors which contribute to the survival of introduced organisms. An example comes from introduction of *Pseudomonas* into soil, where pre-adaptation is necessary. There may not yet be many full-scale applications but there are a number of small-scale successes, with the introduction of organisms in activated sludge and soil.

Heavy metals form an important class of contaminants. Perhaps 50 per cent of all contaminated sites contain metallic pollutants in addition to the organic materials. Soil characteristics may influence the chemical form of the metal, which in turn determines the possibilities for leaching. Metal-contaminated sites require additional considerations such as recycling and ultimate disposal. A process which recovers heavy metals in a form in which they can be immobilised and which leaves "clean" soil, was described. Work on *in situ* bioleaching with Thiobacilli is continuing.

Aliphatic chlorinated compounds can be degraded anaerobically, but only TCE (tri-chloroethylene) can be degraded aerobically. In practice, anaerobic degradation of TCE was observed on one site, but not at another due to complete absence of the required anaerobic organisms. Column studies were also used to determine the possibilities for aerobic co-metabolic degradation of contaminants, especially TCE. Field tests with toluene addition to stimulate TCE-degrading micro-organisms are under way and methane has already been used for this purpose at SRL. Anaerobic degradation of TCE at one site was successful, but the rate of degradation was limited by the electron donor supply. Using microcosm studies, benzoate appeared to be the best electron donor.

There is increasing interest in "difficult" pollutants -- heavy metals and highly chlorinated compounds (requiring co-metabolism). In the latter case considerable effort, but with little success to date, is going into the incorporation of genes for metabolism into usable organisms to avoid the requirement for co-metabolites.

Predictability is a key factor and there is still room for improvement. The main problems include: site-to-site variation and site heterogeneity (both on a micro- and a macro- scale). Some knowledge cannot be transferred from one site to another, e.g. community structure, bioavailability, pollutant toxicity, etc. Site characterisation is necessary - both regarding the hydrogeology and microbial community. *Ex situ* treatment is manageable and a reasonable number of tests will allow predictability (provided good models are available). There is a clear need for statistically validated models and hence for more data and this is important for gaining trust in the reproducibility, predictability and reliability of these techniques.

Three new developments of potential for the future were mentioned. Horizontal transfer of genes occurs naturally under field conditions and may be an alternative to maintaining introduced organisms. Protein-engineering might be used to develop enzymes to degrade specific contaminants.

Finally, the difference in microbial physiology at extremely low substrate concentrations from that at the high substrate levels normally used during microbiological laboratory studies may be exploited.

Session A2.2: Efficacy, reliability and predictability (air/off-gas)

The use of biotreatment technologies (e.g. biofilters and biotrickling filters) for air pollution applications has been increasing in popularity as they are usually sustainable and do not shift a problem from one environmental compartment to another. It has been demonstrated that biotreatment technologies will remove gaseous air pollutants generated by several industrial units located within the middle of a densely populated industrial zone.

Biofiltration is a cheap and reliable technique for the removal of odour and specific biodegradable compounds from waste gases and these techniques have been applied successfully in many fields. A possible risk was thought to be the emission of (pathogenic) micro-organisms, but it has been shown that the levels of such emissions are negligible. For contaminants which produce acids when oxidised, biotrickling filtration is a superior technology, since a re-circulating liquid phase washes the inhibitory reaction products out of the biofilm.

The removal of sulphur from petroleum leads to cleaner burning fuels and lower sulphur emissions. A strain of bacteria that survives in organic solvents has been created which breaks down organic sulphur compounds not removed by conventional processes. A combination of this technology and thermo-chemical desulfurization could provide an effective method for removing most organic sulphur compounds from petroleum. An effective biotreatment process for the two-step removal of sulphur dioxide from flue gases was also described.

A closed biological air filter has been developed for the European Space Agency in order to determine whether such a technology could remove trace gas contaminants in space capsules. A support material was colonised by selected micro-organisms in a near resting state. Biological air filters are expected to have several advantages over physico-chemical systems, such as adaptability to unexpected contaminants. More down-to-earth applications might include systems for domestic and commercial buildings.

Off-gas treatment is now relatively reliable, robust and has specific abilities which cannot be duplicated by chemical/physical treatment methods. It is also cost-effective on a capital and operating costs basis. This technology is particularly appropriate for low concentrations, volatile solvents and organic odours. Emission of pathogens is negligible (confirming the experience with waste water treatment). Few if any models of off-gas or soil treatment have, however, been fully validated statistically and this validation will be developed only when more data becomes available.

Session B1.1: Focus and trends in R&D

In artificially contaminated soil the pollutant is bioavailable, but in reality, mass transfer rate is usually the limiting factor. It is pointless addressing the microbiology alone when improved process engineering is needed to obtain better mass transfer. Addition of PAH-mineralising bacteria may induce a fast and apparently complete PAH degradation in artificially contaminated soil. The apparent disappearance is, however, only partly due to mineralisation -- part becomes non-extractable by a process similar to the natural humification. The possibility exists to intensify and steer this process intentionally and to use humification as a second option for pollutant removal. With soils from contaminated sites, mineralisation may be severely hampered by limited bioavailability of the contaminants which become trapped in soil, clay or tar particles. Pollutants sorb onto and into the soil particles, and the desorption process, depending on the age of the pollution, may be quite slow so that a contaminated soil can be an almost ‘everlasting’ sources of pollutant. Although pollutants may be extractable by organic solvents, they are not necessarily available for biotransformation processes. This is relevant to the question, what is meant by “clean”? If PAHs, for example, can be adsorbed to soil particles and thus not be ecotoxic, do they represent a problem?

Bioremediation of sites polluted with toxic chemicals is one case where the use of recombinant bacteria may eventually complement the activities of naturally-occurring micro-organisms. Molecular techniques are already available which allow construction of strains with novel phenotypes for various aerobic biotransformations. A strategy to deal with gene flow in uncontained applications has been to develop novel tools that resemble to a large extent natural processes. Novel vectors bring into question the somewhat arbitrary distinction between “recombinant” and “natural” bacteria, since their production mimics faithfully the natural processes of DNA shuffling between distant locations of the bacterial genome. The question was therefore raised whether there should be special legislation for these types of GMO which are no different from naturally occurring microbes.

Many molecular techniques have been described that have practical applications in environmental monitoring. Ideal strategies should offer high sensitivity, robustness and reproducibility. Bioluminescent reporter gene technology has been explored for use in on-line measurement of the microbial populations associated with degradation of hazardous chemicals. Large-scale field applications to test the efficacy of these strains in process monitoring and control are in development.

New biological processes are being developed, including: reductive dechlorination, use of thermophilic methanotrophs, co-metabolism, phyto-remediation (in the rhizosphere), the use of recombinant DNA organisms, protein engineering, constructed consortia and inocula adaptation.

Session B1.2: Information transfer and dissemination

Bioremediation is uniquely well positioned to enjoy public support. It is a technology which can develop hand in hand with public confidence. Anticipating public attitudes is important but it should be remembered that public awareness does not equate with public acceptance.

Five routes for information dissemination can be identified:

- focus on the media via workshops for journalists and news editors to explain the principles using case histories;
- briefing papers to provide background information;
- talk direct to the public -- consensus conferences, for example;
- schools are important -- young people are very environmentally aware;
- the Internet -- a powerful way of distributing information and data.

The public is not, in general, aware of the scientific method -- hypothesis testing which allows the elimination of what is not true but does not prove that something is true. People are forever seeking certainty. Another confusion is that between micro-organisms and "germs". Release of a micro-organism is not the same as the release of a disease. There is a lack of good scientific journalists who can translate to the public, in a jargon-free and clear manner. Oversimplification by poor journalists can be as big a problem as an article full of jargon.

There is a real need for the scientific community to communicate both the risks and benefits of genetically engineered micro-organisms based upon what has been learned over the past decades of research on recombinant organisms. Where there are real differences between organisms created by recombinant DNA technology and those that naturally evolve or have been selected as mutants in the laboratory, these must be highlighted.

Liability is a main driving force for remediation, whose definition and attribution depend upon legislation, regulation and operational enforcement. Environmental liability is a "hidden" issue -- it is not obvious until it becomes a legal problem -- which benefits neither customers, workers or shareholders. After asbestos, polluted soil is the next largest global liability issue. In any economic transactions involving the transfer of contaminated land and related liabilities, an important role can be played by banks and insurance companies, who for this reason should be well-informed about the potential of remediation technologies. To transfer liabilities successfully, the business vectors, e.g. banks, insurers, etc., should be used. Companies are affected not just by legislation, but by other issues such as business interruption, loss of good image, for example, during a take-over bid for the company, insurance claims, loss of assets by contamination, cost of legal procedures, etc.

It is most important to have harmonized standards for remediation and also BS or ISO standards for environmental management systems.

"Government", "policy", "regulation", "legislation" are terms often loosely used. When communicating with government it is important to be aware of the differences. It should be remembered that governments require information to operate and to formulate policies, and the quality of the policies also depends on the quality of the information on which they are based.

The OECD assists governments by information transfer in the expectation that if they have the same information they are more likely to develop and implement similar policies. The diversity of experience available in the OECD area is thus made available for the benefit of Member governments.

Session B2: Standardisation and best practice

"Best practice" in soil bioremediation requires the implementation of a series of protocols and procedures at all stages of the treatment process. A proper site investigation, which establishes the nature and delineates the extent of contamination, should be used to establish feasibility. A careful assessment of site geology, hydrogeology and contaminant distribution is essential since effective mass transfer of chemical reagents (nutrients, electron acceptors) to the contaminated zone is a prerequisite for successful bioremediation. During and after the treatment process, a statistically-defined monitoring programme is required to establish that targets are met.

Best practice involves all aspects of the approach to contamination, including intensive treatment to reduce high levels of pollutants, natural remediation, slow removal, and risk-reduction. Although biogeochemical factors have been identified, their influence on treatment strategy is still empirical and more general rules are needed. Standardisation and documentation are important for creating industrial confidence.

The lack of bioavailability of pollutants can be a major drawback in the application of remediation technologies. Rapid remediation of polluted soil requires a mechanical homogenisation of soil particles to liberate the contaminants trapped inside.

Novel detection technologies are becoming available: PCR (polymerase chain reaction), immunological testing, light emission, CCD cameras. Monitoring and tracking degradative organisms and genes is vital in order to demonstrate whether organisms and communities persist and are active. Marker genes can be used as probes for biodegradative potential and fluorescent markers permit the identification and localisation of specific cells in complex environments. Monoclonal antibodies specific to surface determinants of cell envelopes are also valuable as a specific tool for tracking microbes.

Agricultural systems are major contributors to the atmospheric pollutant burden. These include nitrous oxide from nitrification-denitrification cycles, methane production from the rumen and ammonia emissions from animal wastes in livestock housing. Best use of nitrogenous wastes from livestock production is likely to be an important part of an integrated fertiliser strategy for crops. Any effective strategy must be based on life-cycle analysis principles but no strategy will meet its overall goal if it affects profitability to an unacceptable degree.

A cocktail of pollutants often generates the formation of an ecosystem of micro-organisms and a realistic bioremediation strategy will utilise the concerted action of many species. There is, however, only limited information on the co-operative effects and interactions among micro-organisms.

In recent years, a number of new metabolic pathways have been constructed and the potential use of these genetically modified micro-organisms in open environments is being considered as a feasible strategy for cleaning up polluted sites.

Conclusions

In summing up the seven sessions of the workshop, the four co-chairs provided a number of answers to the questions raised by the Chairman in the opening plenary. The answers to the first four are given below and the answers to the fifth question are presented in the "Recommendations".

1) What are the key environmental problems?

No new major issues appear to be emerging although there are always new pollutants. The main problem is technical complexity -- mixed contaminants under complex environmental and site conditions. The mobility of organisms, genes and pollutants is not well understood; the importance of seeking a better understanding of bioavailability and the factors influencing it are beginning to become clear.

Problems for solution providers include lack of data and lack of demonstration opportunities (the chance to do comparative trials). Problem owners have the problem of insufficient guidance on liability and the rules governing degree of clean-up.

Research activities in academe are often not directed at questions of interest to problem owners (there is a need for better communication from industry and closer co-operation with industry and government).

2) *What is the potential of bioremediation with respect to the state-of-the-art?*

Environmental biotechnology can now be regarded as a business rather than just a collection of technologies. Evidence of the health of the market is provided by the number of bioremediation activities already generating cash-flow. There is strong competition for lucrative business, particularly in biofiltration. There is a large and global remediation market but there are many reasons why penetration is slow including the lack of demonstration data. The industry is nevertheless more positive than a year ago in Tokyo regarding the possibilities for bioremediation, chiefly because it has more experience. It is widely accepted that the strength of the bioremediation industry is directly related to the existence of bioremediation standards.

3) *What are the bottlenecks to wider application?*

A) Technical

Technical bottlenecks include lack of both fundamental understanding and field-based R&D data, bearing in mind the complexity of specific locations. There is frequently a lack of the integrated (interdisciplinary) approach which is necessary to achieve optimisation and predictability. A "start from scratch" case-by-case approach is often taken with little progress in establishing ground rules and strategies, the questions being asked differing between industrialists and academics.

In bioremediation, one is dealing with mixed contaminants, conditions of complex geology, and sites poorly characterised with respect to distribution of contaminants and micro-organisms. There is a poor understanding of bioavailability and contaminant mass transfer in a range of environmental conditions.

Another gap is in the ability to measure and monitor the progress of a treatment or remediation process although technology is now becoming available for sophisticated, near real-time, *in situ* monitoring of bioremediation activities. Further understanding is required into all the natural processes, physical, chemical and biological, so that the most promising may be stimulated.

There is much new input from genetic manipulation but outstanding questions remain, for example under what circumstances is it needed, and what should be appropriate targets (enhanced catabolic potential, altered gene regulation, reporter functions).

Scale-up from laboratory to field is regularly seen as a problem: models, both physical and mathematical, are essential -- better and more data equates to better design and engineering.

B) Economic

The essential missing element is comparative cost-effectiveness data. Where that has become available, as in the case of biofiltration, biotreatment has often been shown to have a clear advantage. The initial capital costs of biotreatment may well be higher than physico-chemical alternatives and therefore return on investment is a better measure.

Environmental liability is a major potential problem which must not be underestimated. Hidden costs may include: the requirement for remediation imposed by an authority or by public demand; business losses caused by interruption or loss of reputation; compensation to victims or financial support agencies (banks etc.,); loss of assets by contamination; discovery of inherited problems from past owners, and the cost of legal procedures. Awareness of these problems, which are becoming threats to industry, is growing.

C) Government policy

Legislation can be both an advantage and a problem. Legislation is the ultimate support for the solution providers but is the source of the bottom-line cost of pollution for problem owners. It is through legislation, regulation (and enforcement) that the liabilities are defined, and the market opportunities for bioremediation are created.

There may still be a lack of understanding, among drafters of regulations, of what is required or appropriate as an end-point. Variable legislation and variations in degree of enforcement are seen as inhibiting uptake of bioremediation technologies. Harmonization of risk-based standards is seen as the ultimate goal.

D) Public perception

Lack of understanding by politicians, civil servants and members of the general public may be an inhibiting factor to the exploitation of bioremediation and biotreatment. Dissemination of information related to the health and safety of workers is important, as is community involvement, in discussion of options, and public perception of health, safety and risk aspects. In all cases there is a need to be proactive in communicating with the public.

It is important, particularly with respect to information for government, to target the recipient of the information carefully -- the right sort of information for the right recipient.

4) How can we facilitate the introduction of bioremediation in industry?

The general perception was for more practical demonstrations, test plots and comparative exercises. The move towards risk-based bioremediation targets (since absolute remediation is an unattainable goal), the development of acceptable (standardized) validation protocols, and the determination of the long-term stability (safety) of remediated sites should be encouraged.

Data from all biotreatment and bioremediation success stories would be invaluable both as a marketing tool and a publicity aid.

Environmental problems should be translated into business problems and communicated to the business sector, including in particular banks and insurance companies, which, through their important intermediary roles, can act as effective vectors for the transfer of information.

Developments since Tokyo

One clear message is that the public may well accept environmental applications of biotechnology and that it is possible to present bioremediation as a green technology for pollutant removal. Participants were very confident about the role of biotechnology -- there are achievements to show which could not be matched by traditional techniques such as "pump and treat". However, incineration should no longer be regarded as the main competitor to bioremediation -- there are more sophisticated non-biological techniques available, such as thermal desorption and electrochemical treatment. This reflects the growing perception that biotechnology should be one of a portfolio of techniques.

The differences between the Amsterdam and Tokyo workshops can be highlighted by their different answers to a recurring question, namely, "When will the technology be reliable and predictable?" The Tokyo answer was, "When there is sufficient scientific study to provide an underpinning"; the Amsterdam answer was, "When sufficient competitive field trials and demonstrations have been carried out and validated models developed." Amsterdam saw more demonstrations of commercial uses of biotreatment and bioremediation. The technology can and does achieve goals in a cost-effective manner and can break even against other techniques over a 1-3 year period. Biotechnology is not the solution to all problems, but is appropriate in a wide range of circumstances and niche markets.

Environmental biotechnology at present lacks the robustness of chemical and physical methods. However, the versatility of the pollutants that can be handled, the flexibility of biological systems and the relatively low maintenance costs make bioremediation a very powerful tool in solving environmental problems. Although bioremediation has shown a tremendous potential, it is in some cases not yet considered a "proven" technology for problem owners and authorities. A hurdle to its wider application therefore seems to be the convincing demonstration to industry that this technology offers economic advantages over the competing options.

Despite engineering and accounting uncertainties, bioremediation is being employed for cleaning up air, water and soil -- with little fanfare and without the use of GMOs. The industry has a message of success to impart although both industrialists and academics have sometimes promised more than they could deliver. Although the robustness may not yet have been fully demonstrated, the versatility of the technology should not be down-played. Special treatment for the bioremediation industry shouldn't be necessary -- biofiltration penetrated the market on its own merits, by being cost-competitive, and is now an established technique for a number of pollutants with many success stories. However, soil bioremediation is not so far advanced -- it also works but the applications are more site and pollutant specific -- it may need short term help while demonstrations are run and validating data gathered. Why are there few questions asked about the biological aspects of water treatment? Because it has had many decades to establish itself as a mature technology.

The Amsterdam workshop demonstrated that industrial goals have moved on from defining the opportunities to specifying what works and the operating guidelines, thus validating industrial practice. With reference to GMOs, these may yet be necessary in certain niche applications and they are now being successfully used at the intermediate stages of scale-up. Some industry re-organisation is perceived, blurring the distinction between vendor and customer, as many companies develop technologies for in-house use.

There is increasing pragmatism in government, influenced by budgetary limitations, in the approach to standards of clean-up. The combination of non-zero regulatory standards and interest in less labour-intensive technologies is working to biotechnology's advantage. The scientific community could invite governments to consider more explicitly the circumstances in which intrinsic bioremediation may be acceptable, and could contribute to such consideration. Participants commented that policy has not always been science-based and there was general agreement that it should be founded on risk-assessment rather than absolute (and arbitrarily set) numbers. Regardless of the technology there must be a logical end-point against which each technology can be compared.

Recommendations

The workshop in its final plenary session discussed and proposed recommendations. These were formulated by the participants in a personal capacity. They were addressed to the OECD and, via the OECD, to Member governments and to industrialists, both solution providers and problem owners. In many instances all three, including academe should be involved co-operatively. While recommendations from the workshop are not binding on OECD Member governments participants pointed out that they nevertheless carry considerable intellectual authority and persuasive force.

Technologies for the more distant future will increasingly be based on renewable resources, the recycling of products and the wastes generated by their manufacture, and the use of renewable energy to drive production processes. In the shorter term, the focus of environmental biotechnology should be on bioprevention, biotreatment and bioremediation.

Government

Bioremediation has been shown to be cost-effective for successful long-term treatment and has further great potential. International collaboration on long-term test-bed studies in relation to risk-based goals is recommended. These initiatives should enhance industry-academia partnerships on a global scale. There is an urgent need for government-supported demonstrations of successful technologies, such as air/off-gas treatment, in order to convince public and private customers that environmental biotechnology is becoming mature. Interdisciplinary task forces in association with these demonstrations should be encouraged.

Environmental problems should be translated into business problems, enabling the scientific community to communicate with the business. Governments should support the transfer of technology from science base to applications, from solution provider to problem owner.

Harmonization of legislation and consistent enforcement are required. Industry should be able to expect harmonization of legislation across national (state) boundaries and consistent enforcement. The goal for target setting should be to move from regulations based on absolute concentrations to those based on assessment of risk, taking account of the intended future use of the land. Risk-based targets need to be defined and these targets and associated policies should be developed from quality information obtained from the scientific community.

The whole field of public relations and public education should be addressed. Joint initiatives between governments, industry and academia are required to promote the positive aspects of bioremediation. These should be clear, jargon-free, aimed at the correct level and be able to anticipate public attitudes. Education at an early age is essential. Demonstrations and case studies should be used to convince public and private customers that environmental biotechnology is becoming a mature technology.

Rules of liability for both solution providers and problem owners need to be established in order to remove uncertainties of responsibility associated with land transactions and clean-up. The insight that historic pollution does not necessarily need immediate clean-up, should be adopted.

Solution providers

Solution providers should work together with academia to address the major generic technical barriers identified at this workshop, including scale-up issues, site-characterisation and bioavailability, in order to narrow the gap between the state-of-the-art in the laboratory and those in field conditions. Studies of the combination of organism and micro-environment should be more task oriented, for example, adapting *inocula* to the environment to give a better chance of survival of a high proportion of organisms. Solution providers should also give high priority to the acquisition of data to validate predictive models of bioremediation and should actively support the rapid development of methods for on-line, *in situ* monitoring of pollutants and microbial activity.

A key area of fundamental research, requiring government, industry and academic collaboration, relates to bioavailability. Not all carbon compounds are mineralised but rather are sorbed onto soil particles or become part of, or are chemically linked to, humic material. What is the significance of slow leaching of low levels of residue remaining after bioremediation? Once a material is not available to bioremediation, is it ecotoxic?

Emphasis should be placed on development of on-line, *in situ* monitoring of pollutant levels, micro-organisms and biological activity. It is essential to gather data from field operations in order to develop and validate predictive models.

Solution providers and problem owners should jointly dedicate certain sites on which to base the long-term research necessary to obtain better field data, to be run co-operatively by multidisciplinary teams who could widely publish their data, by, for example, putting their results up on the Internet so that everyone can benefit.

Environmental biotechnology is an interdisciplinary technology in which co-operation between several fields of expertise (from microbiologists to engineers), are a necessity. Interdisciplinary task forces should be encouraged.

Problem owners

Problem owners should be encouraged to consider bioremediation as their first choice, taking advantage of its public acceptance, possible economic advantages and positive image as compared with competing physico-chemical methods. Successful case studies should be reported, and companies should be proactive in disseminating information to the public.

Industry needs to assess its business activities with the goal of better understanding itself, its needs and how it operates in the marketplace, to determine whether radical improvements can be made using biotechnology. Can and should, for example, molecules be designed to become bio-unavailable by being immobilised onto humic material? What long-term effects could such modifications exert on the soil organic fraction, and under what conditions could these compounds become mobilised and thence polluting again?

Closer industrial interaction with academia is necessary in order to stimulate the transfer of technology from science base to applications, from solution provider to problem owner.

Information dissemination should be improved by holding workshops for the media, preparing briefing documents, utilising multimedia and television and involving the community (town and consensus meetings). Schools should be a priority target with a wide use of case studies (success stories).

Since liability is a main source of remediation business, there is a need to translate environmental problems into business opportunities via due diligence. Liability is based on both legislation and public perception. Banks and insurance companies are significant intermediaries in the management of risks and costs, and should be used as "information vectors" for successful technologies. Equally, it is important to know who to target in government, and to give better industrial guidance for academic research; this could be facilitated by the development of focused networks linking individuals having interests, responsibilities, needs and capabilities related to bioremediation.

RÉSUME ET RECOMMANDATIONS

Introduction

En 1994, l'OCDE a publié un rapport intitulé "La biotechnologie pour un environnement propre – Prévention, détection, dépollution". Ce rapport soulevait un certain nombre de questions, d'ordre scientifique, industriel et économique, relatives aux blocages qui s'opposent à une utilisation élargie du biotraitement et de la biodépollution.

Une importante réunion de travail de l'OCDE sur la biodépollution, accueillie par le Gouvernement japonais, s'est tenue à Tokyo en novembre 1994, pour traiter de certains problèmes scientifiques, notamment de ceux relatifs à la réduction efficace et sûre des dangers associés aux produits polluants. Les minutes de cette réunion de travail ont été publiées sous le titre : "Bioremediation : The Tokyo 94 Workshop" (Réunion de travail Tokyo '94 sur la biodépollution).

Les pays Membres de l'OCDE ont décidé de tenir une deuxième réunion de travail, accueillie cette fois par le Gouvernement des Pays-Bas, au cours de laquelle la question de la biodépollution serait abordée dans la perspective d'une diffusion et d'une application plus larges au niveau industriel. Cette réunion de travail, prévue à Amsterdam, devait avoir pour thème la biodépollution du sol, de l'air et des rejets gazeux, réservant la question du traitement, de l'utilisation et de la préservation de l'eau pour la réunion de travail devant se tenir à Mexico en 1996. La réunion de travail des Pays-Bas examinerait les débouchés et les blocages professionnels ; l'efficacité, la fiabilité et la prévisibilité, l'orientation générale et les tendances en matière de R-D ; la normalisation et les meilleures pratiques appliquées, et enfin, le transfert et la diffusion de l'information. Les conclusions et les recommandations de cette réunion de travail sont présentées dans le présent rapport.

Plus d'une centaine de participants, d'industriels, de scientifiques et de gestionnaires venus de nombreux pays ont assisté à la réunion qui a duré deux jours et demi, du 19 au 21 novembre 1995.

Contexte général

Le Président de la réunion, M. Wim Harder (TNO, Pays-Bas), a rappelé le contexte général au début de la séance plénière.

Dans un avenir très lointain, les technologies se fonderont sur des ressources renouvelables, sur le recyclage des produits et des déchets résultant de leur fabrication, et sur l'utilisation d'énergies renouvelables pour actionner les processus de production. A court terme cependant, les biotechnologies d'environnement porteront essentiellement sur la bioprévention, le biotraitement et la biodépollution.

La biodépollution se fonde sur l'aptitude de processus naturels à dégrader les molécules organiques. Les microbes jouent un rôle déterminant dans ces processus et l'ont fait même avant que la photosynthèse n'apparaisse sur la planète. Avec le temps, les microbes ont évolué et sont devenus remarquablement efficaces au point qu'ils peuvent minéraliser la plupart des substances organiques et, ce faisant, refermer ou fermer en partie de nombreux cycles de matières.

Certaines applications industrielles requièrent la maîtrise des activités microbiologiques concernées, ainsi que leur incorporation à des applications techniques et à des bioréacteurs avec lesquels elles constituent des systèmes commerciaux viables. Une application et une diffusion élargies de ces technologies supposent une meilleure compréhension, tant de la base scientifique que des aspects économiques de leur application. Les politiques et les cadres de l'action des gouvernements jouent, et joueront, un rôle majeur dans la diffusion de ces technologies, et l'OCDE est une tribune efficace pour l'examen des politiques et le transfert d'informations aux gouvernements concernés et d'un gouvernement à l'autre.

Pour orienter la réunion de travail, le Président pose cinq questions :

1. Quels sont les problèmes majeurs d'environnement ?

2. Quelles possibilités la biodépollution offre-t-elle par rapport à l'état des connaissances ?

3. Qu'est-ce qui bloque une plus large application – les politiques techniques, les politiques économiques, les politiques gouvernementales ou la perception du public ?

4. Comment pouvons-nous faciliter la pratique de la biodépollution dans l'industrie ?

5. Quelles sont les actions qui peuvent être recommandées aux pouvoirs publics et aux industries ?

Les orateurs de la séance plénière reprennent les questions du Président en abordant toute une gamme de problèmes du point de vue industriel et gouvernemental. Ils soulignent notamment les débouchés professionnels et l'importance des règlements en vigueur pour ceux qui offrent des solutions ; ils établissent des liens entre la biodépollution et l'avenir à plus long terme que représente la bioprévention et retiennent l'efficacité par rapport au coût comme l'élément motivant une utilisation plus étendue des systèmes de biotraitement. Enfin, ils soulignent la nécessité d'une diffusion des informations relatives aux possibilités offertes par ces technologies, à leurs implications et à leurs dangers potentiels au sens large, en même temps que l'intérêt de faire participer la collectivité dans son ensemble.

Il existe de nombreux créneaux spécialisés dans le biotraitement et la biodépollution – prendre soin de l'environnement peut être une occupation profitable. Le problème, c'est que l'autre technologie actuelle peut consister simplement à extraire et à transporter des sols contaminés pour les épandre dans des décharges. Le passage à l'échelle supérieure pour accéder au niveau commercial est l'un des principaux obstacles techniques à surmonter. Les données s'accumulent au sujet des dépenses en capital et des coûts d'exploitation des techniques de biodépollution. Dans un certain nombre de cas, ces dépenses et ces coûts présentent l'avantage d'un certain amortissement par rapport à d'autres traitements qui durent un à deux ans.

Si les investissements initiaux requis pour des processus biologiques sont élevés, les coûts de fonctionnement sont en général assez faibles. Cependant, la biotechnologie apporte des solutions satisfaisantes à bien des problèmes mais, dans la pratique, ses performances restent encore souvent aléatoires. Le marché du traitement est induit par les réglementations gouvernementales, mais seulement lorsque ces réglementations sont effectivement mises en oeuvre. Les facteurs positifs, tels qu'ils sont perçus par l'ensemble de ceux qui offrent des solutions, tiennent entre autres à la législation de l'environnement et aux règlements financiers, à la rigueur des contrôles, aux nouvelles technologies et à une meilleure sensibilisation du public. Les facteurs négatifs comprennent le caractère imprévisible des politiques, la récession économique et la faible rentabilité des investissements. Le soutien des pouvoirs publics à la R-D est important et cette R-D doit toujours être menée en étroite collaboration avec les utilisateurs finals. La coopération est également importante pour le passage des processus à l'échelle supérieure, pour l'intégration des applications à l'ensemble du processus de production, et pour les stratégies de commercialisation globale et de diversification professionnelle.

La dégradation in situ doit être le choix prioritaire, pour des raisons économiques ou autres, comme la réduction de la perturbation des sites. On commence à mieux comprendre les facteurs affectant la durée de survie des organismes utilisés pour augmenter et induire des populations naturelles. Les solutions pratiques associées au cométabolisme sont en cours de mise au point et des progrès considérables sont réalisés en matière de combinaison d'activités métaboliques par manipulation génétique. Cependant, la demande de dépollution évolue constamment. Par exemple, l'oxyde de méthyle et de tertio-butyle se situe maintenant, par la quantité, au troisième rang des substances chimiques produites aux États-Unis. Bien qu'il s'agisse d'une toute nouvelle cible pour la dépollution, une solution biologique a déjà été proposée pour remédier à la pollution causée par ce produit.

Les gouvernements adoptent une attitude plus pragmatique au fur et à mesure qu'ils réalisent que les coûts encourus pour respecter les normes strictes de pollution initialement fixées excéderont tout budget raisonnable en vigueur. Pour la dépollution intrinsèque (situation par défaut), il faut des lignes directrices de façon à définir les résultats socialement acceptables. Une collaboration internationale est nécessaire pour aider les pays qui n'ont pas assez de ressources à faire face à la détérioration de leur environnement. Des initiatives davantage orientées vers la synergie s'imposent puisqu'une action conjointe permet d'attaquer plus efficacement les problèmes.

Les traitements environnementaux doivent posséder trois caractéristiques essentielles : ils doivent être écologiquement acceptables, pas radicalement nouveaux et de nature à pouvoir être recommandés au public. La biodépollution répond à l'évidence à ces trois critères et convient, en outre, particulièrement bien à des pays en développement ayant des traditions culturelles et religieuses variées. Néanmoins, l'acceptation par le public n'est pas garantie. Il est déraisonnable de conclure que le public est d'autant mieux disposé qu'il est bien informé. Son information, ses besoins et son acceptation sont à la base des nouvelles réglementations, en outre, les dires de la presse et la façon dont les politiciens répondent sont étroitement liés. Prévoir les réactions du public est absolument essentiel.

Séance A.1.1. : Débouchés et blocages professionnels (rejets gazeux dans l'air)

Les années 70 et 80 ont vu s'édifier les bases scientifiques du traitement biologique industriel des polluants gazeux. Les années 90 ont été celles de la commercialisation à grande échelle de ces procédés. Si les filtres biologiques restent le dispositif le plus largement utilisé, on est passé de ceux-ci aux filtres percolateurs (où l'eau circule sur le biofilm – voir aussi la séance A.2.2) en tant que technologie actuellement étudiée en vue d'une application éventuelle. Les procédés biologiques d'épuration de l'air et des rejets gazeux présentent un intérêt considérable du point de vue écologique car, au lieu de transférer de l'air vers d'autres compartiments des composés dangereux pour l'environnement, ils les dégraderont en dioxyde de carbone et en eau.

Une large variété de résidus gazeux ont été reconnus comme susceptibles d'être traités par des moyens biotechnologiques pour la plupart desquels des procédés sont déjà commercialisés. Les équipements de biofiltration peuvent traiter des résidus gazeux provenant de nombreuses sources industrielles sous charge variable. Cependant, des biofiltres ont été souvent conçus et construits par les entreprises utilisatrices elles-mêmes, c'est donc une technologie difficile à évaluer. Cette situation n'est pas propice au développement de produit. Le recours au savoir-faire est la véritable condition indispensable au succès – quelqu'un qui offre des solutions peut passer de nombreuses années à réunir le savoir-faire qui lui permettrait de s'imposer avec succès sur le marché. Les aspects économiques de ces procédés biologiques sont très encourageants par rapport aux modes de traitement non-biologiques les plus compétitifs.

Il y a une différence considérable entre le niveau des connaissances mises en oeuvre lors de recherches effectuées en laboratoire et sur le terrain, en outre, le passage à l'échelle supérieure est complexe. Par exemple, la première étape de commercialisation peut consister en une démonstration d'un procédé dans une installation expérimentale, mais une installation à l'échelle industrielle pourrait nécessiter d'importantes modifications de la conception et des moyens techniques, ne laissant inchangée que l'étape du traitement biologique.

Séance A.1.2 : Débouchés et blocages professionnels (sols)

La biodépollution in situ doit être plus efficace par rapport au coût que d'autres méthodes de dépollution puisqu'elle évite l'excavation, nécessite moins d'énergie, et ne génère pas de flux de déchets. Mais jusqu'à présent, la “dépollution” a surtout consisté en transfert vers des décharges. Le coût et la disponibilité des décharges ont rendu cette solution séduisante, notamment au Royaume-Uni, mais les pressions accrues en faveur de l'environnement montrent que cette voie devient de moins en moins acceptable. Des normes fixées par les pouvoirs publics sont essentielles pour susciter de véritables marchés de dépollution des sols.

L'analyse de sols contaminés traitables par des méthodes in situ en Europe a révélé de grands marchés potentiels dans les pays suivants (en milliards de dollars) : Allemagne (16), Italie (10), Fédération russe (9), Roumanie (6), France (6), Pays-Bas (5), Grande-Bretagne (2), Espagne (2), Pologne (2), République tchèque (1) et Belgique (1). La densité démographique est un bon indicateur des possibilités commerciales.

La biodépollution pourrait devenir plus avantageuse puisqu'on trouve dans la nature des organismes ayant une grande aptitude à dégrader des composés fortement chlorés. Au cours des 20 dernières années, les concentrations de PCB (biphényles polycholorés) ont diminué dans la plupart des compartiments de l'environnement grâce à quatre sortes de procédés : transport physique, biodégradation microbienne aérobie, déchloration/dégradation par photolyse (solaire) et déchloration microbienne anaérobie. Les contributions des divers processus de dépollution peuvent être évaluées, quelque soit le cas considéré et ces évaluations servent à apprécier les autres stratégies de dépollution et à prendre les risques en compte.

Un certain nombre d'enseignements peuvent être tirés des traitements à l'échelle industrielle. En particulier, il est absolument indispensable de valider les traitements de façon indépendante et d'étudier avec un sens critique les arguments et les performances des vendeurs. Pour que cette technologie survive et prospère globalement, il est indispensable que les arguments concernant le traitement soient réalistes et que ceux qui ont des problèmes puissent se fier à la réalité de tels arguments.

D'un point de vue écologique, des techniques biologiques pourraient être plus avantageuses que leurs remplaçantes physico-chimiques car elles mettent en oeuvre des processus naturels. De nouvelles stratégies sont néanmoins nécessaires pour traiter des problèmes que pose la disponibilité (ou la non disponibilité) de polluants dans le sol de façon à élargir le champ des applications. Ces stratégies doivent découler de recherches portant sur les interactions entre micro-organismes, produits contaminants et sols, et aussi de la mise au point d'agents tels qu'enzymes ou émulsifiants capables d'accélérer le processus de dégradation, de façon à élargir le domaine d'application. L'utilisation de modes de traitement combinés, c'est-à-dire physico-chimiques et biologiques, doit se répandre ; il faudra de nouvelles techniques de surveillance et d'échantillonnage. La variabilité des données pose un problème majeur. Les méthodes de statistiques et d'échantillonnage sont indispensables, tout comme de nouvelles techniques analytiques.

Séance A.2.1 : Efficacité, fiabilité et prévisibilité (sols)

Il est essentiel de connaître la forme sous laquelle se trouvent les polluants et de poser des questions élémentaires, comme celle de la plus faible concentration possible d'un polluant dans un sol donné. La faisabilité économique dépend de prévisions précises. On a mis au point des essais qui permettent d'évaluer à la fois la concentration finale d'un polluant, après biodépollution, et le temps nécessaire pour y parvenir. Il a été conçu un système mobile qui sert à déterminer l'hétérogénéité à grande échelle. Ce système d'essai peut servir à prévoir l'évolution des zones de diffusion des polluants, à condition d'utiliser des modèles adéquats. Un facteur déterminant qui conditionne la prévisibilité du procédé de traitement réside dans les différences entre divers sites, et la "biogéochimie" des sites, tout comme la nature des polluants et leur disponibilité en vue de leur traitement biologique doivent être soigneusement prises en compte lors du choix d'une méthode de traitement.

Dans les sols, les produits contaminants se répartissent de façon très diverse, que ce soit dans des fûts pleins de polluants mis en décharge ou en fines particules dispersées. Ces produits contaminants peuvent persister dans l'environnement pour diverses raisons, notamment leur structure moléculaire artificielle, leur forte affinité pour le sol et l'absence d'une population de micro-organismes capables de les dégrader. Chaque site contaminé possède ses propres conditions hydrogéologiques qui doivent être évaluées dans chaque cas, en même temps que seront analysées les propriétés caractéristiques du sol, les micro-organismes qui y vivent et l'histoire du site.

Un sol peut contenir en moyenne deux tonnes d'organismes vivants à l'hectare. De nombreuses espèces jouent un rôle déterminant dans la biodépollution mais on ne sait pas vraiment comment influencer leur performance. Des interactions avec des organismes supérieurs peuvent se répercuter sur la structure de la communauté et la question est de savoir si l'on peut manipuler la composition des populations microbiennes. A l'avenir, il faudra composer des communautés microbiennes fonctionnelles et fiables et améliorer les associations de ces communautés grâce à des flux horizontaux de gènes. On pourrait, en outre, s'efforcer activement de diminuer le niveau général de "disponibilité biologique" et donc, d'écotoxicité.

L'inoculation avec des micro-organismes a connu un certain succès dans un nombre limité de cas jusqu'à présent, mais la technique pourrait éventuellement accroître les possibilités de dégradation. Il vaudrait peut-être mieux introduire des gènes cataboliques et promoteurs dans des organismes locaux, tout en recherchant en même temps les facteurs qui contribuent à la survie des organismes manipulés. On peut citer à titre d'exemple l'introduction de *Pseudomonas* dans le sol, où une adaptation préalable est nécessaire. Il n'existe probablement pas encore beaucoup d'applications en vraie grandeur mais on recense un certain nombre de réussites à petite échelle, grâce à l'introduction d'organismes dans des boues et des sols activés.

Les métaux lourds constituent une classe importante de contaminants ; à la différence des polluants organiques, ils ne peuvent être dégradés biologiquement. Environ 50 pour cent de tous les sites touchés subissent une pollution combinée, c'est-à-dire qu'ils contiennent à la fois des polluants organiques et des métaux lourds toxiques. Les caractéristiques pédologiques exerceront une influence sur la forme chimique des métaux qui, à son tour, aura des incidences sur le rendement de l'élimination des métaux. Les sites contaminés par des métaux doivent subir des traitements complémentaires, comme le recyclage des lessives utilisées pour extraire le métal du sol et l'élimination définitive des boues métalliques récupérées. Des participants décrivent un procédé qui récupère les métaux lourds dans une forme permettant leur immobilisation définitive, et qui aboutit à des sols "propres". Sur place, on poursuit un travail de lixiviation biologique avec Thiobacilli tout en modifiant la technologie du procédé de façon à traiter les polluants organiques.

Des composés aliphatiques chlorés peuvent être dégradés en anaérobiose et l'on a trouvé des lignées microbiennes qui dégradent quelques-uns de ces composés en aérobiose. Sur le terrain, on a constaté une dégradation anaérobie du trichloroéthylène sur un seul site, mais pas sur d'autres en raison de l'absence totale des organismes anaérobies nécessaires. Le rythme de la dégradation en anaérobiose dépend de la concentration en donneur d'électrons. D'après des études sur microcosme, le benzoate apparaîtrait comme le meilleur donneur d'électrons. En outre, des études sur colonne ont servi à établir les possibilités de dégradation cométabolique de certains contaminants, notamment le trichloroéthylène. Des essais sur le terrain, avec addition de toluène pour stimuler les micro-organismes qui dégradent le trichloroéthylène, sont en cours et le méthane a déjà été utilisé à cette fin au Stanford Research Laboratory.

On s'intéresse de plus en plus aux polluants "difficiles" – métaux lourds et composés fortement chlorés (cométabolisme indispensable). Dans ce dernier cas, des efforts considérables visent, mais sans grand succès jusqu'à présent, à l'incorporation de gènes responsables du métabolisme dans des organismes adéquats de façon à éviter d'avoir à utiliser des cométabolites.

La prévisibilité est un élément déterminant toujours susceptible d'amélioration. Parmi les principales difficultés, figurent : la variation entre les sites et leur hétérogénéité (à échelle tant microscopique que macroscopique). Certaines connaissances peuvent être transférées d'un site à l'autre, comme la structure de la communauté, la biodisponibilité, la toxicité des polluants, etc. Il est nécessaire de caractériser un site – s'agissant à la fois de son hydrogéologie et de sa communauté microbienne. Le traitement ex situ est envisageable et un nombre raisonnable d'essais permettront d'en prévoir les résultats (à condition d'avoir de bons modèles). Il faut, à l'évidence, disposer de modèles statistiquement validés et donc de plus de données et cela est essentiel pour faire fond sur la reproductibilité, la prévisibilité et la fiabilité de ces techniques.

Trois nouveaux éléments présentant des débouchés pour l'avenir sont évoqués. Le transfert horizontal de gènes s'effectue naturellement sur le terrain et serait une autre façon d'assurer le maintien d'organismes manipulés. L'ingénierie des protéines pourrait servir à la mise au point d'enzymes capables de dégrader des contaminants précis. Enfin, on pourrait exploiter les différences intervenant dans la physiologie des microbes à des concentrations de substrats extrêmement faibles par rapport à leur comportement aux concentrations utilisées normalement lors d'études microbiologiques en laboratoire.

Séance A.2.2 : Efficacité, fiabilité et prévisibilité (rejets gazeux dans l'air)

L'usage de technologies de biotraitement (par exemple, biofiltres et filtres percolateurs) dans des dispositifs de lutte contre la pollution de l'air se répand de plus en plus puisque ces technologies sont ordinairement durables et ne transfèrent pas de problèmes d'un compartiment de l'environnement dans un autre. Il a été prouvé (comme au Mexique) que des technologies de biotraitement éliminent des polluants gazeux atmosphériques produits par plusieurs installations industrielles situées au milieu d'une zone industrielle densément peuplée.

La biofiltration est une technique peu onéreuse et fiable pour la suppression d'odeurs et de composés biodégradables précis à partir de résidus gazeux et ces techniques ont été appliquées avec succès dans de nombreux domaines. On avait craint un risque éventuel d'émissions de micro-organismes (pathogènes) mais il a été démontré que les niveaux de telles émissions sont négligeables. S'agissant de contaminants qui produisent des acides une fois oxydés, la filtration par percolation s'avère supérieure, puisqu'une phase liquide recirculante élimine du film biologique les produits de la réaction d'inhibition.

L'élimination du soufre des hydrocarbures pétroliers conduit à des carburants à combustion moins polluante et émettant moins de soufre. On a créé une lignée de bactéries qui survivent dans des solvants organiques et sont capables de dissocier des composés organiques soufrés résistants aux méthodes classiques. La combinaison de cette technologie avec la désulfuration thermochimique pourrait fournir une méthode efficace pour éliminer la plupart des composés organiques du soufre présents dans les hydrocarbures pétroliers. Des participants décrivent, en outre, un traitement biologique efficace permettant de supprimer en deux étapes le dioxyde de soufre des gaz de combustion.

Un filtre à air biologique fermé a été mis au point pour l'Agence spatiale européenne en vue d'établir si une telle technologie était capable de supprimer des contaminants gazeux à l'état de traces dans les capsules spatiales. Un matériel de support a été ensemencé avec des micro-organismes choisis dans un état proche du repos. Ces filtres à air biologiques devraient présenter divers avantages par rapport aux systèmes physico-chimiques, comme celui de pouvoir s'adapter à des contaminants imprévus. Des applications plus terre à terre pourraient comporter des systèmes utilisables dans des édifices à usage résidentiel et commercial.

Le traitement des rejets gazeux est maintenant assez fiable, bien établi et remplit des fonctions spécifiques qui ne peuvent être remplacées par des méthodes de traitement chimico-physiques. Eu égard aux capitaux investis et aux dépenses de fonctionnement, il présente également des avantages par rapport au coût. Cette technologie est particulièrement appropriée aux solvants volatils à faibles concentrations et aux odeurs organiques. L'émission d'organismes pathogènes est négligeable (confirmant ainsi l'expérience acquise avec le traitement d'eaux usées). Les quelques modèles qui existent pour le traitement des rejets gazeux ou des sols ont cependant été entièrement validés statistiquement et cette validation ne se poursuivra que si davantage de données deviennent disponibles.

Séance B.1.1 : Principale orientation et tendances de la R-D

Dans des sols artificiellement contaminés, les polluants sont biodisponibles mais en réalité, le taux de transfert de matière constitue d'ordinaire le facteur limitant. Il est vain de s'intéresser à la seule microbiologie lorsque l'ingénierie du processus doit être améliorée pour obtenir un meilleur transfert de masse. L'addition de bactéries capables de minéraliser les hydrocarbures aromatiques polycycliques (HAP) peut induire une dégradation des HAP rapide et apparemment complète dans des sols artificiellement contaminés. Cette disparition apparente n'est cependant due qu'en partie à la minéralisation – une fraction devient non extractible par un procédé similaire à l'humidification naturelle. Il est possible d'intensifier et de stimuler délibérément ce procédé et de recourir à l'humidification en tant que deuxième solution pour l'élimination des polluants. Avec des sols provenant de sites contaminés, la minéralisation peut être fortement freinée par la biodisponibilité limitée des contaminants qui se fixent alors sur des particules du sol, de boue et de goudron. Les polluants se sorbent sur et dans les particules du sol et le processus de désorption peut, selon l'ancienneté de la pollution, être assez lent de sorte qu'un sol contaminé peut s'avérer être une source presque “éternelle” de polluants. Si des polluants peuvent être extraits à l'aide de solvants organiques, ils ne se prêtent pas nécessairement à des processus de transformation biologique. Ce qui ramène à la question, qu'entend-on par “propre” ? Si des HAP peuvent s'absorber sur des particules du sol, et donc ne plus être écotoxiques, constituent-ils un problème ?

La biodépollution de sites contaminés par des produits chimiques toxiques est un cas où l'utilisation de bactéries recombinantes viendrait éventuellement compléter les activités de micro-organismes se trouvant naturellement sur place. On dispose déjà de techniques moléculaires qui permettent de réaliser des lignées possédant des phénotypes originaux adaptés à diverses biotransformations aérobies. En matière de contrôle des flux de gènes dans des applications non confinées, une stratégie a consisté à mettre au point des outils nouveaux qui imitent dans une large mesure les processus naturels. De nouveaux vecteurs amènent à s'interroger sur la distinction quelque peu arbitraire entre bactéries “recombinantes” et “naturelles”, puisque leur production imite fidèlement les processus spontanés d'échange d'ADN entre des points éloignés sur le génome bactérien. La question se pose donc de l'opportunité d'une législation spéciale pour ces sortes d'organismes génétiquement modifiés qui ne sont pas différents de microbes d'origine naturelle.

On a décrit de nombreuses techniques moléculaires qui ont des applications pratiques dans la surveillance de l'environnement. Des stratégies idéales doivent offrir une sensibilité, une robustesse et une reproductivité élevées. La technologie de la bioluminescent appliquée au gène reporteur a été explorée en vue de son utilisation lors de mesures en ligne de populations microbiennes associées à la dégradation de produits chimiques dangereux. On est en train de mettre au point des applications à grande échelle sur le terrain pour tester l'efficacité de ces lignées dans la surveillance et le contrôle des procédés.

La mise au point de nouveaux procédés biologiques se poursuit et porte notamment sur la déchloruration réductrice, l'utilisation d'organismes méthanotrophes thermophyles, le cométabolisme, la dépollution végétale (dans la rhizosphère), l'utilisation d'organismes à ADN recombinant, l'ingénierie des protéines, l'adaptation des cultures mixtes et d'inoculums modifiés.

Séance B1.2 : Transfert et dissémination des informations

La biodépollution est remarquablement bien placée pour jouir du soutien du public. C'est une technologie qui peut évoluer en gardant sa confiance. Prévoir les réactions de ce public est important mais on ne saurait oublier que la vigilance du public n'équivaut pas à une acceptation sans réserve.

On peut distinguer cinq voies pour la diffusion des informations.

– essentiellement via les médias lors de réunions de travail à l'intention de journalistes et de rédacteurs pour expliquer les principes à l'aide d'études de cas ;

– documents de synthèse pour donner des informations de référence ;

– échange direct avec le public – lors de conférences de consensus, par exemple ;

– écoles, voie importante – les jeunes étant très sensibilisés à l'environnement ;

– réseau Internet – puissant moyen de diffusion d'informations et de données.

En général, le public n'est pas au courant de la méthode scientifique – mise à l'épreuve d'une hypothèse qui permet d'éliminer ce qui est faux mais ne prouve pas que quelque chose soit vrai. Les gens recherchent toujours la certitude. En outre, on confond souvent micro-organismes et "germes". La libération d'un micro-organisme n'est pas la même chose que la dissémination d'une maladie. Il n'y a pas assez de bons journalistes scientifiques capables de présenter les faits au public dans un langage clair dénué de jargon. La simplification excessive que pratiquent des journalistes médiocres peut poser autant de difficultés qu'un article plein de jargon.

Il est réellement indispensable que la communauté scientifique fasse part aussi bien des risques que des avantages associés aux micro-organismes modifiés génétiquement à la lumière de ce qui a été appris au cours des dernières décennies de recherche sur les organismes recombinants. Il faut préciser en quoi consistent les différences réelles entre des organismes obtenus par la technologie de l'ADN recombinant et ceux qui évoluent naturellement ou ont été sélectionnés comme mutants en laboratoire.

La responsabilité est l'élément moteur de la dépollution dont la définition et les attributions dépendent de la législation, de la réglementation, et des moyens d'application fonctionnels. La responsabilité en matière d'environnement est un problème "caché" – elle n'apparaît que lorsqu'elle devient un problème juridique – ce qui n'avantage ni les clients, ni les travailleurs, ni les parties prenantes. Les sols pollués posent le plus vaste problème de responsabilité après celui de l'amiante. Dans toute transaction économique impliquant le transfert de sols contaminés et des responsabilités connexes, un rôle important peut être rempli par les banques et les compagnies d'assurance, qui doivent pour cette raison être bien informées des perspectives offertes par les technologies de dépollution. Pour que les responsabilités en question soient correctement transférées, il faut recourir aux vecteurs professionnels, c'est-à-dire les banques, les assureurs, etc. Les entreprises ne sont pas seulement concernées par la législation mais aussi par d'autres problèmes comme l'interruption d'activité, la dégradation d'une bonne image, par exemple, lors d'une offre publique d'achat pour l'entreprise, les déclarations de sinistre, la perte de patrimoine à cause de la contamination, le coût de procédures juridiques, etc.

Il est de la plus haute importance d'avoir des normes harmonisées pour la dépollution et aussi des normes du type British Standard ou ISO pour les systèmes de gestion de l'environnement.

Les termes de "gouvernement", "politique", "réglementation", "législation" sont souvent utilisés de façon approximative. Lors d'entretiens avec des pouvoirs publics, il importe d'être attentif aux différences. On se souviendra que les gouvernements ont besoin d'informations pour conduire et élaborer des politiques et que la qualité de ces politiques dépend, en outre, de la qualité des informations sur lesquelles elles sont fondées.

L'OCDE aide les gouvernements en leur communiquant des informations dans l'attente que, disposant des mêmes informations, ils auront davantage tendance à élaborer et appliquer des politiques similaires. La diversité des expériences acquises dans la zone de l'OCDE est donc disponible pour le profit des gouvernements des pays Membres.

Séance B2 : Normalisation et pratiques les meilleures

S'agissant de biodépollution des sols, les pratiques les meilleures supposent la mise en oeuvre d'une série de protocoles et de procédures à tous les stades du déroulement du traitement. Un examen adéquat du site, qui établit la nature et l'ampleur de la contamination, doit servir à déterminer la faisabilité. Une évaluation soigneuse de la géologie et de l'hydrogéologie du site, ainsi que de la répartition du contaminant, est essentielle puisqu'un transfert efficace massif de réactifs chimiques (éléments nutritifs, accepteurs d'électrons) vers la zone contaminée est une condition indispensable à la réussite de la biodépollution. Pendant et après le déroulement du traitement, un programme de surveillance statistiquement déterminé est nécessaire pour établir que les objectifs sont atteints.

Les pratiques les meilleures touchent tous les aspects du traitement de la contamination, notamment le traitement intensif pour réduire les fortes concentrations de polluants, la dépollution naturelle, l'évacuation lente et la réduction des risques. Bien que les facteurs "biogéochimiques" aient été identifiés, leur influence sur le mode de traitement reste empirique et des règles plus générales sont nécessaires. Normalisation et documentation sont essentielles pour instaurer la confiance de l'industrie.

L'absence de polluants disponibles peut être un obstacle majeur à l'application de technologies de dépollution. Une dépollution rapide de sols pollués nécessite une homogénéisation mécanique des particules du sol de façon à libérer les contaminants captés à l'intérieur.

De nouvelles technologies de détection deviennent utilisables : réaction en chaîne de la polymérase, essais immunologiques, émissions de lumière, caméras CCD. La surveillance et le repérage d'organismes et de gènes dégradateurs sont essentiels pour montrer si des organismes et des communautés survivent et sont actifs. Des gènes marqueurs peuvent servir à sonder les possibilités de biodégradation et des marqueurs fluorescents permettent de déterminer et de localiser des cellules spécifiques dans des environnements complexes. Des anticorps monoclonaux spécifiques de déterminants de surface situés sur les membranes cellulaires sont également utiles en tant qu'instruments spécifiques du dépistage de microbes.

Les systèmes agricoles contribuent largement à la charge polluante dans l'atmosphère. Ces polluants comprennent l'hémioxyde d'azote provenant des cycles de nitrification-dénitrification, le méthane produit par les ruminants et les émissions ammoniacales provenant des déjections animales dans les bâtiments d'élevage. La meilleure façon d'utiliser des déchets nitrés provenant de l'élevage va probablement constituer un aspect important d'une stratégie de fertilisation intégrée pour les cultures. Toute stratégie efficace doit se fonder sur les principes d'analyse du cycle de vie mais aucune stratégie n'atteindra son objectif global si elle lèse la rentabilité à un degré inacceptable.

Un mélange de polluants suscite souvent la formation d'un écosystème de micro-organismes et une stratégie réaliste de biodépollution recourra à l'action concertée de plusieurs espèces. Cependant, il n'existe que des informations limitées sur les effets et interactions synergiques entre micro-organismes.

Ces dernières années, un certain nombre de nouvelles voies métaboliques ont été établies et l'utilisation éventuelle en milieu non confiné de ces micro-organismes génétiquement modifiés est considérée comme une stratégie envisageable pour l'épuration de sites pollués.

Conclusions

En résumant les sept séances de la réunion de travail, les quatre co-présidents ont apporté un certain nombre de réponses aux questions soulevées par le Président lors de la séance plénière d'ouverture. Les réponses aux quatre premières questions figurent ci-après et les réponses à la cinquième sont présentées sous le titre “Recommandations”.

1) Quels sont les problèmes majeurs d'environnement ?

Aucun problème majeur nouveau ne semble se manifester bien qu'il y ait toujours de nouveaux polluants. Le problème essentiel tient à la complexité technique – mélanges de contaminants dans des conditions complexes, qu'il s'agisse de l'environnement ou du site. La mobilité des organismes, des gènes et des polluants n'est pas bien comprise ; il commence à devenir évident que la recherche d'une meilleure connaissance de la biodisponibilité et des facteurs qui l'influencent est indispensable.

Ceux qui offrent des solutions se heurtent à des problèmes comme l'absence de données et le manque de possibilités de démonstration (l'occasion de faire des essais comparatifs). Ceux qui ont des problèmes sont confrontés à celui de l'insuffisance des orientations en matière de responsabilité et aux règlements fixant le degré d'épuration requis.

Les activités de recherche poursuivies au niveau académique ne portent pas souvent sur des questions intéressant ceux qui ont des problèmes (il faut améliorer les communications provenant de l'industrie et resserrer la coopération avec l'industrie et les pouvoirs publics).

2) Quelles possibilités la biodépollution offre-t-elle, par rapport à l'état des connaissances ?

La biotechnologie appliquée à l'environnement peut maintenant être considérée comme un secteur d'activité plutôt qu'un simple ensemble de technologies. La vigueur du marché se manifeste par le nombre d'activités de biodépollution générant déjà des recettes. La concurrence est vive pour les entreprises lucratives, spécialisées en particulier dans la biofiltration. Il existe un marché vaste et mondial pour la dépollution mais la lenteur de sa pénétration tient à de nombreuses raisons, notamment le manque de données de démonstration. L'industrie est néanmoins plus optimiste qu'il y a un an, à Tokyo, au sujet des débouchés de la biodépollution, essentiellement parce qu'elle a acquis davantage d'expérience. On s'accorde plus généralement à reconnaître que la vigueur de l'industrie de la biodépollution est directement liée à l'existence de normes en la matière.

3) Qu'est ce qui bloque une plus large application ?

A) Les politiques techniques

Les blocages techniques comprennent l'absence à la fois de compréhension fondamentale et de données de R-D établies sur le terrain, sans oublier la complexité d'emplacements précis. On constate fréquemment un manque d'approche intégrée (pluridisciplinaire) qui est indispensable pour établir optimisation et prévisibilité. Une stratégie au cas par cas à partir de zéro est souvent adoptée sans grand résultat en fixant des règles du jeu et des méthodes, alors que les questions posées sont différentes, selon qu'elles le sont par des industriels ou des universitaires.

En biodépollution, on rencontre des mélanges de contaminants, des conditions géologiques complexes et des sites insuffisamment décrits pour ce qui est de la répartition des contaminants et des micro-organismes. Pour toute une gamme de situations environnementales, la biodisponibilité et les transferts de masse de contaminants sont mal connus.

Une autre lacune réside dans l'aptitude à mesurer et à suivre le déroulement d'un traitement ou d'un processus de dépollution, bien que la technologie permette maintenant un suivi perfectionné, presque en temps réel et sur place, des activités de biodépollution. Il reste encore beaucoup à élucider dans tous les processus naturels, qu'ils soient physiques, chimiques ou biologiques, pour que les plus intéressants puissent être stimulés.

Les manipulations génétiques apportent beaucoup d'éléments nouveaux mais de grandes questions restent en suspens, par exemple, dans quels cas faut-il y recourir, et quels doivent être les objectifs appropriés (renforcement des possibilités cataboliques, régulation des gènes modifiés, fonctions “reporteur”).

Le passage de l'échelle du laboratoire à celle du plein champ est souvent considéré comme un problème : des modèles, tant physiques que mathématiques sont indispensables – disposer de données meilleures et plus nombreuses équivaut à une amélioration de la conception et de l'ingénierie.

B) Les politiques économiques

Ce qui manque le plus, ce sont des données comparatives sur l'efficacité par rapport aux coûts. Partout où on a pu en disposer, comme dans le cas de la biofiltration, le biotraitement s'est souvent avéré présenter un avantage évident. Les coûts initiaux en capital du biotraitement peuvent être bien supérieurs à ceux d'autres méthodes physiques ou chimiques et la rentabilité des investissements constitue donc une meilleure mesure.

La rentabilité en matière d'environnement constitue un problème potentiel majeur qui ne saurait être sous-estimé. Les coûts cachés peuvent comprendre : l'obligation de dépollution imposée par une autorité ou à la demande du public ; les pertes commerciales dues à l'interruption d'activité ou à la détérioration de la réputation ; les dédommagements accordés à des victimes ou à des organismes de soutien financier (banques, etc.) ; la perte de patrimoine par suite de contamination ; la découverte de problèmes hérités de propriétaires précédents et les coûts des procédures juridiques. La sensibilisation à ces problèmes, qui deviennent des menaces pour l'industrie, est en progression.

C) Les politiques gouvernementales

Une législation cohérente prévoyant des mesures d'application et d'harmonisation internationale est tout à fait nécessaire. La législation peut constituer à la fois un avantage et un problème. Elle est l'ultime soutien de ceux qui offrent des solutions mais aussi l'origine des coûts plancher de la pollution pour ceux qui ont ce problème. C'est grâce à la législation, à la réglementation (et aux mesures d'application) que les responsabilités sont définies et que des débouchés commerciaux sont créés pour la biodépollution.

Parmi les rédacteurs de réglementations, il peut encore subsister une certaine incompréhension de ce qui est requis ou approprié en tant qu'effet recherché. Des variations dans la législation et dans son niveau d'application sont considérées comme freinant l'adoption de technologies de biodépollution. On estime que l'objectif ultime est l'harmonisation de normes fondées sur le risque.

D) La perception du public

L'insuffisance de compréhension des politiciens, des fonctionnaires et du grand public peut constituer un facteur freinant l'exploitation de la biodépollution et du biotraitement. La diffusion d'informations concernant la santé et la sécurité des travailleurs est essentielle, tout comme la participation de la collectivité, lors de l'examen des solutions envisageables, ainsi que la façon dont le public perçoit les aspects relatifs à la santé, à la sécurité et au risque. Dans tous les cas, il convient de prendre l'initiative pour les communications avec le public.

Il importe, en particulier pour les informations destinées aux pouvoirs publics, de définir soigneusement le destinataire de l'information – la bonne information allant au bon destinataire.

4) Comment pouvons nous faciliter l'adoption de la biodépollution par l'industrie ?

Le sentiment général est en faveur d'une augmentation des démonstrations pratiques, des parcelles expérimentales et des exercices comparatifs. Il convient d'encourager la progression vers des objectifs de biodépollution fondés sur le risque (puisque la dépollution absolue est hors d'atteinte), l'élaboration de protocoles de validation acceptables (normalisés) et la détermination de la stabilité (sécurité) à long terme des sites dépollués.

Des données extraites de tous les comptes rendus de réussite de biotraitement et de biodépollution seraient fort utiles, tant comme argument commercial que comme soutien publicitaire.

Les problèmes d'environnement doivent être traduits en termes de problèmes professionnels et communiqués au secteur des entreprises, notamment aux banques et aux compagnies d'assurance, lesquelles peuvent, du fait de leur important rôle d'intermédiaires, transférer efficacement ces informations.

Événements intervenus depuis la réunion de Tokyo

Le message est clair : le public pourrait fort bien accepter des applications de la biotechnologie à l'environnement et il est possible de présenter la biodépollution comme une technologie écologique pour l'élimination des polluants. Les participants croient beaucoup au rôle de la biotechnologie – elle a obtenu des résultats qui ne sauraient être égalés par des méthodes classiques se limitant, par exemple, au pompage et au traitement. Cependant, l'incinération ne doit plus être considérée comme le principal concurrent de la biodépollution – il existe maintenant des techniques non biologiques plus perfectionnées, comme la désorption thermique et le traitement électrochimique. Ceci illustre le sentiment grandissant que la biotechnologie doit être un recueil de techniques et qu'il y a place pour des applications hybrides fondées sur des techniques à la fois physico-chimiques et biologiques.

Les différences entre les réunions de travail d'Amsterdam et de Tokyo peuvent être mises en évidence par les réponses différentes qu'elles ont apportées à une question rituelle, à savoir : “Quand la technologie sera-t-elle fiable et prévisible ?”. A Tokyo, la réponse a été : “Lorsqu'il y aura assez d'études scientifiques pour l'étayer” ; et celle d'Amsterdam : “Lorsque l'on aura effectué suffisamment d'essais et de démonstrations compétitifs sur le terrain et mis au point assez de modèles validés”. Il y a eu à Amsterdam davantage de démonstrations d'utilisations commerciales du biotraitement et de la biodépollution. Cette technologie peut remplir, et remplit en fait, les objectifs de façon efficace par rapport aux coûts et pourrait s'imposer par rapport aux autres techniques d'ici un à trois ans. La biotechnologie n'est pas la panacée universelle mais elle convient à une large gamme de circonstances et de créneaux spécialisés.

La biotechnologie appliquée à l'environnement n'a pas actuellement la robustesse des méthodes chimiques et physiques. Cependant, la polyvalence des polluants qui peuvent être traités, la flexibilité des systèmes biologiques et les coûts d'entretien relativement faibles font de la biodépollution un outil très performant pour la prise en charge de problèmes d'environnement. Si la biodépollution a révélé des possibilités considérables, elle n'est pas encore reconnue dans certains cas comme une technologie confirmée par ceux qui ont des problèmes et par les autorités concernées. Pour être plus largement appliquée, il semble donc que cette technologie doive démontrer de façon convaincante à l'industrie qu'elle offre des avantages économiques par rapport aux autres solutions.

Malgré certaines incertitudes au niveau technique et comptable, la biodépollution sert actuellement à épurer l'air, l'eau et le sol – sans grand tapage et sans le recours à des OGM. L'industrie a des succès à annoncer même si les industriels comme les universitaires ont parfois promis davantage qu'ils n'ont pu tenir. Même si la solidité de cette technologie peut n'avoir pas encore été totalement démontrée, sa polyvalence ne saurait être sous-estimée. L'industrie de la biodépollution n'a pas à bénéficier d'un traitement spécial – la biofiltration a pénétré le marché grâce à ses seuls mérites, à ses coûts compétitifs, et c'est maintenant une technique reconnue pour un certain nombre de polluants avec beaucoup de réussites à son actif. Pourtant, la biodépollution des sols n'a pas autant progressé – elle donne aussi des résultats mais les applications dépendent davantage du site et du polluant – il faudrait peut-être la soutenir à court terme en attendant que des démonstrations soient faites et que des données de validation soient réunies. Pourquoi pose-t-on si peu de questions au sujet des aspects biologiques du traitement de l'eau ? Parce que ce traitement a eu de nombreuses décennies pour s'imposer comme une technologie mature.

La réunion d'Amsterdam a montré que l'objectif de l'industrie n'est plus de déterminer des possibilités mais de décrire de ce qui marche, avec les directives d'utilisation appropriées, donc de valider des pratiques industrielles. S'agissant des OGM, ceux-ci peuvent encore être nécessaires pour des applications dans certains créneaux et ils sont maintenant utilisés avec succès lors d'étapes intermédiaires du passage à l'échelle supérieure. Une certaine forme de réorganisation industrielle se fait jour, en estompant la distinction entre vendeur et client, au fur et à mesure que de nombreuses entreprises mettent au point des technologies à leur propre usage.

Sous le coup de limitations budgétaires, les pouvoirs publics deviennent de plus en plus pragmatiques dans leur conception de normes d'épuration. La combinaison de normes réglementaires non nulles et l'intérêt accordé à des technologies nécessitant moins de main-d'oeuvre sont à l'avantage de la biotechnologie. Le monde scientifique pourrait inviter les pouvoirs publics à examiner de plus près les conditions dans lesquelles une biodépollution intrinsèque peut être acceptable, et pourrait contribuer à cet examen. Des participants rappellent que la politique ne s'est pas toujours appuyée sur la science et sont en général d'avis qu'elle devrait se fonder sur une évaluation des risques plutôt que sur des chiffres absolus (et établis arbitrairement). Quelles que soient les technologies, il doit exister un paramètre logique auquel chacune d'elles puisse être comparée.

Recommandations

Lors de leur dernière séance plénière, les participants à la réunion de travail étudient et proposent des recommandations. Ils les formulent à titre personnel. Ces recommandations sont adressées à l'OCDE et, par l'intermédiaire de celle-ci, aux pouvoirs publics et aux industriels des pays Membres, les uns et les autres offrant des solutions et étant confrontés à divers problèmes. Dans bien des cas, ces trois partenaires, université comprise, doivent joindre leurs efforts. Si les recommandations émanant de la réunion de travail ne sont pas contraignantes pour les gouvernements des pays Membres de l'OCDE, des participants rappellent que ceux-ci sont néanmoins investis d'une autorité intellectuelle et d'une force de persuasion considérables.

Les technologies destinées à un avenir très lointain devront s'appuyer de plus en plus sur des ressources renouvelables, sur le recyclage des produits et des déchets générés lors de leur fabrication, et sur l'utilisation de l'énergie solaire et d'énergies renouvelables pour alimenter les processus de production. A plus court terme, les biotechnologies d'environnement devront être axées sur la prévention, le traitement et la dépollution biologiques.

Les politiques gouvernementales

La biodépollution s'est révélée efficace par rapport aux coûts pour des traitements à long terme réussis et elle a encore de grandes possibilités. La collaboration internationale pour des études à long terme sur banc d'essais en fonction d'objectifs fondés sur le risque est recommandée. Ces initiatives doivent renforcer le partenariat entre industries et universités à une échelle globale. Il est urgent que les pouvoirs publics encouragent des démonstrations de technologies performantes, comme le traitement des rejets gazeux dans l'air, afin de convaincre le public et la clientèle privée que les biotechnologies d'environnement arrivent à maturité. A l'appui de ces démonstrations, il convient d'encourager des groupes d'études pluridisciplinaires.

Les problèmes d'environnement doivent être exprimés en problèmes professionnels, ce qui permet au monde scientifique de communiquer avec celui des affaires. Les gouvernements doivent encourager le transfert des technologies depuis leurs fondements scientifiques jusqu'à leurs applications, c'est-à-dire de celui qui offre la solution à celui qui a le problème à résoudre.

Il est indispensable d'harmoniser la législation et de l'appliquer de façon cohérente. L'industrie doit se préparer à voir la législation s'harmoniser par-delà les frontières nationales (ou des États) et être appliquée de façon cohérente. Dans le choix des objectifs, on s'efforcera de passer de réglementations fondées sur des concentrations absolues à des règles basées sur l'évaluation du risque, compte tenu de l'utilisation prévue à l'avenir pour le terrain considéré. Il faut définir des objectifs en fonction des risques et faire évoluer objectifs et politiques connexes en fonction d'informations de qualité provenant de la communauté scientifique.

La question des relations avec le public et de sa sensibilisation doit être globalement prise en compte. Des initiatives conjointes entre pouvoirs publics, industriels et scientifiques sont indispensables pour promouvoir les aspects positifs de la biodépollution. Ces initiatives doivent être exprimées clairement, sans jargon, appliquées au niveau convenable et capables d'anticiper les réactions du public. La sensibilisation précoce est essentielle. Démonstrations et études de cas doivent servir à convaincre le public et la clientèle privée que les biotechnologies d'environnement sont en train de parvenir à maturité.

Il est nécessaire d'établir des règles de responsabilité applicables aussi bien à ceux qui offrent des solutions qu'à ceux qui ont des problèmes, de façon à supprimer les incertitudes en matière de responsabilité liées aux transactions et à l'épuration des terrains. Il faut accepter la notion selon laquelle une pollution ancienne ne doit pas nécessairement être épurée immédiatement.

Offrir des solutions ...

Ceux qui offrent des solutions doivent s'efforcer, avec les universitaires, de surmonter les principaux obstacles techniques génériques évoqués lors de la présente réunion de travail, notamment les problèmes de passage à l'échelle supérieure, de caractérisation des sites et de biodisponibilité, de façon à réduire l'écart entre l'état des connaissances au laboratoire et celles utilisables sur le terrain. Les études portant sur la combinaison organisme et "micro-environnement" doivent être davantage orientées sur les tâches à remplir, par exemple, l'adaptation de cultures mixtes à l'environnement pour améliorer les chances de survie d'une fraction importante d'organismes. Ceux qui offrent des solutions doivent également accorder une haute priorité à l'acquisition de données servant à valider des modèles prévisonnels de biodépollution et doivent encourager activement la mise au point rapide de méthodes pour la surveillance en ligne, sur place, des polluants et de l'activité microbienne.

Un domaine essentiel de recherche fondamentale, nécessitant la collaboration des pouvoirs publics, des industriels et des universitaires concerne la biodisponibilité. Les composés carbonés ne sont pas tous minéralisés mais sont plutôt sorbés sur des particules du sol ou intégrés ou encore liés chimiquement à des matières humiques. Que signifie une lixiviation lente de faibles concentrations de résidus subsistant après biodépollution ? Dès lors qu'une matière ne se prête pas à la biodépollution, est-elle écotoxique ?

On insistera sur l'organisation d'une surveillance en ligne, in situ, des concentrations de polluants, des micro-organismes et de l'activité biologique. Il est essentiel de réunir des données provenant d'opérations sur le terrain de façon à établir et à valider des modèles prévisionnels.

Ceux qui offrent des solutions se joindront à ceux qui ont des problèmes pour réserver certains sites sur lesquels se fonderont les recherches à long terme nécessaires pour réunir de meilleures données sur le terrain, ces sites devant être gérésen coopération par des équipes pluridisciplinaires qui pourraient diffuser largement leurs données en mettant, par exemple, leurs résultats sur Internet de façon à ce que chacun puisse en bénéficier.

Les biotechnologies d'environnement constituent un domaine pluridisciplinaire au sein duquel une coopération entre plusieurs domaines de compétence (du microbiologiste à l'ingénieur) est une nécessité. Les groupes d'études pluridisciplinaires doivent être encouragés.

... À ceux qui ont des problèmes

Ceux qui ont des problèmes seront amenés à considérer la biodépollution comme la première solution à choisir, compte tenu de son acceptation par le public, des avantages économiques qui peuvent l'accompagner et de son image positive par rapport à des méthodes physico-chimiques concurrentes. On signalera les études de cas faisant état de réussites et des entreprises prendront les devants pour en informer largement le public.

L'industrie doit analyser ses activités professionnelles de façon à se faire une idée plus précise d'elle-même, de ses besoins et de la façon dont elle fonctionne sur le marché, pour déterminer quelles améliorations radicales peuvent être obtenues grâce à la biotechnologie. Par exemple, des molécules peuvent-elles et doivent-elles être conçues de façon à ne plus être biodisponibles une fois immobilisées sur des matières humiques ? Quels effets à long terme de telles modifications pourraient exercer sur la fraction organique du sol, et dans quelles conditions ces composés pourraient-ils être mobilisés et donc polluer à nouveau ?

Un resserrement des relations industrie/université est nécessaire pour stimuler le transfert de technologie à partir de bases scientifiques vers des applications, de celui qui offre des solutions vers celui qui a un problème à résoudre.

On améliorera la diffusion de l'information en organisant des réunions de travail pour les médias, en préparant des documents d'information, en recourant au multimédia et à la télévision et en faisant participer la collectivité (réunions urbanisme et consensus). Les écoles seront une cible prioritaire où on utilisera largement des études de cas (relatant des réussites).

La responsabilité suscitant d'importantes activités de dépollution, il convient de traduire avec toute la diligence requise les problèmes d'environnement en débouchés professionnels. La responsabilité se fonde à la fois sur la législation et sur la perception du public. Les banques et les compagnies d'assurance sont des intermédiaires non négligeables dans la gestion des risques et des coûts et doivent être utilisées pour propager l'annonce de technologies prometteuses. De façon similaire, il importe de savoir à qui s'adresser parmi les pouvoirs publics et de donner de meilleures orientations industrielles pour la recherche universitaire ; la chose pourrait être facilitée par l'établissement de réseaux ciblés reliant ceux qui ont des intérêts, des responsabilités, des besoins et des aptitudes en matière de biodépollution.

OPENING REMARKS

OPENING REMARKS

by

W. Harder
TNO-Institute of Environmental Sciences, Energy Research and Process Innovation,
Delft, The Netherlands

It is a great pleasure to welcome you to the OECD Workshop Amsterdam '95. When the publication "Biotechnology for a clean environment" (OECD, 1994) was being completed, a discussion in various OECD fora was initiated, regarding the exploitation of the potential of biotechnology for the prevention, detection and remediation of environmental pollution. These discussions led to the suggestion to organise a series of workshops on different specific issues and, at the invitation of the Government of Japan, the first Workshop on Bioremediation was held in Tokyo in November 1994. The Proceedings of this Workshop have just been published and they provide ample proof (for those who did not attend) of the success of this gathering.

Our present workshop in Amsterdam is the second in the series and is specifically devoted to Wider Application and Diffusion of Bioremediation Technologies. It is held at the invitation of The Dutch Government and I kindly ask your co-operation to make this workshop as great a success as our predecessor.

Our workshop focuses on bioremediation of off-gases and soil. As in the previous workshop, we shall use the word "bioremediation" in a broad sense. In addition to biological clean-up of sites contaminated with hazardous chemicals and the treatment of industrial off-gases, it may also include other ways in which the application of biotechnology can help to improve the quality of our environment.

The organizers or the workshop have decided to follow the example set in Tokyo and have asked the chair to formulate a number of questions for you to consider during the next two days. These questions are:

1. What are the key environmental problems that bioremediation should address?
2. What is the potential of bioremediation with reference to the state of the art?
3. What are the bottle necks to wider application with respect to the following issues: technical, economic, government policy, public perception?
4. How can we facilitate the introduction of bioremediation in industry?
5. Which actions can be recommended to governments and industry?

I invite you all to consider these questions in the discussions during the various sessions and hope that we shall come to specific answers and recommendations at the end of our workshop. Although we must realise that our recommendations cannot be binding to any of the OECD Member countries, our reviews do carry considerable persuasive force and intellectual authority.

Finally, it is a privilege to serve as your Chairman for this workshop and I thank the organisers for the invitation. Together with the co-Chairs and speakers, we shall endeavour to provide you with a stimulating programme with ample time for discussions. I trust that we shall have a memorable workshop.

WELCOMING SPEECH

by

Dr. P.G. Winters
Deputy Director-General of Industry and Services, Dutch Ministry of Economic Affairs

Introduction

On behalf of the Ministry of Economic Affairs, which is hosting this OECD Workshop on Bioremediation Technologies, I would like to welcome you to Amsterdam. This workshop will focus on the contamination of two of our natural resources: soil and air. Clean water is also a problem, of course, even in the Netherlands with its rivers, channels, and rainy climate. There was in fact a limit on water use here this past summer. However, water will be the subject of another OECD workshop, which will take place next year in Mexico.

Before introducing today's topic, I would like briefly to present Dutch industrial and technology policy. Broadly speaking, Dutch industrial policy aims to create conditions favourable to industrial competitiveness. To achieve this aim, the government takes measures in the areas of infrastructure, tax relief, and deregulation, among others. More specifically, industrial and technology policy focuses on technological innovation.

Convinced that knowledge is a key to international competition, not only for companies but also for countries seeking to attract new investors, the government stimulates the development of technology by enterprises and the public research institutes, as well as the commercial exploitation of this knowledge. It recently published an ambitious White Paper on technology policy, entitled "Knowledge in action", which defines two main points of concern. One is the need to stimulate private and public investments in research, as total Dutch investments have dropped to well below the OECD average, an indication that Dutch firms spend less on R&D than their competitors in other countries. The second is the need to improve the output from R&D expenditure; this mainly concerns the economic output resulting from public research. In this respect, the government pays special attention to the interaction between public research (universities, TNO) and private research.

In addition, the White Paper announces policy efforts in a number of areas: costs of R&D (tax deduction for R&D activities); technological co-operation (between public and private research, but also between enterprises); diffusion of knowledge; technical education; trends of importance to the Dutch economy, such as the electronic highways (with government and enterprises collaborating to define projects that will stimulate their use) and environmental technology, with its high market and ecological potential. This last is a topic of special concern to this workshop as well.

The Ministry of Economic Affairs therefore strongly supports the objectives of this workshop: to ensure the international diffusion of experience with bioremediation technologies and to encourage their industrial application. Today, all countries have to deal with environmental problems, and governments and industry have to deal with the globalisation of markets when searching for viable solutions. The challenge to this workshop is to help define how co-operation on and diffusion of bioremediation technologies can foster new economic activities. Let me therefore draw attention to some aspects of the economic challenges and the role of co-operation and diffusion as they relate to bioremediation technologies.

Problems

The past decades have witnessed world-wide growth of industry, with great benefit to the world economy. However, this growth has considerably increased the stress on the environment, to the point where nature alone can no longer restore our severely contaminated soil, water, and air. Today, we are facing two major problems. The first one is financial. According to recent market studies, it is estimated that it would cost a minimum of Gld 100 billion to fully restore soil quality in a small country like Holland using proven, conventional technology. It is obvious that the funds to solve this problem are not readily available. The second problem is technical complexity, as pollution is caused by different substances and combinations of substances, and soil composition varies greatly.

Yet solutions must be found, and technology can be used to find them. The challenge for science, industry, and governments is to create economically and technologically viable conditions for implementing these solutions.

Opportunities

Bioremediation offers promising ways to meet this challenge. Good results have been obtained under laboratory conditions for difficult situations where other technologies have failed. This promising start is described in OECD's *Biotechnology for a Clean Environment,* which contains a worthwhile review of these technological developments. However, technology is not the only factor involved; a further challenge is to solve problems in an economically viable way, at less cost, and more rapidly. The estimated size of the environmental problem in the Netherlands makes it clear that there are real economic opportunities available if industry invests in the creation of economically sound solutions. Indeed, according to the same OECD study, the total environmental market in the OECD Member countries could reach $300 billion by the year 2000, and the market for bioremediation could attain at least $50 billion a year.

This presents a great challenge to the supply side, but bioremediation also creates opportunities for the demand side. According to studies carried out by a recently started Dutch programme, NOBIS, it may be possible to reduce costs by up to 50 per cent by using bioremediation rather than classical remediation technologies. This presents a very great opportunity, and the challenge for all the parties involved is to transform these opportunities into real economic activities.

Co-operation

There are many actors concerned by bioremediation: research institutes and universities, industry and local authorities with problems that need to be solved, and of course the environment industries. By co-operating and sharing their expertise, they may be able to reduce the time required to achieve market breakthroughs. For instance, universities and institutes could focus their R&D

activities on existing problems in industry, and they could make use of the capabilities and practical experience of environment industries as they seek to transform basic science into practical developments. There is, for example, growing co-operation between industry and R&D institutes within the framework of the Dutch programme for stimulating industrial technology (PBTS). Last year, 12 proposals involving such co-operation were funded in the area of soil remediation, and nine in the area of air cleaning.

Diffusion

Many other countries are also developing experience with different situations. Taken together, their varied experiences can create a valuable knowledge base. Diffusion of this information encourages acceptance on the demand side and removes obstacles such as questions of predictability, risk, and costs. It can also avoid wasting time, effort and money. And last but not least, it may help governments set in place the conditions under which these new technologies can be implemented most effectively.

Policies

Governments play an important role in the success of developments such as bioremediation, and this is especially true for environmental technologies which can be the key to reconciling ecological sustainability and economic growth. The challenge for governments is to set guidelines that stimulate rather than constrain industry's innovative potential. In the Netherlands, the government has been encouraging the development of environmental technologies for several years, for example, by the PBTS programme mentioned above.

This year, the Ministry of Economic Affairs initiated NOBIS, a new five-year programme on *in situ* bioremediation. This programme has been set up as a public-private partnership: in addition to the Ministry, the partners are institutes, bioremediation companies, and those with environmental problems to solve. The Ministry of Environment and Housing is involved as well. Through this programme, there is interaction between policy making and practice. The objectives of NOBIS are two-fold. In the first place, it seeks to develop innovative and, above all, practical solutions for many different situations. In the second place, it aims to create a solid knowledge infrastructure of institutes and industries as a basis for long-term co-operation. With financial support of Gld 25 million from the government, this programme will generate industrial activities worth an estimated Gld 65 million and even more over the longer term.

Recently, the Dutch Ministry of Economic Affairs and the Ministry of Education, Culture and Science also launched a new R&D initiative entitled Economy, Ecology, and Technology. This much broader initiative focuses on the development of promising technologies with high environmental effectiveness and high economic potential. The ultimate goal is to create sustainable industry and prevent future environmental contamination. Biotechnology is certainly an area with such potential.

Conclusion

Bioremediation is a technology that offers many opportunities to improve the environment and to develop economic activities. It is in fact a technology that can help nature do its work. This workshop is addressing the conditions needed to turn these opportunities into reality. I am convinced that the discussion will be fruitful and that some answers and recommendations will be forthcoming. I hope you will keep in mind the importance of industrial applications and diffusion, which are the basis of innovative solutions and ensuring the economic growth of bioremediation.

OECD'S BIOREMEDIATION WORK

by

R. Nezu
Director, Directorate for Science, Technology and Industry, OECD

Good evening, Dr. Harder, Ladies and Gentlemen,

On behalf of the OECD and its Secretary-General, it is my pleasure to open the Amsterdam Workshop on the Wider Application and Diffusion of Bioremediation Technologies. Our greetings and our thanks to all of you: first of all, to the Dutch sponsors of this meeting, in particular the Ministry of Economic Affairs, whose generosity has made this meeting possible, to Dr. Wim Harder from TNO, who kindly accepted to chair the workshop, to the four Co-chairs and the Rapporteurs. I am confident that you will do excellent work during the next few days.

Our appreciation also goes to the important scientific speakers who have come from 14 different countries around the globe and, last but not least, to the many participants and national delegates.

We are meeting today at this beautiful site almost exactly one year after the Tokyo workshop on Bioremediation, the OECD's first international meeting on the subject. Many of you have seen the Proceedings of that event, which have just been published. Let me recall and share with you what my predecessor, Mr. Tanaka, said in his opening remarks at last year's conference. He said:

> "We do not want our work on the science and technology of bioremediation to end with the Tokyo workshop; we would like this to be a beginning, a starting signal, rather than an end".

I thank everybody here for responding so positively to that wish. This year, the Netherlands is the sponsoring country, next year it will be Mexico. You will be pleased to hear that the OECD Working Party on Biotechnology – the oversight body for our work – met ten days ago and agreed to give high priority to bioremediation and related activities. It is not overstating the case to say that your efforts in preparing this workshop are already bearing fruit at policy levels.

The Tokyo workshop covered many subjects and focused often on the scientific aspects of bioremediation. This meeting centres on the bioremediation of air and soil, particularly in terms of the wider application and industrial diffusion of bioremediation technologies. Here, I am tempted to make an analogy to a track and field event called hop, skip and jump. Some years ago, I met a top-ranking Japanese athlete in this event, and, out of curiosity, I asked him: "What is the right balance between the distances to be gained by the individual hops, skips and jumps?" His answer was interesting. He said that the most appropriate balance was 40 per cent for hop, 25 per cent for skip, and the remaining 35 per cent for jump. I asked why the second step made such a small contribution.

He replied that, on the contrary, it was the most important part, because it was the means of connecting the first and the third. He explained that many athletes spend too much energy on the first and second and thus lose their momentum or balance.

Today I think we are at the stage where we too need to maintain momentum and balance. Therefore, our Dutch friends have rightly opted for a subject that is both concise and practical. Few countries know as well as the Netherlands how to combine first-class research and scholarship with pragmatic applications and with a healthy sense for business and profits. Many other countries could learn from our Dutch colleagues, and I venture to say that I include my own country, Japan. This ability is one of the reasons why this relatively small country, threatened by the sea and devoid of natural resources, has become a major country, in fact a world leader in many skills and technologies, including bioremediation. Looking through the programme of this workshop, I was truly impressed by the number of scientific and technological experts from virtually every field of bioremediation that you have been able to bring together.

Here I would like to make another analogy. As an industrial policy maker in Tokyo and as Director of Science, Technology and Industry at the OECD, I have been very interested in technology development and the contribution of technology to competitive industries. Why, for example, is the Netherlands so strong and competitive in growing tulips? I do not think any of you would say that the country is particularly well-suited in terms of weather, soil or water. Yet, while there is no reason to believe that the natural environment is good for growing tulips, this country has led the world for three centuries. Sometimes, great sums of money were invested in tulips and when the price of tulips collapsed, so did the international financial market. This is a very good example of establishing a competitive position through hard work and conscious efforts to develop on technology and know-how.

Now, let me point to four reasons why bioremediation is of particular interest to the OECD:

First, bioremediation is an emerging field. Many scientific questions remain unsolved, even if there are already many profitable applications. Bioremediation relates both to science and R&D and to technology application and diffusion, the priority issues in the overall programme of work of my Directorate. Neither should be neglected.

Second, government policies and the related framework conditions play a major role in the development and diffusion of bioremediation technologies. Not only R&D policy, but also pollution prevention and control policies can affect bioremediation, positively as well as negatively. Many consider the OECD an appropriate platform for its Member countries to discuss such policies and policy differences – this workshop can be useful in this respect.

Third, information for policy makers and the public at large is one of the main problems and challenges of bioremediation and of biotechnology in general. I believe that the OECD has been doing a good job in this area. I hope this workshop will again contribute to better understanding and acceptance of this technology.

Fourth, allow me to end with a personal note. I confess that biotechnology and bioremediation are new to me. My last responsibility before joining the OECD, was as Deputy Director-General for trade negotiations in the Japanese Ministry of International Trade and Industry (MITI). The sectors with which I am most familiar are colour TVs, steel, textiles, semiconductors, and automobiles, but the relatively long list of issues that I have dealt with does not include biotechnologies. However, I have recently been fascinated to see biotechnology surfacing in all kinds of trade discussions, including those within the OECD. Biotechnology is a key area, along with micro-electronics and new material developments, with great scientific and technological potential and where Europe, the United States and Japan are competing head to head. Without any doubt, there is the possibility of conflicts in these areas. At the same time, opportunities for co-operation are even greater, and the OECD wishes to play an active role in minimising the risk of conflict and maximising the benefit of co-operation. Let me assure you that I shall keep my eyes on the economic and trade significance of bioremediation.

Ladies and Gentlemen, with your help, I hope that in a year's time I will no no longer say that bioremediation is new to me. Thank you.

SETTING THE SCENE

THE MICROBE'S CONTRIBUTION TO SUSTAINABLE DEVELOPMENT

by

W. Harder
TNO-Institute of Environmental Sciences, Energy Research and Process Innovation, Delft, The Netherlands

Introduction

The past decades have witnessed an increasing awareness that human activities, in particular intensive agriculture and industrial technologies, must be brought in harmony with the global material cycles in the biosphere. In other words, we shall have to make a transition from exploitation of our natural resources towards a partnership with the global ecosystem. Although in recent years much emphasis has been placed on the need to reduce the contamination of soils, ground water, sediments, surface water and air with hazardous chemicals, the issue is in fact much broader. Technologies of the more distant future should be based on renewable resources, they should use "mild" production processes, the resulting products and services should be environmentally compatible and any waste they generate should be recyclable (OECD, 1994).

In the years following the publication of the Brundtland report in 1987, several attempts have been made to translate its basic philosophy and recommendations into an operational approach for the immediate future. In the case of industrial technologies, a working group of the World Business Council for Sustainable Development (WBCSD, 1995) recently considered the following elements essential:

1. De-materialise: reduce the amount of raw materials used for delivering the desired product function;

2. Increase energy efficiency: this not only applies to the energy required to react raw materials and assemble products, but also to energy consumed during product use and disposal;

3. Eliminate negative environmental impact of processes and products: reduce diffuse emissions and by-products formed during the production process, but also during product use;

4. Close material cycles: design for recyclability, but not at any cost;

5. Borrow from natural cycles, particularly where renewable resources and recycling are concerned;

6. Extend the durability and service life of products, particularly in the usage phase.

I shall adopt this approach in an attempt to identify where the major challenges are for bioremediation technologies for the next decade.

Despite the rapidly developing interest in the use of these technologies, bioremediation is not yet universally understood nor trusted by those who must approve of its use (National Research Council, 1993). In the context of the present workshop it is therefore necessary to evaluate the hesitation associated with a wider application and diffusion of bioremediation technologies. This hesitation is, at least in part, due to insufficient understanding of the basic principles of bioremediation. In view of this we need not only to advocate further research to improve our ability to apply logical strategies for bioremediation, but should also stress the need for a better education of decision-makers with respect to the potential and limitations of the technology. Against this perspective, this paper evaluates the role of microbes in bioremediation in the context of a sustained development of the earth's ecosystem. It also attempts to identify key environmental problems and emerging opportunities for biotreatment and re-enforces a recent plea to emphasize the role education can play in helping scientists, engineers and decision-makers to understand the inherent complexity of bioremediation processes (Verstraete *et al.*, 1995).

Principles of sustainable development: lessons from nature

Life on earth is confined to a shell with a mean thickness of approximately 10 kilometres at the surface of our planet. Of this surface shell, which is in contact with an atmosphere of about 50 kilometres thickness, about two thirds is covered with water, the rest is land. There is virtually no exchange of materials between the planet and outer space and very little exchange between the surface and the interior of the earth. This means that the biosphere must be regarded as a closed system with respect to materials and it thus resembles a closed glass jar with a lid, screwed firmly on. However, it is an open system with respect to energy; our planet receives electromagnetic energy from the sun and radiates energy back into space. In this environment of water, sand, rock and a reducing atmosphere, life developed about 3.5×10^9 years ago. Initially the energy required for the synthesis of complex macromolecules that are essential for living creatures was derived from chemical reactions. Somewhat later early prokaryotes acquired the ability to use the sun as an energy source for biosynthetic purposes. This type of photosynthesis is dependent upon an organic or inorganic electron-donor other than water and could not serve to ensure sustainable development of life on our planet because the available electron-donors were in limited supply.

Around 2×10^9 years ago oxygenic photosynthesis arose and this "photosynthetic revolution" has changed the face of the earth, both in terms of numbers of living creatures, biological complexity and chemistry. The main reason for this is that this type of photosynthesis makes use of water as an electron-donor; since life itself is dependent on liquid water, the development of the ability to use this very compound as an electron-donor was a most significant event in evolution. Biological production processes use carbon dioxide and water as raw materials (and smaller quantities of other inorganic compounds) and produce oxygen as a by product. Carbon dioxide was initially an abundant feed stock, but the amount of this compound progressively held in biomass and fossil fuel has led to a drop in its concentration in the atmosphere by about a thousandfold; in exchange oxygen is now present.

Figure 1. Input/output relationships in biological processes in the biosphere

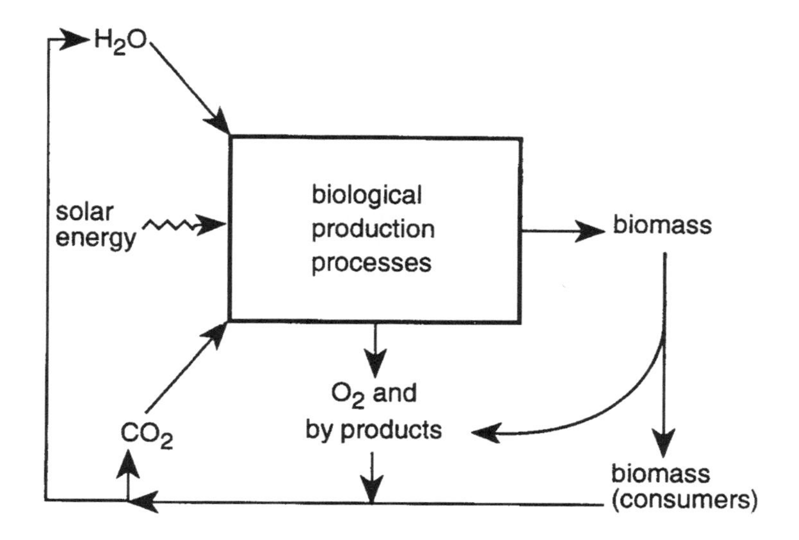

Source: Author.

Since the biosphere is essentially a closed system with respect to materials, sustained production and consumption processes in the biosphere have been critically dependent (and still are) on closed material cycles of essential elements (Figure 1). Microbes have fulfilled this role ever since the advent of oxygenic photosynthesis and they have become remarkably effective. They inhabit all possible locations where life exists and make use of virtually all types of chemical reactions that generate energy (Schlegel and Jannasch, 1992). This ubiquity of micro-organisms is based on three major properties: their small size for easy dispersal by air and water, their metabolic versatility and flexibility and their ability to tolerate unfavourable conditions. Their role in ensuring sustainable development in the biosphere has been studied during the last century and although there are still major gaps in our understanding, there are a number of lessons to be learned.

1. Sustainability does not mean "keep everything as it is".

2. Biodegradation essentially relies on chemistry; the often complex chemical reactions are catalysed by biological catalysts in the form of micro-organisms.

3. The biosphere has a tremendous experience with bioremediation but is not infallible with respect to man-made chemicals. The biodegradative potential of (unnatural) compounds should therefore be clearly established before they are discharged into the biosphere.

4. Biodegradation processes are generally carried out by complex communities of micro-organisms, whose composition is a reflection of the environment in which they exist. Introduction of specialised laboratory strains into such communities without paying due attention to so-called field application vectors (Lajoie *et al.*, 1992) is hardly ever successful, while super inocula generally serve no other function than that of a source of nutrients for the indigenous microbial population.

5. Biodegradation processes can be quite slow and do not always lead to complete mineralisation (viz. humic materials in the soil). This is not necessarily unacceptable, but the environmental compatibility of products formed must be demonstrated.

6. The rate of biodegradative processes is critically dependent on the presence of appropriate micro-organisms as catalysts and provision of appropriate micro-environmental conditions for optimal and sustained activity. Despite the notion that "everything is everywhere and the environment selects", it may be advantageous to reduce the start-up phase of (new) processes by introducing appropriate communities, organisms or even genes (but see 4). Provision of the required set of micro-environmental conditions to support the activities of the organisms requires a full understanding of this interaction and a close collaboration between microbial ecologists, microbial physiologists and bioprocess engineers.

7. In order to speed up biodegradative processes in case natural processes are too slow to meet our needs or desires, appropriate technological support of the microbial populations must be provided.

Advances in our understanding of biodegradative processes in the biosphere have made it clear that the central question is not whether biodegradation is successful, but rather how we can improve the rate of these processes in case they are too slow to cope with the high rate of discharge of waste from our industrialised society. This requires continued improvements in the technology of existing bioremediation processes and the development of novel concepts and technologies. Technologies of the more distant future (i.e. 30-50 years) should be based on the principles that ensured sustainable development in the biosphere. They should make use of solar energy, should be based on renewable resources and any (waste) product that they generate should be recyclable.

Figure 2. Input/output relationships in (industrial) processes

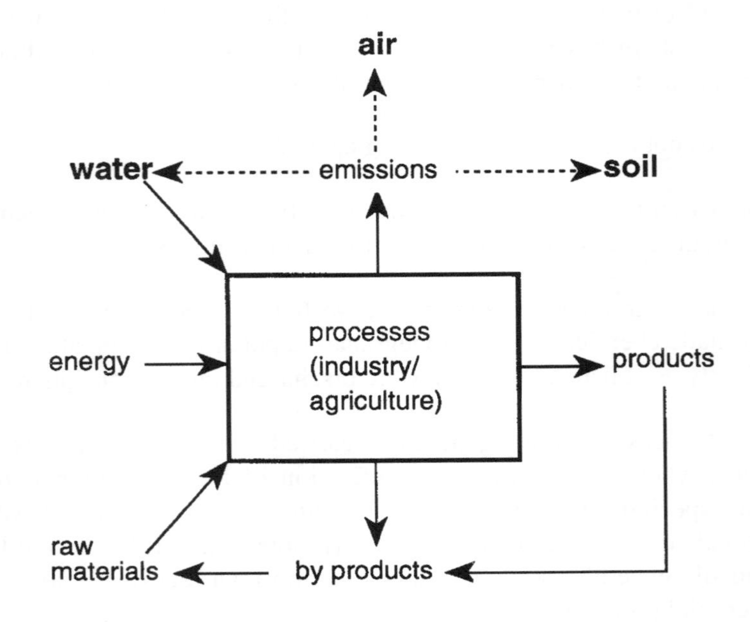

Source: Author.

Challenges for bioremediation technologies

For the immediate future environmental biotechnology has an important role to play in the prevention, detection and remediation of environmental damage (OECD, 1994). In the context of the present workshop it is appropriate to restrict our discussion to industrial processes (Figure 2). In view of the upcoming OECD workshop on water, we shall leave this important environmental compartment outside our discussion.

Following the list of essential elements put forward by WBCSD (loc.cit.) the following observations can be made:

1. De-materialise. Although there is clearly scope for bioprevention and bioremediation technologies in the conversion of natural resources to feed stocks for the process industry, the overall effect of this recommendation will be to reduce the stream of byproducts and emissions from industrial processes. This does not abolish the need for process-integrated and add-on biotechnologies, but will lead to a need to develop more diverse and smaller scale units.

2. Increase energy efficiency. This recommendation calls for more energy efficient processes and offers potential for biotechnological alternatives to current technologies. However, a discussion of this potentials falls outside the scope of this paper.

3. Eliminate negative environmental impact. This relates to diffuse emissions and by-products formed during the production process, but also during product use. Traditionally emphasis has been given to waste products and liquid waste streams (OECD, 1994), but recent studies have made clear that we should pay more attention to emissions into air (Baart and Diederen, 1991). Particularly in the case of volatile organic compounds, they constitute a significant source of diffuse pollution of water and soil following deposition from air and we shall examine the application of bioprocesses to reduce these emissions in the course of our workshop. However, there is perhaps a more immediate threat to our environment and this relates to an increase in the concentration of ozone in the troposphere (note that this is different from the well-known stratospheric ozone problem; Van Aalst, 1993). In the troposphere (the first 10 kilometres of our atmosphere) ozone is formed by photochemical reactions of nitrogen oxides and volatile organic compounds. The average ozone concentration at ground level has risen from 10 ppb at the end of last century to around 30 ppb in recent years. Also, the seasonal behaviour of ozone has changed and this indicates enhanced photochemical production of ozone from increased anthropogenic emissions. During the summer, in periods of sunny and warm weather, ozone (and other pollutants) builds up to concentrations far exceeding the WHO Air Quality Guideline of 60 ppb. This so-called summer smog episode extends over large parts of Europe and it may last for periods of up to several days to a week. Ozone starts to affect human health from concentration exceeding 80 ppb (hourly average), while plants are already affected by concentrations as low as 40 ppb (growing season average). It is clear that the tendency for the average ozone concentration to increase and the build up during summer smog periods constitutes a severe environmental problem. Since summer smog results from photochemical reactions of nitrogen oxides (emitted by traffic, industry and power plants) and volatile organic compounds (from biological sources, traffic and industry), it is important to reduce these emissions. In order to be effective, reductions of (European) emissions of NO_x and VOC of at least 70 per cent are needed. This not only requires a wide introduction of improved 3-way catalysts in cars and power plants, but also the development of (bio)filtration technology to reduce VOC emissions. For the future we should also reconsider the current practice of burning

much of our solid waste. Since this occurs at temperatures whereby NO_x is formed, alternative technologies including gasification and biotechnologies should be developed.

4. Close material cycles. This recommendation is primarily directed at the recycling of products and by-products of industrial processes. For products an important aspect is design for recyclability and here is scope for the development of new biomaterials by biotechnological processes. For waste streams added-value processes which convert a waste stream into useful products are to be preferred over end of pipe treatment, particularly if they have economic advantages. There is clearly scope for biotreatment and bioremediation in this context (OECD, 1994).

5. Borrow from natural cycles. In the context of our discussion of principles of sustainable development, this recommendation is self-evident. Where possible we should not only base our future technologies on renewable resources, we should also develop production processes, products and services that are compatible with material cycles as they occur in natural environments. This does not constitute a licence to load everything that is "natural" on to our environment. The rate at which we add compounds to natural cycles must be carefully balanced to avoid overloading, should be continuously monitored to ensure compatibility and must be supported by appropriate technology in case overloading is unavoidable.

6. Extend the durability and service life of products. This recommendation has technological, psychological and sociological impacts and may, when effective, lead to a reduced throughput of materials in the industrialised countries. It does not constitute an immediate challenge in the context of our present workshop.

A few remarks must be made with respect to the other topic of our workshop, namely bioremediation of soil. The experience so far has clearly demonstrated that the soil and its inhabitants are intrinsically heterogeneous. At different locations soil systems may differ greatly with respect to physical and chemical properties and therefore generalised information is of limited value. It is essential that quantitative information on these properties and on the nature and distribution of pollutants in a specific location is generated before a plan for bioremediation is made and put into effect. In order to learn from our experience for future applications, the progress of the remediation process should be carefully monitored and the results used to construct predictive models. Fortunately a number of well-controlled large-scale field experiments are currently being conducted or planned and the experience gained should help us to turn bioremediation into a less-controversial technology. These field experiments should ideally be accompanied by laboratory experiments to address bottlenecks in our knowledge relating to: i) how microbes behave in the field; ii) how to provide them with appropriate nutrients and the correct micro-environment, and iii) how to make the pollutants available for the microbial cells.

For the future, the priority must be to prevent further contamination of soils by those chemicals and products which pollute the soil and which are persistent or difficult to biodegrade once they have reached the soil environment. These chemicals must be banned by (international) legislation. Remediation of existing polluted sites is a tremendous task and can only be handled when priorities are set based upon a carefully conducted risk assessment. Wherever possible an initial approach may be based on the ICM principle (isolation-control-monitoring) which may be linked to natural bioattenuation or *in situ* bioremediation. If necessary intensive curative methods must be chosen and bioremediation is then one of the available technologies. A special point of concern is at which concentration of pollutants the soil may be considered "clean". This is particularly relevant for compounds that are not or have become hardly bio-available. Perhaps, as pointed out by

Verstraete *et al.*, (1995), bioremediation in this case should focus on engineered biological humus-binding processes rather than on complete mineralisation.

Conclusions

The principles of bioremediation as they function in the biosphere have secured sustainable development of life on our planet for more than 2 x 10^9 years. The lessons that can be learned from the way in which natural ecosystems function point the way towards technologies for the more distant future (> 50 years). They should make use of solar energy, should be based on renewable resources and (waste) products that they generate should be readily recyclable. It is evident that biological processes will be an integral part of these future technologies and that micro-organisms will be major active ingredients in closing the material cycles as they have done so successfully in natural ecosystems.

For the more immediate future it is expected that bioremediation shall enjoy a wider application, particularly in the prevention of further environmental damage and in the remediation of problems that are part of our legacy. In order for this to happen it is necessary that bioremediation is more universally understood and trusted by those who must approve of its use. I am confident that the present workshop will contribute to achieve this aim.

REFERENCES

BAART, A.C. and H.S.M.A. DIEDEREN (1991), “Calculation of the atmospheric deposition of 29 contaminants to the Rhine catchment area”, Report R91/219 to the Interministerial Working Group for Atmospheric Deposition, TNO-MW, Delft.

LAJOIE, C.A., S.Y. CHEN, K.C. OH and P.F. STROM (1992), “Development and use of field application vectors to express non-adaptive foreign genes in competitive environments”, *Appl.Environ. Microbiol.* 58, pp. 655-663.

NATIONAL RESEARCH COUNCIL (1993), *In situ bioremediation: When does it work?*, National Academy Press, Washington, D.C.

OECD (1994), *Biotechnology for a clean environment: Prevention, detection, remediation*, OECD, Paris.

SCHLEGEL, H.G. and H.W. JANNASCH (1992), "Prokaryotes and their habitats" in *The Prokaryotes*, A. Balows, H.G. Trüper, M. Dworkin, W. Harder and K.H. Schleifer (eds.), 2nd Ed., pp. 75-125, Springer-Verlag, New York Inc.

VAN AALST, R.M. (1993), "Atmospheric pollution a European concern", in *The European Common Garden*, T. Ribeiro (ed.), pp. 87-104, Group of Sesimbra, Brussels.

VERSTRAETE, W., E. TOP, P. VANNECK, M. DE RORE and G. GENOUW (1995), "Lessons from the soil", in *Contaminated soil '95*, W.J. van den Brink, R. Bosman and F. Arend (eds.), pp. 15-24, Kluwer Academic Publishers, The Netherlands.

WBCSD (1995), “The six dimensions of Eco-efficiency”, *Tomorrow* 5, p. 40.

ENVIRONMENTAL BIOTECHNOLOGY: BUSINESS OPPORTUNITIES AND BOTTLENECKS

by

J.H.J. Pâques
PAQUES B.V., Balk, the Netherlands

The market

The environmental market is very diversified. As a result of world-wide changes in thinking, it has thousands of niche markets. It is quite clear that industries have to be concerned about the environment in order to ensure continuing operation of their business, but it is also clear that care for the environment offers business opportunities.

The environmental market is a growing industrial market, estimated at US $200 billion world-wide, and is predicted to grow at a rate of 5.5 per cent per year to the year 2000.

The so-called "20/80-rule" applies to the environment industry as to others: 20 per cent of firms account for about 80 per cent of output in the individual market segments. The main segments of the environment industry (see Table 1) are described below.

Table 1. The main segments of the environment industry
(percentage)

	North America	Europe	Japan
Equipment and related services:	74	76	79
Water and effluent treatment	24	34	22
Waste management	25	15	22
Air quality control	12	17	25
Other (land remediation, noise)	13	10	10
General services	26	24	21
Total	100	100	100

Source: OECD estimates.

Overall, equipment for water and effluent treatment accounts for the largest share of industry output, although there are regional differences, depending on local environmental concerns as well as environmental legislation. In the Netherlands, this market segment accounted for Gld 712 million in

1992, an increase of 13.2 per cent over 1991. The production of equipment for water and effluent treatment is largely a mature market; the technologies are well-established. Relatively speaking, this market is projected to have the slowest growth.

Products for collecting and transporting solid wastes are the largest component of the waste management segment. This segment is expected to increase in importance, owing in part to stricter controls on waste disposal.

Air quality control equipment removes pollutants from a gaseous stream or converts pollutants to a non-polluting or less polluting form. Biological scrubbers and filters to treat flue gas are becoming more important.

The most important other categories of environmental equipment are products for land reclamation and noise reduction.

General environmental services largely involve engineering and consulting services to solve specific problems. This segment is dominated by large engineering firms, which provide technical engineering and construction services. These services are expected to increase in importance, as the move towards cleaner process technologies requires greater engineering and analytical expertise.

A substantial percentage of this market will concern the biotechnology segment. Environmental market revenues will increase in the coming years as the overall environmental market grows, and biotechnology will take a large share of this market.

The environmental biotechnology market

Biotechnology often works in closed loops and therefore offers a good solution to environmental problems. However, the environmental biotechnology market is not a simple one. It depends, on the one hand, on the results of research and development within universities, research centres and companies. On the other hand, it is strongly influenced by uncontrollable environmental factors.

The environmental market is mainly driven by governmental regulation. As politicians often look at the short-term, its development is unpredictable. In the Netherlands, the unpredictability of government environmental policy is regarded as the most negative factor in the development of such a market. On the other hand, new legislation is viewed as offering the most new opportunities (Table 2).

Table2. Positive and negative factors in the development of the environmental market

Positive factors	Negative factors
New environmental legislation – 30%	Unpredictable environmental policy – 26%
New financial regulations – 18%	Economic recession – 25%
Stricter control – 16%	Low return on investment – 21%
New environmental technologies – 14%	
Public awareness – 11%	
Other – 11%	

Source: Author.

Effective and consistent environmental regulation can create markets for new products and technologies and stimulate the demand for different types of pollution control equipment. It also stimulates the development of advanced techniques and technologies. Government funding of basic environmental research, particularly R&D, can be an important source of support and an incentive to its growth. In most countries, government R&D spending for environmental objectives has increased in the past five years, but only a small share of total government expenditure goes directly for the development of environmental technologies. However, in the context of environmental R&D programmes, government support to technology development is now increasing.

In the Netherlands, with the highest relative share of government spending on environment-related R&D, the Dutch National Environmental Policy Plan has allocated US$ 65 million over four years for research on environmental technologies.

In recent years, increasing public environmental awareness has become an indirect driver for the environmental goods and services market. For many companies, a positive attitude towards the environment is an important aspect of their corporate image.

How to succeed

In order to be successful, several elements are important. First of all, suppliers have to invest a considerable amount in R&D, as the market is becoming more and more technology-driven. Given the fact that the environmental market is very unpredictable, it is difficult to develop long-term R&D programmes.

Another factor, which is often underestimated but is very important, is scale-up. In the field of biotechnology, interesting laboratory developments are frequently announced. However, it is often difficult to scale these developments up into a workable and reliable installation. When new processes are designed and constructed for the first time, experience and frames of reference are missing. The problems of scale-up can be solved if there is very close co-operation between biotechnology developers, hardware application engineers, and the end users.

Related to basic research strength is the capability to integrate environmental technologies into total productive systems. Environmental biotechnology will increasingly be integrated in production processes. Owing to the increasing emphasis given to the development of clean technologies by governments, it is expected that integrated process approaches will come to play a more important role in the environment industry.

Furthermore, success in the environmental biotechnology market requires a global marketing strategy and the ability to offer a diversified range of products. A world-wide focus is extremely important, and makes it possible to compensate for economic recession in one geographic area by growth in another. However, a global strategy will only succeed if the owner of the technology co-operates with local firms in their respective markets.

For example, in the winter of 1990 our company was approached by Budelco, one of Europe's largest zinc smelters. Based in the southern part of the Netherlands, it has produced zinc for approximately 100 years and has polluted hundreds of hectares with SO_4 and heavy metals. In the early 1960s, when the processes used in the mill were changed, pollution ceased. Today, Budelco produces zinc essentially emission-free, but the heritage of the first 60 years of zinc production remains to be dealt with.

In the mid-1980s, together with Shell Sittingbourne, Budelco started a research programme based on a groundwater control strategy. By pumping several hundred cubic meters of polluted water per hour, it was possible to prevent extension of the polluted area. Of course, this groundwater had to be treated so that it could be discharged directly into surface water.

Following initial contacts in December 1990, we started a laboratory study programme in our laboratories; within a few months we determined that biological treatment of groundwater contaminated with SO_4 and heavy metals was technically and economically feasible. On the basis of this information, Budelco asked us to perform a large-scale pilot study on site. It started in early 1991, and within a few months, we achieved the same results as on laboratory scale. Budelco then asked us to produce the basic design with a cost estimate of plus or minus 15 per cent.

When this design was presented in the summer of 1991, we were invited to quote for a turn-key plant to our design, and a firm order was given in autumn 1991. We were able to design and build the treatment plant within 12 months. The start-up of the overall system took place in May 1992, that is, within two and a half years of our initial contacts.

During this time we had developed a new technology, executed pilot studies, made a design, built a plant, and started it up successfully. Treatment of this type of wastewater is extremely important for companies like Budelco but also for the mining industry. At this moment we are, for example, carrying out a pilot study for a large copper mine in Salt Lake City, Utah, where both the amount of groundwater and the concentration of heavy metals are many times greater than at the Budelco site.

Why do I tell this story? To stress the fact that R&D on new technologies should always involve close co-operation with the end user. Without co-operation with Budelco, our researchers would not have focused on the final goal and been able to solve the problem in the most economical way within a very short span of time.

I would like to stress that scale-up is the most important step in the development of environmental biotechnology. At the moment, there are no significant government subsidies or programmes to facilitate this work.

Summary

It is important to look at environmental biotechnology from the angle of an integrated production process. We still tend to view environmental biotechnology processes in terms of end-of-pipe solutions. It is clear that we will see a trend towards incorporating environmental measures into the production process, and environmental biotechnology can play an important role. There must be close co-operation between manufacturers of process equipment and suppliers of environmental technology, and this will result in fundamental changes in existing communication patterns.

The shift towards the use of clean technologies and away from end-of-pipe approaches also has economic implications. Because clean technologies involve fundamental changes in processing, they will expose companies to greater economic and market risks and will have greater impact on their competitiveness than end-of-pipe solutions. The environment industry can therefore be considered a strategic sector, given its effects on the competitiveness of other industrial sectors.

The environmental biotechnology market has thousands of niche markets which can be successfully explored only if companies have very specific know-how, proven successes, the capability to work world-wide, a wide range of products and services, and a willingness to co-operate with clients at the local level in solving problems.

Trade in environmental equipment and services and the globalisation of the environmental industry would obviously be facilitated by greater harmonisation of environmental standards.

Questions, comments, and answers

Q: What was the nature of the technology that was applied successfully in the Budelco case?

A: Anaerobic reduction of sulphate in the groundwater gives a heavy metals sulphide which is returned to the mill. This is followed by an aerobic oxidation of sulphide to sulphur for the sulphuric acid plant.

Q: In the OECD we are talking about reform of regulations. Do you know of specific areas of technology where you want changes in the form of or removal of regulations which would be beneficial?

A: In our own experience it is not that legislation is not good - on the whole legislation is a good thing, but it is often not applied. Many markets are fantastic from an academic point of view, but from an entrepreneur's viewpoint there is no market because there is no legislation to enforce it.

LONG-TERM OPPORTUNITIES AND SUSTAINABLE TECHNOLOGY IN BIOREMEDIATION AND BIOPREVENTION

by

Kunio Nakajima
Ministry of International Trade and Industry, Tokyo, Japan

Introduction

Biotechnology has been applied to various fields, such as chemicals, medicine, agriculture, and foods. In the future, its applications will be even broader. Among other fields, the environment is an area where biotechnology can be important and make a significant contribution to saving resources and energy and to reducing waste. The OECD's *Biotechnology for a Clean Environment,* which appeared in 1994 (the so-called "Green Book"), estimated that the environment industry's potential market, in terms of conservation or remediation, would reach US$ 300 billion in the year 2000, compared to US$ 200 billion in 1990. The environment-related market is expected to grow rapidly in Japan; it is expected to reach US$ 200 billion overall in the year 2000 (Table 1), and environmental applications of biotechnology are expected to reach US$ 1 billion (Table 2). The size of the potential market is one reason why environmental applications of biotechnology are of interest to the relevant industries.

Table 1. Forecasts of market trends for the environment related industry in Japan (US$ billion)

	Present	2000	2010
Supporting service	13.4	20.0	34.8
Waste treatment/Recycling	109.3	161.7	228.0
Remediation	8.7	14.5	24.3
Energy related	19.4	31.3	40.2
New products	2.3	5.5	23.2
TOTAL	152.9	232.8	350.2

Source: Industrial Structure Council, MITI, Japan, June 1994.

In most cases, industry adopts biotechnology if it sees the merits of biotechnology. Thus, biotechnology is applied to production processes only when, compared to conventional technologies, it improves the quality of products or the cost-effectiveness of processes for chemicals, medicine, or foods. This is a prerequisite for realising viable commercial applications of biotechnology to the environment.

Table 2. Forecasts of biotechnology related market trends in Japan

	1994	2000
Medicine	3.8	15
Foods	0.1	10
Chemicals	1.8	4
Agriculture and fisheries	0.2	3
Environment		1
Supporting instruments	0.6	1
Others	1.5	
TOTAL	8.0	34

Source: MITI, Japan, 1993.

MITI's policy to promote environmental application of biotechnology

Bioremediation

The OECD Tokyo '94 Workshop on Bioremediation concluded that biotechnology can be used effectively to solve environmental problems. Various recommendations for establishing scientific methods of evaluating efficacy and safety were identified, as were various problems to be solved through R&D. Concerning the safety of transgenic organisms, OECD Member countries do not perceive any specific risks. Thus, it appears feasible to use transgenic organisms in bioremediation as long as they are appropriately managed to ensure safety. This is one of the conclusions of the Tokyo '94 Workshop. Another is that it is essential to demonstrate both efficacy and safety in order to put bioremediation to practical use. These conclusions have affected our policy for the promotion of bioremediation.

Safety evaluation

When the OECD published *Recombinant DNA Safety Considerations* in 1986 (the so-called "Blue Book"), the relevant authorities in Japan used this report to establish guidelines for the application of biotechnology in industry. These guidelines are mainly for contained use. The next step involved study of the safety evaluation of deliberate release of transgenic organisms. MITI discussed this issue in terms of model cases, using sewage processing and clean-up of hazardous material in soil. At the outset, it was assumed that the potential effect of transgenic organisms on the environment could be evaluated by evaluating potential risk first in the laboratory, then in a closed chamber, and finally under field conditions.

In 1993, the OECD issued *Safety Considerations for Biotechnology*, which described how to evaluate the environmental safety of modified crop plants and biofertilizers during scale-up. Taking this report into consideration, we are now updating procedures to evaluate the safety of deliberate release of transgenic organisms, including those used for bioremediation.

On the basis of this report, we now focus on a general evaluation of bioremediation operations, because it is expected that the practical and primary application of bioremediation would involve augmentation of native organisms through nutrition. MITI continues to collect data on case studies of safety evaluation from experiments on organisms, including transgenic organisms able to degrade TCE or collect phosphates from waste water.

Efficacy measurement

Efficacy measurement is essential to encourage the diffusion of bioremediation technology because, without it, no one could evaluate the cost-performance of this technology and make a decision to choose one from alternative technologies. To establish efficacy measurement, it is necessary to collect, accumulate and share experiences of field application. The remediation of environmental pollution is also an important issue in Japan, but there is limited experience of bioremediation. The lack of experience makes it difficult for the industries to be convinced of its effectiveness and to be able to estimate the cost of remediation, as well as making it difficult to win public confidence. MITI is now trying to accumulate experience and to form the foundation for the wider application and diffusion of bioremediation (Table 3).

Table 3. Projects for safety and efficacy evaluation of bioremediation by MITI

Study on safety measures for using transgenic organisms under natural environment	Budget: Period:	US$ 500 thousand for 1995 FY 1993-1996 FY
R&D project on soil bioremediation	Budget: Period:	US$ 500 thousand for 1995FY 1995-1999 FY

Source: Author.

R&D projects of bioprevention by MITI

Bioremediation is a technology for restoring polluted environments. Over the long-term, however, preventive measures are more important and effective in terms of overall cost.

Green products

Biodegradable plastics are a typical green product. Waste plastics present a very serious environmental problem. Biodegradable plastics can be degraded into water and carbon dioxide by micro-organisms in the environment. Therefore, they have less environmental impact and help reduce total wastes. Recently, private Japanese companies have been promoting efforts to commercialise biodegradable plastics (Table 4). MITI is encouraging and supporting these efforts by its R&D projects (Table 5) and some other policy measures.

Yet, the commercialisation of biodegradable plastics presents a number of problems. These include the definition of biodegradability and methods of testing it, labelling to distinguish biodegradable plastics from conventional plastics, the development of wider fields of application. The most important issue is cost reduction.

Table 4. Trends in development of biodegradable plastics by private companies in Japan

Product name	Company	Composition
Chemical synthesis:		
Polylactic acid	Shimadzu Corporation	Polylactic acid
	Mitsui Toatsu Chemicals	Polylactic acid
	Mitsubishi Resin	Polylactic acid
	Dainippon Ink & Chemicals	Polylactic acid
Praccel	Daicel Chemical Industries	Polycaprolactone
Bionolle	Showa Highpolymer	Aliphatic polymer
Natural polymer:		
Novon	Chisso Corporation	Starch and additive
Amipole	Nippon Corn Starch	Starch

Source: Author.

Table 5. Status of MITI projects on technical development of biodegradable plastics

	(expected total budget)
R&D of biodegradable plastics	1990-1997 FY (US$ 17 million)
Development of film manufacturing technology	1993-1996 FY (US$ 6.5 million)
Development of testing and assessment method	1989-1996 FY US$ 3.7 million)

Source: Author.

Government commitment is important to the promotion of the commercialisation of biodegradable plastics, in terms of development and standardisation of testing methods, creation of initial demand, and international harmonisation of regulatory requirements. At the same time, industries and consumers also have an important role to play. Industries need to verify the safety of the by-products of the degradation of biodegradable plastics, and consumers need to be educated about the merits of biodegradable products.

Green processes

An example of a green process is the use of biotechnology to produce hydrogen. The oil-refining process requires large amounts of hydrogen, which are usually produced from fossil fuel, thereby also producing carbon dioxide. However, if we were able, through biotechnology, to control photosynthesis by micro-organisms, hydrogen could be produced using sunlight as the energy source. This would make it possible to reduce the consumption of fossil fuel and carbon dioxide emissions in the oil-refining process. This technology would also be reproducible by culturing the micro-organisms.

In many cases, biotechnological processes could replace present chemical processes that require large amounts of energy and fossil fuel. In order for our economies to be more in harmony with the environment, conventional technologies will have to be transformed. Organisms can produce various materials essential to their survival, while adapting themselves to the environment. If we can reproduce such processes in industrial processes, significant amounts of energy and resources will be saved and waste will be reduced. We can expect biotechnological processes to be a key to making the global environment from pollution.

Conclusion

Technologies can be divided according to whether they are marketable or not marketable. Technologies applied to environmental problems, including bioremediation, are basically not marketable. Technology diffusion is a logical extension of research and development, but basic research in itself will not promote the commercialisation of non-marketable technology. Regulations that affect taxation or financing can also affect the promotion of technology and help make non-marketable technology more marketable. In order to promote the diffusion or commercialisation of technology, it is very important to link basic research efforts to the social system (Figure 1).

Figure 1. Links between basic research efforts and the social system

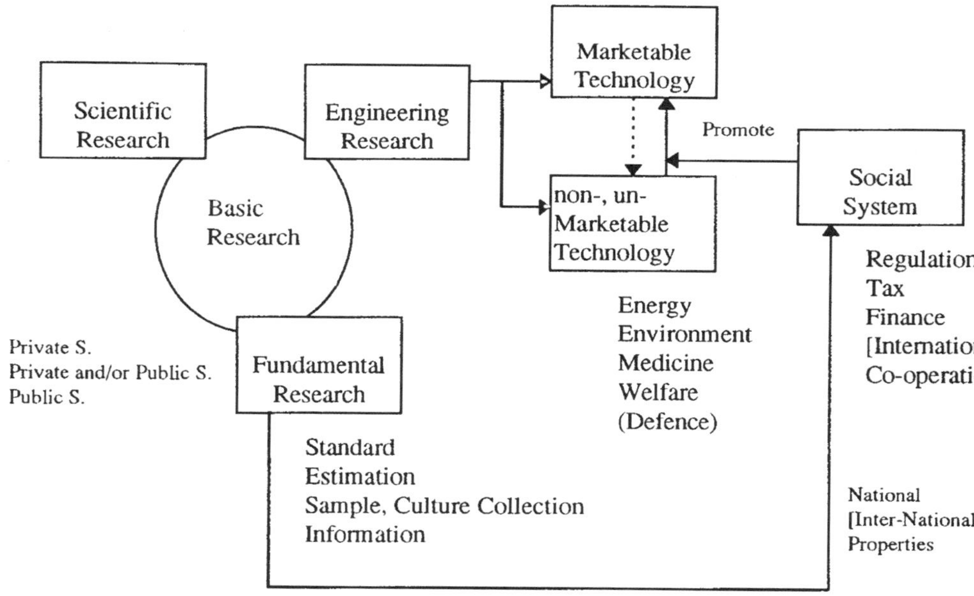

Source: Author.

The third session of the OECD Working Party on Biotechnology has just agreed to start a new activity to study and evaluate the potential of biotechnology for clean industrial products and processes. The activity is expected to lead to a second "Green Book" which, it is hoped, will encourage the co-ordination of R&D policies within the social system of Member countries. Japan positively supports this new activity.

Conclusion

R&D IN THE FIELD OF BIOREMEDIATION TECHNOLOGIES

by

Ronald Unterman
Envirogen Inc., Lawrenceville, United States

Introduction

As we increasingly confront our pollution problems, including the production, release, escape, and disposal of compounds previously thought to be recalcitrant, we have seen the development and implementation of new remediation and pollution control technologies. Among these technologies are traditional biotreatment systems as well as new, advanced biodegradation processes. The ultimate driving force for the development and use of biotreatment systems is their cost effectiveness as compared to chemical and physical treatment technologies and their efficacy, especially for dilute contaminants. The technical foundation for these systems are the biological processes that have naturally evolved over time at ambient temperatures and with surprisingly broad capabilities. It is the harnessing of these microbiological activities and their implementation in advanced engineered applications that forms the basis for the increasing use of biotreatment systems for biodegradation of even the most difficult hazardous chemicals.

Research and development in the field of bioremediation has generally encompassed the traditional areas of biocatalysis and systems development. However, in addition to these two fundamental areas of bioremediation, i) microbiology and ii) process engineering, two other regulatory and applications areas need to be further studied, these being iii) the establishment of reasonable cleanup targets and iv) the monitoring, demonstration and documentation of effectiveness.

Biocatalyst development (bioadsorption/bioconversion)

In terms of biocatalysis development, the major R&D thrust continues to be the isolation and characterisation of new and better catalytic activities for the degradation of xenobiotics, and in particular the halogenated organics. Additional target chemicals include other difficult to degrade chemicals such as: higher molecular weight polyaromatic hydrocarbons (PAHs), methyl-tertiary-butyl-ether (MTBE), and chlorofluorohydrocarbons (CFCs). In addition, other work is focusing on the bioadsorption/bioconversion of metals using both micro-organisms and plants. Less traditional microbial approaches are now using anaerobic cultures and fungi and have shown excellent promise in recent years. An unresolved issue that is still being studied is whether the use of pure cultures can be efficacious as compared to the use consortia, and ultimately whether and how genetically engineered micro-organisms can be used for bioremediation and pollution control.

Although indigenous organisms are generally the first choice for biocatalysts in biotreatment systems, there are increasing examples of the use of added micro-organisms.

Systems development

In terms of systems development, biological treatment can be applied either *in situ* or *ex situ* depending on the site and matrix characteristics and nature of the target chemicals. *In situ* treatments are ultimately the most cost effective approach for treatment of soils and ground water. Within this framework, one can choose between alternatives such as intrinsic remediation, biostimulation of indigenous organisms, or where necessary, bioaugmentation (the use of added organisms). For *ex situ* treatment one can chose among bioreactor systems for treatment of solids, liquids, and gases using both traditional and new bioreactor designs. For example, the biological treatment of industrial air emissions has expanded beyond the initial use of biofilters for odour control, to their wider use for the treatment of volatile organic compounds (VOCs). Furthermore, new reactor designs such as biotrickling filters, bioscrubbers, and membrane bioreactors are now being evaluated and commercialised for other process streams which are not readily treated with traditional packed bed biofilters. Many key process parameters are under development for these new systems and include such factors as: packing bed life, degradation of chlorinated organics with concomitant pH control, biomass control, introduction of gaseous nutrients such as ammonia and triethylphosphate, simultaneous biodegradation of mixtures of insoluble and soluble compounds, and generally the mass transfer limitations which can dramatically affect contact time and therefore capital costs.

One of the most difficult challenges has been the development and application of biotreatment systems for cometabolic targets. These systems pose distinct problems as compared to the biodegradation of organic chemicals that serve as the sole source of carbon and energy for microbes. In these systems there are issues of catalyst stability, negative selection of the biocatalysts and the need for exogenous reductant due to the energy requirements of these systems. The most studied example of cometabolic biotreatment has been for the biodegradation of TCE and the first commercial systems for this chemical are now just being applied.

Establishing reasonable regulatory targets

Another important problem requiring further research relates to the question of determining reasonable clean-up endpoints. Too often our approach to remediation has been to target an absolute contaminant concentration with little information as to the true risk posed by the starting and final concentrations. Thus, further work needs to be conducted to develop and refine effective risk assessment tools and models. This effort must eventually yield a technical foundation for risk-based decision-making and include a cost/benefit analysis for making these decisions. Factors such as bioavailability, humification and leachability must be included in the process. Is our clean-up goal the complete mineralisation of an organic chemical or is bioimmobilisation/humification a sufficient endpoint for such targets as the amino byproducts from TNT and the oxidised metabolites from PAHs? What are environmentally acceptable endpoints for our clean-ups? Can one biodegrade the lower molecular weight, more soluble chemicals at a site and leave unchanged the essentially insoluble and immobile, higher molecular weight targets? This last concept is being articulated as "biostabilisation".

Monitoring, demonstration and documentation

Another area that requires further development is the use and application of new monitoring tools. Extensive work is ongoing with approaches such as: reporter genes, biosensors, DNA probes, DNA fingerprinting, monoclonal antibodies, fatty acid analysis, subsurface respiration, and carbon isotope ratios. These new tools will allow us to better understand the mechanism and ultimate results of our bioremediation systems. Further work is also required to develop more efficient and unequivocal treatability protocols which can clearly distinguish between biodegradation results and abiotic losses. Continued development of analytical protocols will also help facilitate the application of bioremediation systems, and includes better leachability and toxicity protocols. Ultimately our goal must be to demonstrate and document the cost effectiveness and reliability of biological treatment systems. In doing so we will prove their efficacy to corporate clients, government agencies and the public, and biological treatment will become a widely accepted remediation and pollution control technology.

Applications and market

From an enzymatic perspective one can view the list of target organic pollutants as a spectrum from the readily biodegradable to the most recalcitrant, essentially non-biodegradable compounds. At the "easy" end are simple hydrocarbons and alcohols representing readily biodegradable chemicals, which can be treated using traditional wastewater and soil treatment technologies. At the other extreme are recalcitrant compounds such as dioxin, chlordane, toxaphene, Mirex, and other complex halogenated compounds, which are essentially nonbiodegradable with today's biotechnology. Within the limits of this spectrum one can then place chemical targets based upon their demonstrated rates of biodegradability. For example, trichloroethene (TCE), and higher molecular weight polychlorinated biphenyls (PCBs) would be at the more difficult end because of the requirement to degrade these chemicals cometabolically, whereas dichloromethane or monochlorobiphenyls would be towards the "easy" end because of the ability of microbes to utilise these compounds as sole carbon and energy sources. As a general rule, branched, complex structural compounds are more difficult to degrade than straight chain molecules (e.g. t-butyl alcohol vs. n-butyl alcohol). Likewise, an increasing degree of halogenation also generally decreases the rate of activity of a series of related compounds. However, within the constraints of this general view it must also be recognised that over the last 20 years it has become clear to environmental microbiologists that most compounds can ultimately be biodegraded by some microbial community. This is especially true in light of the more recent discoveries over the last 10-15 years of extensive biodegradative activity by anaerobic microbial communities, and in particular, the ability of both anaerobic and aerobic microbes to dechlorinate more highly chlorinated organic chemicals.

In addition to the inherent biodegradability of target chemicals, the limited solubility of many of these chemicals also has a significant impact on their observed rates and extent of biodegradation. In the case of many high molecular weight and hydrophobic compounds, their recalcitrance can be attributed to both low rates of enzymatic activity, as well as very limited bioavailability. Thus, a greater understanding of the limiting factors of bioavailability are being brought into play as advanced biological systems address the issue of increased mass transfer and increased bioavailability.

From an applications perspective one needs to view the biotreatment target not only in terms of the organic chemical being biodegraded, but also in terms of the matrix containing the target. Thus, we have come to view the application of biotreatment systems in terms of the four environmental "states of matter":

1) solids (soil, sediments, sludges);

2) liquids (groundwater, industrial wastewater);

3) gases (industrial air emissions, soil vent gas);

4) the subsurface (saturated and vadose zones).

These biotreatment systems encompass both our historical problems, as addressed by bioremediation (soils, sediments and aquifers), as well as our currently generated wastes requiring pollution control (air and water industrial effluents). The use of biological treatment systems for both of these market segments is growing dramatically as new technologies are being applied (Table 1; asterisks indicate degree of market emphasis).

Table 1. Applications for biotreatment systems

	Remediation	Pollution Control
1. Solids (soil, sediment, sludge)	***	*
a) PCBs		
b) PAH		
c) fuels		
d) TNT, nitro-pesticides		
2. Liquids (groundwater, wastewater)	***	***
a) PCE, TCE, TCA, etc.		
b) dichloromethane		
c) nitrobenzene, aniline		
d) alcohols		
3. Gases (industrial VOCs, soil vent gas)	*	***
a) odours (H_2S, CS_2)		
b) styrene, acrylonitrile		
c) BTEX, isopentane		
d) dichloromethane, TCE, etc.		
4. Subsurface/*In situ*	***	
a) intrinsic bioremediation		
b) bioventing/biosparging (biostimulation)		
c) bioaugmentation		
d) "biocurtain" / "funnel and gate"		

Source: Author.

The challenges and successes that we have seen over the last years for the implementation of advanced biodegradation systems encompass both microbiological, as well as engineering developments. The examples below illustrate the progress being made in these areas. Continued developments in both the microbiology and engineered applications will result in an increased use of biodegradation systems over the coming years, thereby bringing safe, natural, and cost-effective solutions for remediation as well as pollution control problems.

Cometabolic targets

Of particular interest and challenge for bioremediation are the halogenated organic compounds which have been widely used industrially because of their combination of chemical properties and thermal stability. Examples include: PCBs as a fire retardent fluid in heat exchangers, electrical equipment, and high temperature hydraulic fluids; TCE and other chlorinated solvents as nonflammable industrial solvents; and CFCs as nonreactive refrigerants and degreasing agents. However, the same chemical properties that confer upon these compounds their thermal and chemical stability, also confer upon them a general recalcitrance for biodegradation. Although these chemicals have been shown to be biochemically stable, research over the past 20 years has shown that PCBs, TCE, and HCFCs are biodegradable.

These halogenated chemicals pose a special challenge to microbial processes because in general they cannot be used as a sole source of carbon and energy for micro-organisms. Thus, organisms must be selected which grow on alternative substrates (co-substrates) in a process designated as cometabolism. This biochemical limitation presents a challenge to process development in that the co-substrate is often a competitive substrate for the enzymes which biodegrade the targeted chemicals. In addition, the application and enrichment of the biodegradative organisms is not easily maintained because there is no natural selection pressure for these organisms. In fact, in many cases there is a selective disadvantage for the micro-organisms which attack these chemicals due to the fact that the chemical or its byproducts are toxic to the organisms. In addition, the biodegradation process often consumes energy in the form of NADH, thereby conferring another selective disadvantage.

The initial approach for treating PCB contaminated soils and sediments has been to apply proper nutrients, and when necessary superior cultured micro-organisms, under aerobic conditions to oxidise the PCBs through ring cleavage. More recent work, however, has focused on the anaerobic microbial transformation of PCBs which results in the anaerobic dechlorination of higher chlorinated PCB congeners to lower chlorinated homologs. These lower chlorinated homologs can then be destroyed with an aerobic biotreatment step. Thus a two stage anaerobic/aerobic biotreatment is currently under development for the bioremediation of higher chlorinated PCB mixtures, as well as higher concentrations of lower chlorinated PCBs.

The extensive use of chlorinated solvents such as TCE, has caused widespread contamination of ground water throughout the United States. However, a broad group of micro-organisms have now been shown to be able to biodegrade TCE using any of several co-substrates including phenol, toluene, methane, ammonia, and ethyl benzene. Because the microbial cultures that degrade these chemicals are only active under cometabolic conditions, extensive work has been conducted on developing the proper process engineering controls to optimise the use of the competitive co-substrates. Additional work has also focused on the genetic engineering of these strains to isolate the structural genes and control them under non-competitive process conditions. To date the use of these cultures has been demonstrated in above-ground bioreactors under long-term (7 months) operation for the destruction of TCE in ground water with influent concentrations of 5-10 ppm and effluent concentrations below 50 ppb. Although field systems have utilised indigenous or superior naturally occurring strains under optimised process conditions, future applications will include the use of genetically engineered micro-organisms in bioreactors and eventually in ground water and surface soil applications.

Another class of difficult chemicals are the chlorofluorocarbons and their newer derivatives the hydrochlorofluorocarbons (HCFCs). However, it has now been shown that certain classes of methanotrophic organisms can biodegrade selected HCFCs. This work can now allow manufacturers to make product choices based in part on the ultimate biological fate of the produced chemicals; thus, the HCFC market could move toward biodegradable alternatives.

In situ biotreatment technologies

In many cases biological approaches for remediating polluted environments provide a significant advantage over alternative technologies in both cost savings and reduction of environmental and human exposure. *In situ* treatment systems provide an added advantage of destroying the contaminant in place, thereby further reducing human exposure. Cost is also reduced because contaminated media do not have to be excavated, relocated, or otherwise disposed of. Furthermore, in the case of deep vadose zone or aquifer contamination, excavation is often impossible.

The application of *in situ* bioremedation techniques for subsurface TCE (and other chlorinated solvents) contamination has been at the forefront. This work has focused on the application of nutrients and co-substrates in the subsurface (phenol or methane), as well as the injection of micro-organisms to facilitate the degradation of sorbed TCE. *In situ* bioremediation offers the distinct advantage that a direct biological attack of the material sorbed to aquifer solids will speed the clean-up time frame as compared to the slow pump and treat approaches that have been tried to date. Some pump and treat remediations are expected to take as much as 50-200 years because of the rate limiting desorption of organic materials under the pump and treat conditions. In contrast, the direct biotreatment of the chemical contaminants in the subsurface allows for the biodegradation of source material which would otherwise leach slowly over the long time frame of the pump and treat operation.

Intrinsic bioremediation

Research in this field has focused on the development and implementation of direct, non-invasive subsurface bioremediation technologies. The simplest approach is the identification of intrinsic bioremediation whereby it can be demonstrated that an indigenous microbial population exists under suitable environmental conditions and that degradation has already occurred and is continuing. This is always the first choice of remediation because it requires no intervention besides monitoring of the natural progress of biodegradation. The potential for this technology to reduce the economic burden of environmental restoration is so substantial that the United States Air Force has drafted a guidance document specifically for intrinsic remediation.

Biostimulation

Another approach which has been used for a number of sites and contaminants, is the stimulation of the degradative activity of indigenous populations to remediate the target chemicals when this alone is sufficient for remediation to occur. Biostimulation, as this process is called, is applied in cases where a degradative population exists within the contaminated zone, but where nutrient or other conditions are insufficient for microbial activity. Often oxygen is the limiting substrate and introducing oxygen either as hydrogen peroxide, pure oxygen, or air, can in many cases induce the natural population to degrade the target chemicals. The application of bioventing and biosparging approaches for introducing atmospheric oxygen to the subsurface is often a very cost effective approach for remediation utilising the indigenous microbial population. Biostimulation can also be accelerated or augmented by physical processes such as heating of the subsurface where lower ambient temperatures are limiting, to help stimulate bacterial activity, or hydro- or pneumatic fracturing to increase permeability and zones of influence.

Appropriate additions of nutrients or cometabolites to the subsurface can stimulate degradation by indigenous micro-organisms for an increasing number of contaminants. Biostimulation has been used extensively as an *in situ* treatment for hydrocarbon contamination, and more recently, biostimulation protocols have been developed for more recalcitrant contaminants. Certain chlorinated organic compounds that are cometabolically degraded by a number of aerobic strains can be degraded *in situ* after the addition of methane. Another recently demonstrated *in situ* treatment technology for VOCs involves stimulating the degradative activity of anaerobic bacterial populations. In the appropriate environment, the addition of specific nutrients and the maintenance of anaerobic conditions can result in the biotransformation of even the most highly chlorinated VOCs to non-toxic products. One of the more recalcitrant halocarbon compounds, tetrachloroethylene (PCE), has been shown to be degraded to ethene under anaerobic conditions after the addition of sodium benzoate and magnesium sulfate to recirculating ground water in an aquifer. Methanogenic bacterial activity stimulated by the addition of a variety of electron donors to aquifer sediments has also been shown to dechlorinate PCE to ethane in laboratory studies.

Bioaugmentation

Despite the fact that naturally-occurring organisms can degrade most contaminants, biostimulation will not suffice in cases where i) the contaminant is a cometabolic target and the necessary nutrients/inducers are not present and cannot be added to the environment, or ii) competent degradative organisms are not present among the indigenous population. In these cases, bioaugmentation (adding bacteria to the subsurface) may be the only effective strategy.

In addition to adding bacteria directly to the vadose zone or aquifer, a number of approaches have been devised for using immobilised bacteria *in situ.* Bacteria have been immobilised on solid substrates and then introduced into the subsurface as a "funnel and gate" or "biocurtain" approach for intercepting the migration of a plume and degradation of the target chemicals. In a similar approach, degradative bacteria have been attached to a cylindrical reactor and placed in a recirculating well to treat ground water *in situ.* The idea of creating subsurface biological treatment zones has also been used in the development of a technology known as the "lasagna" process. A combination of hydrofracturing and electro-osmosis is being used to create parallel aerobic treatment zones within very tight clay formations. The electro-osmosis process is then used to move contaminants from the clays into these sand filled treatment zones which can be amended with nutrients and oxygen.

Constitutive variants of degradative micro-organisms

In situ bioremediation of TCE (and related VOCs) has been limited by the inability of naturally-occurring TCE-degrading micro-organisms to degrade TCE in the absence of inducing co-substrates, some of which are themselves toxic chemicals. All known naturally occurring

TCE-degrading micro-organisms require the presence of inducing chemicals to stimulate TCE biodegradation. However, bacteria capable of constitutive expression of TCE-degrading genes/enzymes have been developed and include constitutive toluene *ortho*-monooxygenase (TOM) variants of *P. cepacia* G4 and toluene monooxygenase (TMO) variants of *P. mendocina* KR-1. Because the constitutive variants continue to express TOM in the absence of inducing substrates, the cells continue to degrade TCE as long as the cells have sufficient energy reserves to catalyze the reaction.

Adhesion-deficient strains

One of the major difficulties when using bioaugmentation for remediation of subsurface contamination is delivering degradative micro-organisms to the zone of contamination. Bacteria often adhere to solid surfaces and plug porous materials rather than pass through them. Thus, *in situ* biotreatment technologies are currently limited by poor migration of bacteria through aquifer solids. In an *in situ* field study involving *P. cepacia* G4 injection into a TCE plume, the data strongly suggested that movement of the micro-organisms was severely retarded by the aquifer material. Although this field pilot demonstrated that *P. cepacia* could sustain TCE-degrading activity under *in situ* conditions, clearly, more aquifer material could be treated if the organisms or the aquifer conditions were altered to reduce adsorption and retardation of bacterial movement.

Thus, adhesion deficient TCE degrading bacterial variants of G4 and KR1 have now been developed and are being field tested. These metabolically proficient, adhesion-deficient strains are more suitable for *in situ* applications than the wild-type organisms because they are less likely to plug the injection well during application, and can travel farther through aquifer sediments, thereby increasing the effective zone of remediation. Because some of these variants have also been selected to constitutively degrade TCE, they can continue to degrade TCE as they penetrate contaminated aquifers.

Enhancing energy reserves in degradative organisms

In addition to problems of gene/enzyme induction and adhesion, microbial biocatalysts for co-metabolic targets usually require the addition of nutrients to stimulate and prolong their activity. However, adding a co-substrate to an aquifer can stimulate the growth of other non-target-degrading organisms and create problems with biofouling and oxygen depletion. Furthermore, the use of co-substrates adds to the cost of the treatment process and may be operationally inefficient if the microbes are transported or migrate away from the co-substrate.

An alternative approach for enhancing and maintaining biological activity *in situ* is to utilise biocatalysts that are enriched in energy reserves. The production of energy storage polymers, most commonly poly-ß-hydroxybutyric acid (PHB), by bacteria is a long-studied phenomenon. PHB is produced by many bacteria, usually under conditions of a nutrient limitation in the presence of excess carbon, and may account for up to 80 per cent of the bacterial cell dry weight. Utilisation of PHB as a source of reducing substrate for increasing the TCE transformation rate and capacity of a methanotrophic mixed culture has been demonstrated.

The primary advantage of using energy-enriched organisms for *in situ* remediation versus feeding indigenous organisms is that it forces more efficient utilisation of added oxygen and inorganic nutrients. The degradative organisms carry an internal food reserve with them into the aquifer. The food reserve is not available to less efficient indigenous organisms, and thus it does not stimulate increased oxygen demand or biofouling associated with feeding of indigenous microbes. The only additional oxygen that may be needed is that required to support oxidation of the target contaminants and low rates of cellular respiration.

Table 2. Advantages of genetically engineered organisms for treating hazardous waste

Limitation	Genetic Engineering Solution	Benefit
Incomplete Degradation	1. Uncoupling metabolism from degradation 2. De-regulation of genetic controls	1. Support activity with inexpensive, non-toxic substrates. 2. Eliminate need for toxic inducing substrates 3. Obtain treatment standard levels
Low Rates of Degradation	1. Selection of high performance host organisms 2. Remove degradive "bottlenecks"	1. Use smaller, less expensive bioreactors 2. Decreased fermentation/treatment costs
Recalcitrant Target Compounds	1. Add substitution-specific functions (e.g. dehalogenation activity) 2. Alter enzyme specificity	1. Increased range of treatable compounds 2. Increased substrate range of single organisms
Formation of Toxic Intermediates	1. Re-route metabolites 2. Add complementary activities/pathways	1. Extended treatment life 2. Extended range of treatable compounds
Chemical Mixtures (e.g. PCBs, mixed organic wastes)	1. Combine metabolic activities 2. Broaden substrate specificity	1. Decreased fermentation costs (single organism) 2. Selection of environmentally robust host

Source: Author.

Genetic engineering for hazardous waste treatment

Genetic engineering represents the future of biological treatment technologies by providing a mechanism for improving the degradative efficiency of micro-organisms selected for treating specific contaminants. Although genetically engineered micro-organisms (GEMs) are not yet used to treat hazardous wastes, recent advances in this area are likely to lead to future application of GEMs to address some of our more difficult hazardous waste problems. A brief outline of some of the potential benefits of genetic engineering in treating hazardous wastes is provided in Table 2. GEMs will probably have their greatest use in treating the most difficult types of wastes such as complex mixtures, extremely recalcitrant chemicals like chloro- and nitroaromatics, or substrates that are only attacked by co-metabolic mechanisms.

Bioavailability

The applicability of bioremediation to insoluble target chemicals is determined in part by the availability of the organic molecules. There are three major physical/chemical characteristics that govern the rate and extent of biodegradation of hydrophobic organic compounds (HOCs) . These include i) affinity of the compound to a sorbent matrix, ii) the solubility of the chemical, and iii) the distribution of the compound between air and water. The most critical of these three factors in developing a biodegradation process for HOCs, such as PCBs, is the rate and extent of sorption/desorption of the HOC to an organic matrix such as soil. It is this sorption of hydrophobic organic compounds to soil organic matter which limits the diffusion of HOCs. In fact, the concentration of organic matter, which includes both natural organic matter (e.g. humic acids) and residual synthetic co-contaminants, greatly influences the degree of sorption and thus desorption from the soil. The bioavailability can be enhanced by any of several physical, chemical, or biological means (e.g. high-shear mixing, and chemical oxidation or biodegradation of cocontaminating chemicals). An alternative approach for increasing HOC bioavailability is to include a surfactant, solvent, or emulsifier to help solubilise the HOC. Several commercially available surfactants have been tested, however, the commercial usefulness of this approach is still unknown. In many cases surfactants have no beneficial effect.

From another viewpoint one can consider the limitations of HOC bioavailability as a potentially positive factor in determining HOC remediation strategies. For example, one can develop a conceptual approach for evaluating the limited range and extent of PCB bioavailability and degradation in terms of a risk-based clean up standard. Experiments have shown that at 50 per cent total PCB biodegradation extensive degradation of the lower chlorinated homologs results, and this in turn results in a substantial decrease in the leachable fraction of PCBs. Additionally, studies have been conducted to assess the biodegradation of water soluble PCBs in the low parts per billion range (µg/L), and these studies have demonstrated that virtually all of the water soluble PCBs can be biodegraded. When these observations are combined one can propose an approach for biodegrading the lower chlorinated PCB homologs to decrease subsequent migration of the residual PCB fraction. In addition, a model can be supported from biodegradation studies with PAHs for a significant rate of degradation for any subsequently solubilised PCBs. This clean-up approach is called "biostabilisation" and is being considered as a viable remediation alternative.

Conclusions

The development and implementation of biotreatment systems for hazardous wastes is of critical importance to both our industrialised and developing societies as we face the high cost of remediating our historical environmental problems, as well as close the effluent loop on our industrial processes. Improvements in both microbial biocatalysts and bioreactor systems are leading the way toward the next generation of biotreatment systems. These technologies are being applied to an ever increasing extent to address our soil, sediment, water, and air pollution problems and will ultimately displace other physical and chemical processes, because of their efficacy, safety and cost-effectiveness.

Questions, comments, and answers

Q: How close is your company to being able to establish ground rules from your own knowledge or are you still reliant on case by case studies, using these aerobic, anaerobic and co-metabolic studies?

A: In terms of process control, some information can be generalised, but on the whole it is not. We still rely on case by case studies.

Q: You touched on the concept of bioavailability being advantageous but different contaminants have different exposure pathways. Can you comment on this?

A: When we look at exposure route, some can be controlled by, for example, a 10 inch clay cap leaving only leaching as a problem. You have to look at exposure routes on a site-specific basis. In the example I gave, we were looking at lagoons but a river would be a quite different problem

Q: What in your opinion are the major constraints to wider application of bioremediation?

A: In soils remediation there are often process limitations rather than microbial limitations. For example, mass transfer is a problem rather than the catalytic activity of the microbes.

Q: May there also be legislative problems?

A: New legislation is being written with flexibility, using risk based goals rather than absolute mineralisation-based goals. I consider the experience base to be the major problem, rather than legislative problems to date. For example, client companies do not wish to be the first to implement a technology; they are much happier to be the second!

INFORMATION TRANSFER AND THE PUBLIC CONTEXT

by

Bernard Dixon
Bio/Technology, Ruislip, United Kingdom

An industrialist, charged with the development of a particular technology, can be viewed as having responsibilities not only to his staff and shareholders, but also to his fellow citizens and to the wider world. What are the three main criteria for that technology – in addition, of course, to the self-evident need to be useful and profitable?

I would propose the following. The technology should be ecologically acceptable; it should neither pollute the biosphere nor consume finite resources, and, ideally, it should actually improve the quality of the natural environment. Second, it should not be radically new, because it would then create uncertainties and unpredictability. Third, it should, for these two reasons, be a technology that can be confidently commended to the public, to politicians, and to regulators.

Bioremediation is such a technology. Indeed, a few moments of reflection and comparison with, say, the nuclear or chemical industry will indicate that bioremediation is as well-positioned for widespread support as any technology is likely to be.

Being "green" in both purpose and methodology, it meets the first criterion, as its only impact on the environment is through its cleansing role. Bioremediation is appropriate, too, for application in both developing and developed countries.

Equally persuasive is the argument that while bioremediation is poised for more vigorous development, it is not intrinsically novel. It therefore meets the second criterion. The transforming and scavenging roles to be played by micro-organisms when used to degrade pollutants in the soil and elsewhere are comparable with those of bacteria and other organisms involved in the world-wide process of sewage disposal. Indeed, it is difficult to draw any hard and fast line between dedicated, scientifically based bioremediation and, for example, the continual recycling of elements in the biosphere, or the sort of natural cleansing that occurred along the Kuwaiti and Saudi Arabian coast following the deliberate leaking of oil during the occupation of Kuwait in 1990-1991 (Sorkhoh *et al.*, 1992).

Here, I would like to recall one aspect of the formation, 50 years ago this year, of the Society for General Microbiology in the United Kingdom. The person who gave the society its name was the Cambridge scientist Marjorie Stephenson whose book, *Bacterial Metabolism*, published in 1930, virtually founded the study of microbial biochemistry. Just before her book appeared, she was studying the anaerobic organisms thriving in sludge containing effluent from a sugar beet factory in the river Ouse near her home. As a result, she played a major role in discovering the hydrogenases produced by bacteria we now recognise as extremely valuable scavengers. I recall this to point out that bioremediation is neither unnatural nor inherently novel. Thus, its practitioners are at least partially immune to the criticism, which can be levelled against any genuinely new technology, that its consequences and risks are unpredictable.

Fulfilment of the first two criteria by no means ensures public acceptance – the third criterion. Nor should it. Nevertheless, there is an enormous difference between the extension and conscious direction of organic processes that fashion the biosphere day by day, year by year, century by century, and the introduction of a technology founded upon techniques and approaches which mark a sharp departure from what has gone before. And public attitudes towards this and any other technology are, in turn, influenced, sometimes very strongly, by information disseminated through the media and other channels.

At this point I must acknowledge that "information transfer and dissemination", as addressed during this OECD workshop, is by no means limited to information provided to the general public – the sort of information we all digest in forming opinions on matters that come into public discussion. We shall consider other types of information and other routes through which it flows. It can be argued, however, that public information or misinformation is of crucial importance in the cat's cradle of links between technology, regulation, political action, and public endorsement or rejection of new and potential developments.

Let us consider a few examples, from other fields, to illustrate the power of public opinion on matters of science and technology. On the positive side, it was public agitation, not initially any action by experts (indeed, quite the reverse), that led to the recognition of Lyme disease in the United States in 1975. Concerned women in the Connecticut village of Old Lyme brought to light an "epidemic" of arthritis and thus played a crucial role in a chain of events that led eventually to the characterisation of the spirochaete responsible for the infection and the tick that carried it from deer to humans. Likewise, during the 1960s, public demand lay behind US government measures to increase funding for cancer and sickle cell anaemia dramatically.

More recently, public anxiety has helped to retard the development of food irradiation. This was graphically illustrated at the 2nd World Congress of Foodborne Infections and Intoxications in Berlin in 1986. An entire session of that meeting had been scheduled to examine the considerable potential for the use of ionising radiation as a means of ensuring that foods are free from dangerous micro-organisms. Given the world-wide increase in food poisoning attributable to *Salmonella* and other pathogens, the new technology was sorely needed (and still is).

The meeting took place, however, just a few weeks after the Chernobyl accident. As a result, virtually all of the speakers agreed that public disquiet about radiation was so strong that politicians and regulatory agencies would find it impossible to authorise food irradiation in the foreseeable future. One participant concluded that the technology was "a dead duck for all time".

Perhaps the clearest way to illustrate the significance of information – or its lack – in shaping public opinion concerns the use of animals in scientific laboratories. My own evidence is drawn from the United Kingdom, but I believe that the same lesson applies elsewhere as well. On the one hand, the scientific community did very little, for many decades, to explain, honestly, accurately, and sensitively, the need for humane animal experimentation. On the other, lobby groups opposed to animal experiments provided more and more information (and in some cases misinformation) that has helped to shape public opinion, especially among the young. Only late in the day did the scientific community realise that its failure to engage in public discussion had contributed to increasing opposition to the use of animals in scientific laboratories.

I have strayed quite some way from the subject of bioremediation, but my purpose has been to emphasise the significance of public information and public attitudes as the background against which all developments in science and technology take place. More than ever before, these developments can proceed only to the extent that scientists and their regulators enjoy public trust. The politicians too, who set the agendas, guidelines and laws under which regulators operate, are heavily influenced by public opinion and debate, and in particular by the way it is reflected in the media.

To cite just one example, the last time there was a national newspaper strike in the United Kingdom, when virtually no newspapers appeared for several weeks, a friend of mine in the House of Commons pointed out that the number of questions asked by members of Parliament had contracted to about a quarter of the normal figure. Certainly, it is not only in the United Kingdom that politicians rely on the media to indicate topics of public interest and concern, and the link is both direct, as a result of parliamentarians reading the press, and indirect, as when topics aired in the media reach them through their constituents.

What is likely to be the public mood concerning the emerging technology of bioremediation? Some reports suggest a very gloomy answer, certainly as regards the use of genetically modified organisms (GMOs) for this purpose. "Public opinion in Germany has largely turned against science in general and against recombinant DNA" was a comment made by a participant in a recent survey of European Molecular Biology Organisation (EMBO) scientists (Rabino, 1992). Likewise, a report in Britain by the Advisory Committee on Science and Technology concluded that public concerns could hamper progress in biotechnology (ACOST, 1990). I have heard many similar comments over recent years by conference speakers from other European countries.

Yet, if one looks at the best available evidence, the picture is rather different. A 1991 Eurobarometer study, conducted in 12 countries (Marlier, 1992), showed that one out of two European citizens believes that biotechnology and genetic engineering will improve our way of life over the next 20 years. Only one in ten believes the opposite. The greatest optimism is shown by men, young people, people with a higher educational level, and those who are "comfortably off".

These findings and those of other surveys certainly do not support the idea that there is strong public opposition to biotechnology/genetic engineering. They do indicate that it is simplistic and misleading to measure the temperature of public opinion along a single scale from enthusiasm to hostility. One of the single most suggestive results in the Eurobarometer survey was that, of all the countries studied, the Danish public showed both the strongest level of support for biotechnology *and* the highest perception of risk. In other words, it is possible to combine high regard for the achievements of science, in areas such as medicine, agricultural and environmental improvement, with concern about potentially adverse consequences of new technology.

Further evidence comes from a survey conducted recently in Britain (Evans and Durant, 1995). The investigators measured general attitudes in the community by asking nine questions, ranging from whether scientists can be trusted to whether science is changing our way of life too quickly. The results showed a generally positive view of science: 70 per cent of respondents believed that science and technology are making our lives healthier and more comfortable, and 80 per cent backed government support for "research which advances the frontiers of knowledge". However, the pattern was not uniform. Some participants agreed with both positive and negative statements.

To determine whether the subjects' overall view of science was reflected in their opinions on individual areas of research, Evans and Durant used new sets of questions. The specific topics included creating new forms of animal life and finding a cure for cancer. Here, the results revealed at best a moderate correlation, which was strongest for useful and basic science and weakest for morally contentious research. In other words, participants' responses to questions that explored their general attitude towards science and technology did not make it possible to predict accurately how they felt about particular types of research.

A third analysis compared the respondents' attitudes to science with their factual knowledge, as assessed from their response to various statements from the natural and medical sciences, such as light travels faster than sound and sunlight causes skin cancer. Here, Evans and Durant discovered that despite statistically significant correlations between attitudes and knowledge, the strength of the link varied considerably.

Factual knowledge correlated moderately well both with the participants' general attitudes and their attitudes towards useful and basic research. On the other hand, knowledge was almost wholly unrelated to attitudes to non-useful research, and there was a strong negative association between knowledge and support for research that may be seen as morally contentious.

Commenting on this last finding, Evans and Durant conclude that "it would be unwise for scientists and science policy makers to presume that a better informed public is automatically a public that is more supportive of any and all forms of scientific research. On occasion, the opinions of a scientifically well-informed public may serve as a check on public and political support for certain areas of research."

This is an important warning. We should, however, consider the results of this survey alongside those of others that have invited comments on specific applications of biotechnology. It is here that we find more concrete evidence that helps to predict the likely public view of the use of micro-organisms in bioremediation.

Consider, for example, a survey on attitudes to biotechnology carried out a few years ago in the United Kingdom (Martin and Tait, 1992). The researchers evaluated attitudes among a population sample that included people chosen at random from electoral registers who lived in two parts of the country where releases of genetically modified organisms had occurred; members of Friends of the Earth and other "green" groups and non-technical staff of companies involved in biotechnology.

Even with such a sample, which might have been expected to bias the results away from a positive endorsement of the technology, support for the use of genetic modification for environmental decontamination was as strong as for its use in medicine. Some 65 per cent of respondents were "comfortable" with the idea of deploying recombinant organisms to clean up oil slicks and to detoxify industrial waste. This compared with 59 per cent in the case of medical research and 57 per cent in the case of "making medicines". On the other hand, the levels of support for using viruses to attack crop pests, cloning prize cattle, and making hybrid animals were 23 per cent, 7.2 per cent and 4.5 per cent, respectively.

The percentages for respondents who were "uncomfortable" with these various applications of biotechnology were almost exact mirror images of the positive figures. Only 13 per cent were uneasy about the use of GMOs to clean up oil slicks or to detoxify industrial waste, whereas 72 per cent were uncomfortable with their use to clone prize cattle, 82 per cent were uncomfortable with the idea of making hybrid animals, and 49 per cent were unhappy about using viruses to attack crop pests.

Martin and Tait drew attention in their report to the unexpectedly wide range of opinion reflected in these overall percentages when they broke down their sample into its seven constituent sub-groups. Even areas such as improving crop yields and making medicines, which are not normally considered controversial, showed a wide range of response. However, the range of people who were comfortable with the idea of using genetic modification for the two types of environmental decontamination were among the narrowest of all.

It would be as dangerous to generalise from these figures, taken from one survey carried out in just one country, as it would be to overlook warnings about the fragility of public opinion and the fact that more information, more widely understood, does not necessarily lead to greater public endorsement of a particular technology. Nevertheless, it seems that the hard evidence does not justify the worst fears of many commentators as regards alleged public hostility towards biotechnology. Of the various applications of biotechnology, bioremediation appears to be one of those most likely to enjoy public confidence and support.

A different type of argument to bolster this conclusion comes from the part of Europe that has seen the most vigorous antagonism towards gene technology, the former West Germany. Opposition has abated over the past five to ten years, and in a recent book entitled *Resistance to New Technology*, Joachim Radkau of the University of Bielefeld attributes this in part to changes associated with reunification (Radkau, 1995). On the one hand, the economic and social problems arising from reunification diverted attention away from the alleged dangers of gene technology. On the other, allegations concerning these risks and hazards were overshadowed by information on the scale of industrial pollution in East Germany, and indeed other eastern states. "The risks of genetic engineering seemed relatively harmless beside the destroyed landscape resulting from East German chemistry and brown coal," Radkau wrote.

It would, of course, be absurd to argue that bioremediation is invariably the optimal choice for cleaning up polluted environments or that it is entirely risk-free. Even its own proponents are not always in agreement as to the most appropriate form of bioremediation to adopt in particular circumstances, whether, for example, to use *ex situ* decontamination of material removed from a site or *in situ* cleansing and whether, in turn, to base *in situ* remediation on stimulation of existing microbial flora (by the introduction of nutrients and oxygen, for example), or on the introduction of dedicated, perhaps genetically modified scavengers.

These issues clearly have implications for regulation and public acceptability, and here again it is clear that no consensus has been reached on why bioremediation efforts are not proceeding more rapidly and more more confidently. Is the state of the technology to blame? Or are regulators needlessly nervous about public acceptability? Last year, an exchange of letters in *Nature* illustrated very clearly one area of disagreement.

"Innovative products of the new biotechnology for bioremediation remain on the drawing board, with researchers and companies intimidated by regulatory barriers and disincentives," wrote Henry Miller of the Hoover Institution Institute for International Studies. Alleging discrimination by the Environmental Protection Agency against "micro-organisms created with high-precision rDNA technology", Miller concluded, "Until governments demonstrate rationality and sensitivity to scientific principles in their regulatory policies towards the new biotechnology, we will be slopping bacterial growth media on the beach instead of putting high technology to work" (Miller, 1994).

In response, three authors from Shell Research (Lethbridge *et al.*, 1994) argued that "stringent regulation is not the issue here". The real problem, they insisted, was the technology itself and the misapprehension that anything as complex as crude oil could be degraded by so-called superbugs. "Even if it were technically possible to incorporate all the necessary genetic information into recombinant micro-organisms, the burden of maintaining all these genes is likely to be so great as to make recombinant strains non-competitive in the natural environment."

I conclude with a paradox. Although opinions vary as to why bioremediation is not being more widely and more vigorously adopted, there is little doubt that nervousness concerning public acceptability is at least one significant element in the current, somewhat hesitant mood. Yet – and despite the fact that more information, better understood, does not guarantee public acceptance – this is a technology that is as well positioned for public endorsement as any technology is likely to be.

To quote last year's *Biotechnology for a Clean Environment* (OECD, 1994), "environmental biotechnology must be seen in the context of a future in which industrial technologies will have to be increasingly in harmony with the global material cycles of the biosphere". The paradox is that bioremediation, which in these terms contrasts starkly with so many other technologies, is not yet being portrayed and commended as vigorously as it could and should be.

Questions, comments, and answers

Q: You put GMOs at the focus of public perception. Perhaps this concentration distorts the approach to the natural technology. Should we not get bioremediation accepted first as, say, waste water treatment ,and fight the GMO battle later?

A: I am not sure that this is the best tactical approach. Separating the two could be more of a problem. The view of biotechnology as new and dangerous has abated and it may be better say that GMOs are not so different and therefore need not be addressed separately.

C: Since there are no uses of GMOs in remediation we should speak of incremental changes, introducing a new catalyst as and when necessary.

Q: I am not comforted by the statistics. Often a small, strongly motivated minority can have a great influence.

A: I agree that that the system is unstable and can be disturbed by one mischievous article or film. We have to make use of the goodwill that is out there.

Q: How can we enhance the supportiveness of the general public?

A: By being more proactive, for example by organising workshops for journalists and bringing the technology and relevant information into the public arena.

Q: On whom should the burden fall for generating this acceptability? Industry? Scientists? Government?

A: A number of initiatives are underway to promote public understanding of science involving all these participants. Companies are generally timorous but can do much. For example, Novo Nordisk has been excellent in this regard. Regulatory bodies should also be involved in order to make their processes more transparent. Public acceptance includes a whole range of activities and different bodies should work more closely together in attaining this.

Q: How far is education linked to public perception, e.g. primary schools?

A: Education is extremely important. We are influenced most at an early age and attitudes are formed in junior school. However, the media often do not want to be 'educators' but see themselves as 'entertainers' through the dissemination of 'exciting news stories'. There is a vacuum here currently being filled by lobbying groups.

REFERENCES

ADVISORY COUNCIL ON SCIENCE AND TECHNOLOGY (1990), *Developments in Biotechnology*, HMSO, London.

EVANS, G. and J. DURANT (1995), "The relationship between knowledge and attitudes in the public understanding of science in Britain", *Public Understanding of Science* 4, pp. 57-74.

LETHBRIDGE, G., H.J.J. VITS and R.J. WATKINSON (1994), "Exxon Valdez and bioremediation", *Nature* 371, p. 97.

MARLIER, E. (1992), "Eurobarometer 35.1: Opinions of Europeans on biotechnology in 1991" in *Biotechnology in Public*, J. Durant (ed.), pp. 52-108, Science Museum, London, for the European Federation of Biotechnology.

MARTIN, S. and J. TAIT (1992), "Attitudes of selected public groups in the UK to biotechnology" in *Biotechnology in Public*, J. Durant (ed.), pp. 28-41, Science Museum, London, for the European Federation of Biotechnology.

MILLER, H.J. (1994), "Regulatory disincentives", *Nature* 370, p. 244.

OECD (1994), *Biotechnology for a Clean Environment – Prevention, Detection, Remediation*, OECD, Paris.

RABINO, I. (1992), "A study of attitudes and concerns of genetic engineering scientists in Western Europe", *Biotech Forum Europe* 10/92, pp. 636-640.

RADKAU, J. (1995), "Learning from Chernobyl for the fight against genetics? Stages and stimuli of German protest movements: A comparative synopsis" in *Resistance to New Technology: Nuclear Power, Information Technology and Biotechnology*, M. Bauer (ed.), Cambridge University Press, Cambridge.

SORKHOH, N., R. AL-HASAN and S. RADWAN (1992), "Self-cleaning of the Gulf", *Nature* 359, p. 109.

Parallel Sessions/A1 and A2

A1.1. BUSINESS OPPORTUNITIES AND BOTTLENECKS: AIR/OFF-GAS

SIMULTANEOUS BIOLOGICAL REMOVAL OF SO_2 AND NO_x FROM FLUE GAS

by

C.J.N. Buisman
PAQUES Environmental Technology, Balk, the Netherlands

Introduction

In 1993, PAQUES B.V. and Hoogovens Technical Services started a co-operation which resulted in a joint venture called "Biostar Development".

Hoogovens Technical Services is a market leader in the Netherlands, France, the Czech Republic and Poland for SO_2 removal from flue gases of coal-fired power stations. It also supplies engineering, contracting and consulting services on a world-wide basis. Hoogovens Technical Services is headquartered in IJmuiden, the Netherlands, and has subsidiary offices in the United States, Canada, Brazil and India.

PAQUES B.V. is specialised in biotechnological environmental technologies and is a world-wide market leader in anaerobic waste water treatment. PAQUES B.V. is headquartered in Balk, the Netherlands, and has subsidiary offices in the United States and Hong Kong.

In the period 1989-1992, a fortunate synergy occurred in the field of biotechnology for gas cleaning. Biotechnological research (Agricultural University Wageningen, the Netherlands), directed to find new micro-organisms and bio-processes to make useful products, revealed new ways and biotechnological mechanisms to extract sulphur components from waste water, while producing elemental sulphur (Buisman *et al.*, 1990).

Biotechnological development work (PAQUES B.V.) was carried out and the first commercial installation to produce sulphur from polluted groundwater (300 m^3/hr) was started up in 1992 (Scheeren *et al.*, 1991).

The importance of this recent research and development in the area of waste water treatment was recognised, as well as its potential applications, in various other industrial processes. Through intensive co-operation between Hoogovens Technical Services and PAQUES, the concept for a new biological flue gas desulphurization process (BIO-FGD) producing sulphur as a by-product was invented. It consists of the combination of a sodium scrubber with two biological reactors resulting in an attractive new concept for a gas-cleaning process.

Attractive aspects are:

- Capital and operating cost are appreciably lower compared to the limestone gypsum process.
- Less chemicals and energy are used compared to the conventional scrubbing processes.

- Applications for this process are not only in the "classical" flue gas desulphurization of power stations, but also in the areas of SO_2 removal from smaller quantities of industrial off-gases and gas-cleaning in incinerator installations, etc.

- Sulphur is produced instead of sulphate salts and can be reused.

Status of the BIO-FGD technology

Laboratory research was carried out and a 2 MW pilot plant was successfully operated in 1993, treating 6 000 Nm^3/hr of flue gas from a coal-fired power station in the Netherlands. On this basis a commercial BIO-FGD system was ordered in 1993. This FGD has now been in operation since July 1994, treating 50 000 Nm^3/hr of flue gas (15 MW size).

A second pilot plant was started up in May 1995, using hydrogen gas as the electron donor and treating the same flue gas as the first pilot plant.

On lab-scale, the simultaneous removal of SO_2 and NO_x has been studied since 1994, and a small pilot plant was started up in November 1995.

Description of the process

The process being applied for SO_2 removal is called the Biological Flue Gas Desulphurization (BIO-FGD) process. Sulphur dioxide (SO_2) is absorbed from the flue gas and is converted into elemental sulphur using a combination of scrubber technology with two biological process steps. Apart from the production of sulphur, another important advantage is that the bleed stream from the system contains little impurities (mainly chloride).

The overall reaction that takes place in the BIO-FGD system is as below:

$$SO_2 + 3H_2 + \frac{1}{2} O_2 \text{ ---> } S^\circ + 3H_2O$$

If desired, NO_x can be removed simultaneously. NO_x is than biologically converted into nitrogen gas (N_2). This item is described at the end of this paper.

As can be seen in Figure 1, the BIO-FGD installation consists of the following four main parts:

- absorber unit;

- anaerobic biological reactor;

- aerobic biological reactor;

- sulphur recovery section.

Figure 1. BIO-FGD process scheme

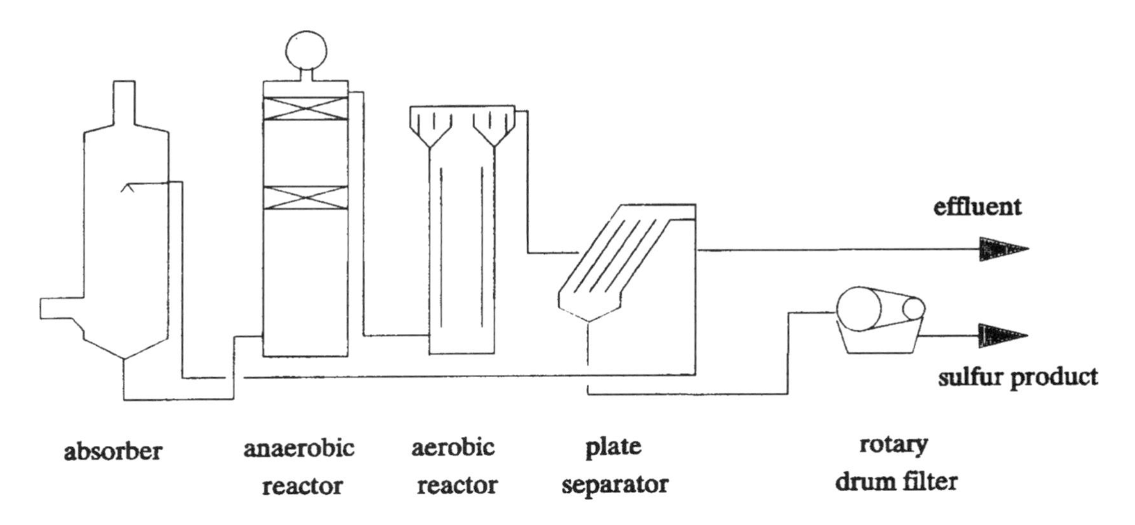

Source: Author.

The absorber unit

In the absorber unit of the BIO-FGD installation, the sulphur dioxide is absorbed in the scrubbing liquid (sodium-based scrubber).

The quantity of liquid used in this new biotechnological process is small compared to the amount used in conventional (lime)stone scrubbing processes (low L/G). This is attractive from an economical point of view. The SO_2 is absorbed as sulphite, according to:

$$NaOH + SO_2 \text{ ---> } NaHSO_3$$

The sulphite containing absorber bleed stream is led to the first biological reactor.

The anaerobic biological reactor

In the first biological reactor, an anaerobic reactor, the sulphites are reduced, forming dissolved hydrogen sulphide according to:

$$NaHSO_3 + 3\ H_2 \text{ ---> } NaHS + 3\ H_2O$$

This process of sulphate/sulphite reduction is well described in literature (Widdel, 1988).

To support this reaction, oxidizable compounds (called "electron donors") have to be added. These compounds are converted to carbon dioxide, water and additional biomass. Organic compounds and hydrogen are, in principle, candidates to serve as electron donors.

In the BIO-FGD process, other impurities from the flue gases are removed together with the biological sludge that is formed in the first biological reactor (the heavy metals that might be dissolved are precipitated in the form of highly insoluble metal sulphides).

The aerobic biological reactor

In the second reactor, hydrogen sulphide is partially oxidised to elemental sulphur, also by a biological process (Buisman *et al.*, 1990). For this partial oxidation, the amount of air injected into the reactor is controlled to prevent complete oxidation, resulting in the unwanted formation of sulphate. The partial oxidation occurs according to the following equation:

$$NaHS + \frac{1}{2} O_2 \longrightarrow S^{\circ} + NaOH$$

Through this process, the alkalinity (NaOH, used in the absorber unit) is recovered, and it shows that overall, no alkalic chemicals are consumed for the removal of sulphur dioxide.

For removal of chlorides and fluorides, still some caustic has to be added.

Sulphur recovery

The last section of the installation is the recovery of the sulphur from the effluent of the second biological reactor.

In the process given in Figure 1, the sulphur suspension from the aerobic reactor is pre-concentrated in a settler and is de-watered by a rotary vacuum drum filter. After this, the sulphur is melted with a heat treatment and recovered as a high quality liquid sulphur.

Costs of the new process

In order to evaluate the economics of the BIO-FGD process on 600 MW scale, capital and operating cost have been compared to the limestone gypsum forced-oxidation process. The results show that the BIO-FGD process capital cost are approximate 30 per cent lower compared to the LSFO system. Also, variable costs are lower resulting in the overall cost of the BIO-FGD process being 30 per cent lower.

Pilot plant results

About a year after research on lab-scale was initiated, a pilot plant at 2 MW-scale was started up in the second half of 1993.

The goal for this pilot plant was to demonstrate the chemistry of the process treating real flue gas instead of the synthetic influent used in the laboratory-scale experiments.

The pilot plant was designed to treat 6 000 m^3/hr of the 2 million m^3/hr flue gas produced at the 600 MW coal-fired power station Amer -8 situated in Geertruidenberg (Netherlands).

For the same situation, a second pilot plant was started up in May 1995. In this unit, the use of hydrogen gas as the electron donor is studied.

Pilot Plant I showed the technical feasibility of the process and the possibility to grow active biomass on hot toxic flue gasses. Pilot Plant II showed, by using hydrogen gas, the economical feasibility of the BIO-FGD process.

Process control parameters

One of the most important parameters to be controlled is the pH of the absorber bleed stream entering the first biological reactor. This pH is controlled by adding small quantities of caustic to the scrubbing liquid.

The pH of the bleed of the absorber is maintained between 7 and 8.

A lower pH will decrease SO_2 removal efficiency of the absorber, and lower the buffering capacity. A higher pH results in too high a pH in the second biological reactor where alkalinity is produced by the partial oxidation of hydrogen sulphide to form sulphur.

Another important parameter to be controlled is the dissolved oxygen concentration in the aerobic reactor in order to partially oxidise hydrogen sulphide to sulphur. Uncontrolled oxidation will lead to the formation of thiosulphate and sulphate, and loss of alkalinity.

Other parameters measured are the inlet and outlet sulphur dioxide concentrations of the flue gas, the flow of the incoming flue gas, temperatures and pH of the biological reactors and the conductivity of the water in the system.

Furthermore, the different sulphur components, organic components, chloride, fluoride, bicarbonate and heavy metals are determined on a regular basis (lab analyses).

Finally the characteristics of the sludge produced in the anaerobic reactor and the sulphur produced in the aerobic reactor are determined.

Electron donor

The choice of electron donor consumed by the bacteria in the anaerobic reactor is very important because of it's impact on the operating cost of the BIO-FGD process. So a low cost and readily available world-wide compound should be used.

The first pilot plant was started up with ethanol as a model electron donor.

Simultaneous economic and laboratory studies were conducted to find lower cost electron donors as replacement for ethanol.

Hydrogen gas, which can be very easily produced by cracking methanol or reforming natural gas (world-wide available), turned out to be cost-effective and very effective in laboratory experiments.

Over 1 000 mg/l of sulphide could very easily be produced in the anaerobic reactor without any loss of hydrogen to methane-producing bacteria.

Result of the start-up of a laboratory reactor using hydrogen is shown in Figure 2.

Figure 2. Effluent sulphide concentration of a laboratory reactor using hydrogen gas as electron donor for sulphate/sulphite reduction (at 35 °C)

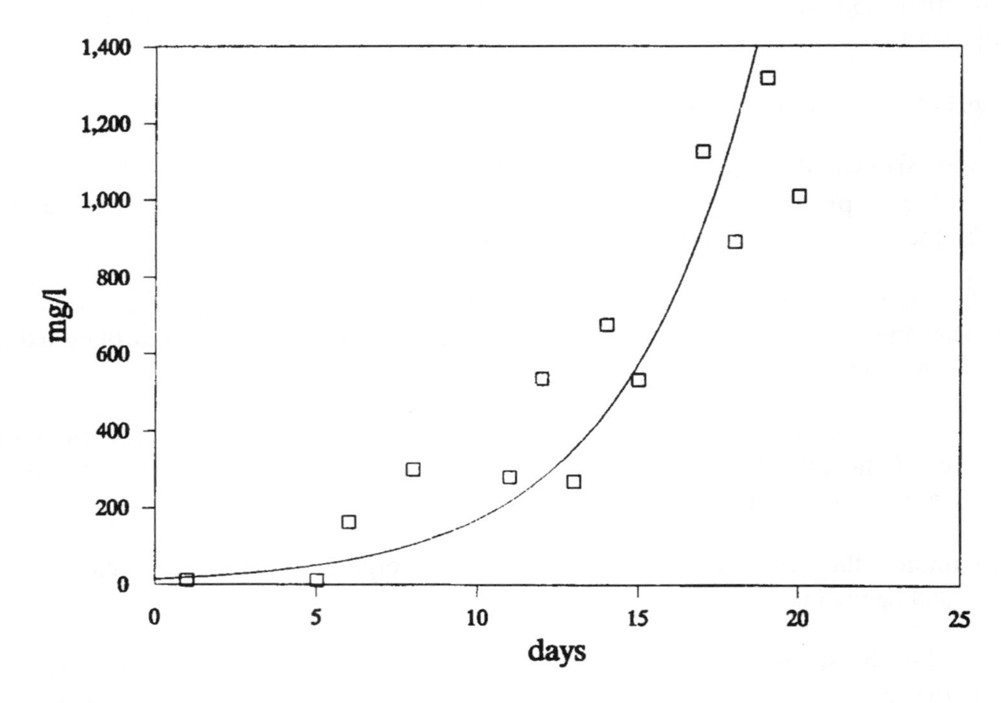

Source: Author.

Therefore it was decided to test hydrogen gas in the pilot plant.

Using an insoluble gas like hydrogen instead of an organic compound like ethanol requires a different type of anaerobic reactor.

A gaslift loop reactor was installed and hydrogen gas was tested since May 1995.

Results of pilot plant operation

Absorber unit

At low amounts of water used (L/G = 3) and a pH value of 7 in the absorber bleed high removal efficiencies for sulphur dioxide (> 95 per cent) could easily be achieved.

Anaerobic reactor

Using ethanol as the model electron donor for sulphate reduction, it was shown that hydrogen sulphide production increased gradually after start-up of the anaerobic reactor. The hydrogen sulphide concentration in the effluent of this reactor increased from zero to 200 mg/l (Figure 3) within one month due to the growth of the sulphate-reducing bacteria. (The initial seed sludge was obtained from an anaerobic reactor from a paper mill.)

Figure 3. Effluent sulphide concentration of the anaerobic reactor on ethanol

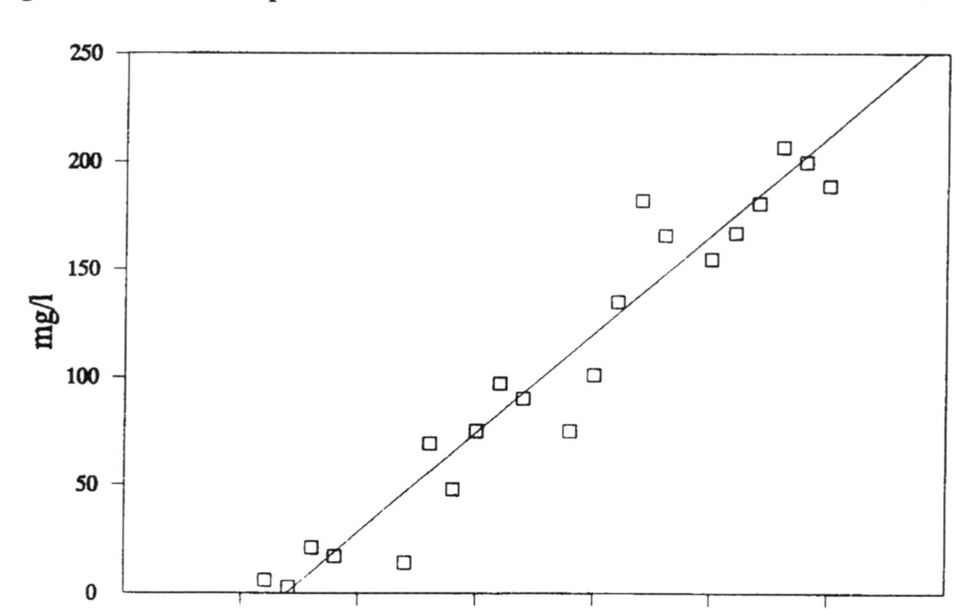

Source: Author.

Production was increased in the months afterwards by increasing the effluent concentration to 350 mg/l and shortening the hydraulic residence time.

It was shown, that at thermophilic conditions (50°C/120°F) and with a salinity of 1 000 up to 8 000 ppm of chloride, new smaller granular sludge occurred. This biomass has a high activity and settles well in the reactor.

It could be concluded that no compounds, absorbed from the flue gas, had no toxic effect on the activity or on the growth of the biomass whatsoever.

Hydrogen reactor

The second pilot plant, using hydrogen gas as electron donor, proved to start up rapidly. In this reactor biomass is grown on a basalt carrier. Within a month a gradual increase in production was observed resulting in hydrogen sulphide effluent concentrations up to 250 mg/l, eventually resulting in sulphide concentrations exceeding 700 mg/l.

It is shown that also with hydrogen as electron donor, no toxic effects were observed from components in the flue gas on the growth of the biomass.

Aerobic reactor

In the second biological reactor, over 99 per cent of the sulphide is removed from the start. Initially, mainly sulphate was produced but with increasing load rates the production shifts to mainly sulphur, reaching 95 per cent sulphur formation after about a month. This is shown in Figure 4.

Figure 4. Sulphur and sulphate production in the air lift loop reactor

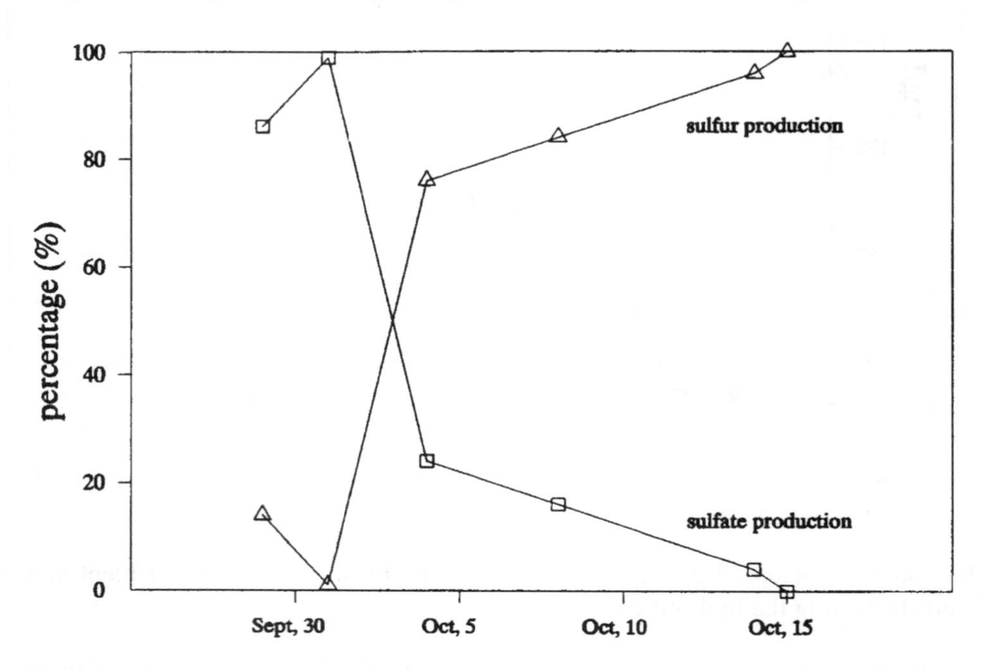

Source: Author.

The overall efficiency over the two biological reactors herewith reached values of 75 per cent of sulphur dioxide converted into biological sulphur within one month. This is shown in Figure 5. The research program will focus on increasing this conversion up to 95 per cent.

Sulphur recovery

In the pilot plant a tilted plate separator was used to recover the produced sulphur from the effluent of the aerobic reactor.

The efficiency of this recovery depends on the particle size distribution of the produced sulphur. During the experiments in the first pilot plant it was shown that the particle size could significantly be increased by increasing the sulphur concentration in the reactor. This can be done by returning part of the sulphur slurry from the tilted plate separator to the aerobic reactor.

Figure 5. Conversion of $S0_2$ into sulphur

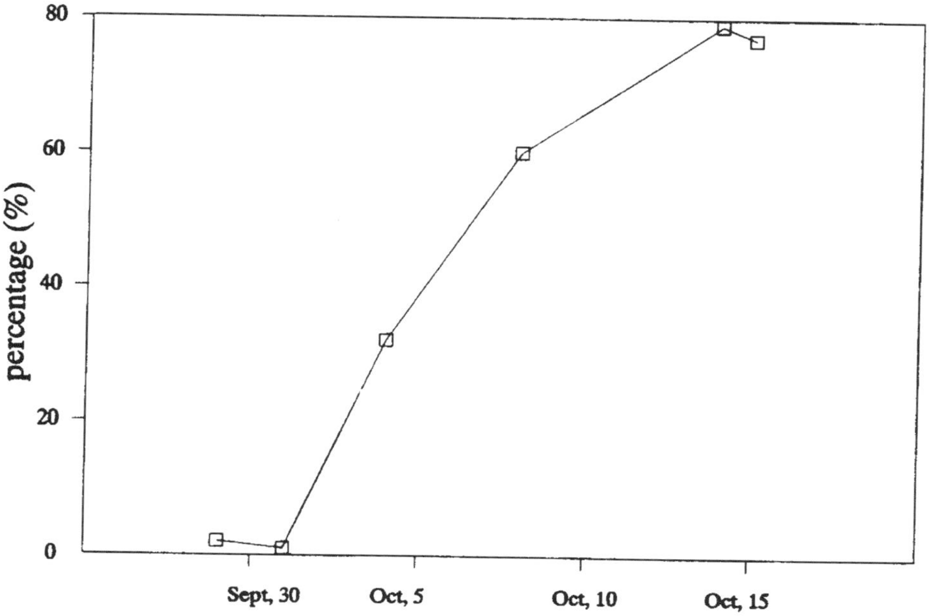

Source: Author.

Simultaneous removal of NO_x and SO_2

Recently Biostar has started a research programme for the removal of NO_x within the BIO-FGD system. To make the BIO-FGD suitable for $DeNO_x$ the scrubber needs to be adjusted. Due to the low solubility of nitric oxide, a metal chelate is used for chemically enhancing the absorption of NO. The study has focused on the use of Fe(EDTA) (EDTA = ethylenediaminetetraacetate). In the scrubber, a $Fe^{II}(EDTA)^{2-}$ solution is brought in contact with the flue gas. Besides the absorption of SO_2, the following reaction takes place:

$$Fe^{II}(EDTA)^{2-} + NO \rightarrow Fe^{II}(EDTA)(NO)^{2-}$$

This reaction results in the production of a so-called nitrosyl-complex. This complex is biologically reduced in the anaerobic reactor using an electron donor, for instance ethanol:

$$6\,Fe^{II}(EDTA)(NO)^{2-} + C_2H_5OH \rightarrow 6\,Fe^{II}(EDTA)^{2-} + 3N_2 + 2CO_2 + 3H_2O$$

The $Fe^{II}(EDTA)^{2-}$ solution is thus regenerated and can again be supplied to scrubber. The nitrosyl complex $Fe^{III}(EDTA)^{-}$ is formed, but this is biologically reduced into $Fe^{II}(EDTA)^{-}$.

Bench scale tests have already proven the feasibility of simultaneous biological removal of SO_2 and NO_x, and in November 1995 a small pilot plant was started up.

Conclusions

It can be concluded from the pilot plant research so far, that it is possible to treat the hot (120 °C) and toxic flue gases from a coal-fired power plant biologically without problems. Ethanol and hydrogen gas can be used as the electron donor. The shift from ethanol to hydrogen gas as electron donor for large-scale applications resulted in a complete new reactor design for the anaerobic reactor. A second pilot plant at 2 MW scale was started up in May 1995 and provided the necessary data concerning the scale-up of this gaslift reactor and the aerobic airlift reactor.

By using hydrogen, it is proven that this BIO-FGD process can work with significant lower operational costs than conventional processes.

Summarising

The BIO-FGD process offers the following advantages:

- 30 per cent lower cost compared to LSFO;
- High SO_2 removal efficiency, up to 99 per cent;
- Sulphur as the product;
- Ideal for retrofit (small absorber);
- Existing FGD's using sodium hydroxide or sodium carbonate can be modified at a low capital cost. A pay back time of less than three years is achieved by savings on operating costs;
- Possible simultaneous NO_x/SO_2-removal.

Questions, comments, and answers

Q: Can you envisage these gas phase reactors avoiding use of a liquid scrubber stage altogether?

A: We know of some experiments on bacteria in very dry conditions but in the case of converting SO_2 to H_2S this would be very difficult because the gases are too hot. In the NO_x case it may be possible.

Q: How do you prevent complete oxidation all the way through to sulphate or sulphuric acid?

A: The answer is to limit the amount of oxygen available. Then only the bacteria which can produce sulphur survive, and the sulphate producers are washed out.

Q: Is there a biological agent for decontamination of NO?

A: We wash out the NO with iron-EDTA and when it is in solution we convert it to nitrogen by biological means. The organism has not yet been isolated and it may be a consortium.

Q: If you are going to use this process on a very large scale, will the cost of electron donors be very high? What type of electron donors are used and what are the economics of this?

A: We are currently operating at 2 tonnes per hour and using hydrogen (from natural gas). Electron donor costs will not be prohibitive even though this is the most influential operating cost.

Q: How do you think this technology can be facilitated? Is the industry too conservative?

A: The commercialisation phase is much more expensive than the research -- perhaps there could be some sort of insurance scheme and government support. A minimum of 50 per cent cost benefit is needed before industry becomes interested in a new technology.

Q: Have you any experience with public perception, and do you anticipate a problem?

A: We come in when the coal-fired plant is already in place but the location of the site is controversial, more so than the actual technology. People don't want anything 'in their own backyard'. Bio-solutions may have advantages in this respect.

REFERENCES

BUISMAN, C.J.N., P. YSPEERT, B.G. GERAATS and A.G. LETTING (1990), "Optimization of sulphur production in a biotechnological sulphide-removing reactor", *Biotechnology and Bioengineering* 35, pp. 50-56.

HEIJNEN, J.J., M.C.M. VAN LOOSDRECHT, R. MULDER, R. WELTEVREDE, A. MULDER (1993), "Development and scale-up of an aerobic biofilm air-lift suspension reactor", *Water science and technology* 27, pp. 253-261.

SCHEEREN, P.J.H., R.O. KOCH, C.J.N. BUISMAN, L.J. BARNES and J.H. VERSTEEGH (1991), "New biological treatment plant for heavy metal contaminated groundwater", Conference Non-Ferrous Metallurgy, Brussels, pp. 403-416.

WIDDEL, F. (1988), "Microbiology and ecology of sulphate and sulphur reducing bacteria", in *Environmental Microbiology of Anaerobes*, A.J.B Zehner (ed.), John Wiley, New York.

YSPEERT, P., T. VEREIJKEN, S. VELLINGA and A. DE VEGT (1993), "The IC reactor for anaerobic treatment of industrial wastewater", proceedings Food Industry Environmental Conference, Atlanta, November 14/16.

WHY INTRODUCE BIOFILTRATION IN INDUSTRIAL PRACTICE?

by

K.-H. Engesser, M. Reiser, T. Plaggemeier and O. Laemmerzahl
Department for Biological Waste Gas Treatment, Institute for Sanitary Engineering, Water Quality and Solid Waste Management, University of Stuttgart, Germany

Introduction

An increasing amount of money is to be spent in order to meet the likewise increasing demands established by clean air acts. A great interest therefore exists in the application and improvement of biological waste air purification techniques as these are usually very cost-effective.

Use of biofiltration for waste gas treatment is state-of-the-art, especially in Germany, as well as in the Netherlands and other European countries (Kirchner, 1995; Ottengraf and Diks, 1992). One of the main fields of application is the treatment of odorous gases emitted, e.g. by waste water treatment facilities as well as the food production industry, and of VOC containing gases emitted by, e.g. the glassfibre producing/handling industry.

A total of about 2.5 Mio t of hydrocarbons/yr was emitted from the area of the old states of Germany (before reunification) during the 1980s. From these, about 250 000 t/yr may be eliminated by methods of biological waste air treatment. Actually, only about 2 to 4 per cent of this amount is subject to purification operations of this type. In the Netherlands, 80 000 t/yr of hydrocarbons were emitted from industrial sources in the year 1992. For the whole of Europe, emissions have been calculated (1990) indicating an annual loss of 1 360 000 t for benzene, 1 650 000 t for toluene and 50 000 t for styrene. Assuming that substantial amounts of these emissions stem from industrial point sources, a huge application potential for techniques of biological waste air purification becomes obvious.

Three principle devices for biological waste gas treatment are used:

1. biofilters, in which immobilised micro-organisms adhering to an organic matrix like compost or bark degrade the gas pollutants;

2. bioscrubbers, in which the pollutants are washed out using a cell suspension and this suspension is regenerated by microbial activity in an aerated tank;

3. biotrickling filters, in which immobilised micro-organisms adhering to an inert matrix degrade the pollutants while they are suspended in a water film and supplied with inorganic nutrients by a medium trickling through the device.

Why use biofiltration?

Biofiltration would not have found such a wide application just because it is a natural and therefore ecologically very reasonable method, but it is also cheap. In fact, for the treatment of appropriate waste gases it is much cheaper than any other applying "conventional" method. The construction of a biofilter system is simple and operational costs are low. Because maintenance needs are usually low, staff costs are minimised. To ensure proper function, the filter material has to be replaced regularly (approximately every two to three years) and the waste gas stream has to be preconditioned (i.e. humidified).

Nowadays a system must not only be reasonable in economic terms in order to be introduced into industrial use, but has also to take ecological aspects into consideration. In fact, waste gas treatment using biological methods is not only a natural process, but also a process not touching environmental compartments other than air. Some abiotic methods simply transform the air problem into another, e.g. a water problem. In a physisorptive scrubber contaminated water arises, while chemisorptive processes often lead to a solid waste problem. Biological processes degrade the air pollutants to water, carbon dioxide and biomass; if at all, the excess biomass is the only remaining of the pollutants that has to be dealt with (Krill and Menig, 1994).

Biological waste gas treatment is also an excellent method from the point of view of energy balance. Abiotic methods like condensation and membrane methods regain the pollutants, but are expensive due to high energy needs. Furthermore, if more than one pollutant is to be removed, these methods yield a corresponding mixture of the pollutants that often cannot be separated in a cost-effective way. Thermal incineration needs additional fuel if the pollutant concentration is too low to give rise to an autothermic process, causing this method to be expensive. Additionally, extra carbon dioxide is produced.

As a biological method, the public acceptance of biological methods usually is high. This strongly eases the approval of such facilities.

When may biological methods be used?

Biological methods are superior over other techniques when the total pollutant concentration is below approximately 1 g/(m^3 of off-gas) and large waste gas streams have to be purified. The pollutants to be removed have to be biodegradable, and the higher their water solubility, the easier the pollutants may be removed from waste gas using biological methods.

Because micro-organisms need to be adjusted to a certain substrate, the method will allow good purification quality, especially when the pollutant composition does not change too much with time. The presence of biotoxic, acidic or alkalinc substances usually affects the performance of biofilters as well as too high gas temperatures. Since preconditioning, i.e. humidifying of the waste gas, is necessary anyway in biofiltration, these problems may be dealt with by treating the waste gas in a scrubber unit before entering the biofilter. When choosing bioscrubbing or biotrickling filtration the humidifying of the waste gas is dispensable. In biotrickling filtration, components giving rise to acidic or alkaline compounds are washed down by the mobile liquid phase so the active biology itself is not affected.

When these conditions are met, biofilters will reliably clean up polluted air over a long period of time in most cases. For example, styrene containing off-gas is purified with good efficiency in a biofilter (Figure 1). In spite of temporary changes in concentration and high filter volume load the styrene content of the crude gas is reduced by more than 75 per cent at a high filter volume load.

Figure 1. Example for a biofilter in industrial operation with good performance

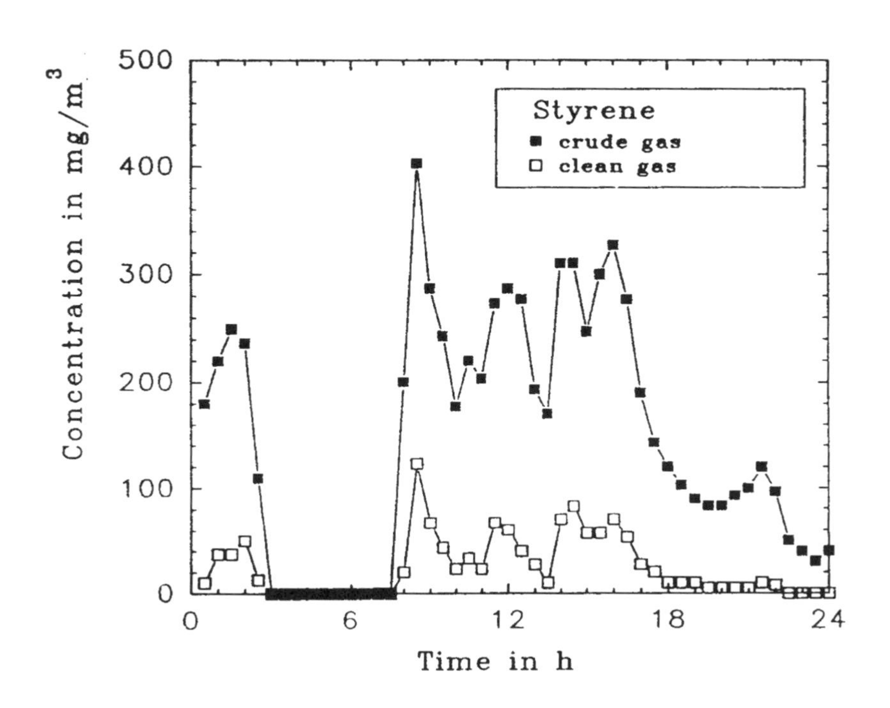

Note:

1. In a two-stage biofilter styrene was degraded with an efficiency of more than 75 per cent at a high filter volume load of 250 $m^3/(m^3\ h)$.

Source: Reiser, unpublished data.

What problems may occur and how can they be solved?

One of the most important running parameters is the maintenance of a constant humdity of about 40 to 50 per cent (w/w) in the filterbed. An excess or shortage of humidity will lead to a decrease in the purification efficiency. In a dried up biofilter the water activity is decreased, which means that the biocatalytic activity is affected and additional gas channels are formed allowing the gas to pass through very easily without sufficient contact with the micro-organisms. A waterlogged biofilter contains anoxic zones not directly participating in the degradation of the pollutant which decreases the active surface of the filter and therefore its performance. Also, the usual specific filter resistance (approximately 1 000 Pa per m filterheight, dependent on the specific filter load) is increased dramatically, affecting the energy demands of the ventilating system.

Figure 2. Crude gas forced through two different biofilters filled with different media

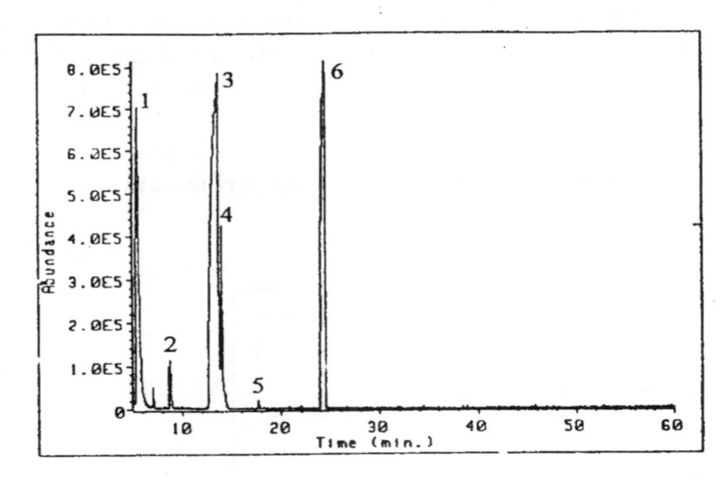

Total ion chromatogram "crude gas biofilter, laquer processing industry"
C_{org}-concentration: 290 mg/m^3

1. Acetone (20 per cent)
2. 2-Butanone (2 per cent)
3. *n*-Butyl alcohol (31 per cent)
4. 1-Methoxy-2-propanone (18 per cent)
5. 5-Methyl-hexanone (0.2 per cent)
6. *n*-Butyl acetate (25 per cent)

The rest 3.8 per cent

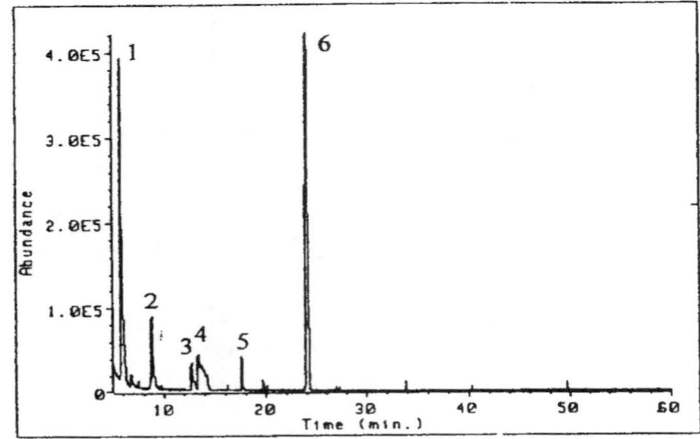

Total ion chromatogram "clean gas biofilter 1"
C_{org}-concentration: 30 mg/m^3
efficiency η = 90 per cent

1. Acetone (35 per cent relating to the crude gas)
2. 2-Butanone
3. *n*-Butyl alcohol (0.5 per cent relating to the crude gas
4. 1-Methoxy-2-propanone (0.6 per cent relating to the crude gas
5. 5-Methyl-2-hexanone
6. *n*-Butyl acetate (18 per cent relating to the crude gas)

Total ion chromatogram "clean gas biofilter 2"
C_{org}-concentration: 20 mg/m^3
efficiency η = 93 per cent

1. Acetone (2 per cent relating to the crude gas)
2. 2-Butanone
3. *n*-Butyl alcohol (1 per cent relating to the crude gas)
4. 1-Methoxy-2-propanone (9 per cent relating to the crude gas)
5. 5-Methyl-2-hexanone
6. *n*-Butyl acetate (0.05 per cent relating to the crude gas)

Note:

1. The same crude gas was forced through two different biofilters filled with different media. The yielding clean gases show a different composition. This shows the existence of two different microbial populations being responsible for the degradation of the pollutants.

Source: Reiser, unpublished data.

Besides humidity, the characteristics of the native filter material and the adhering microbial flora play an important role in filter performance (Figure 2). In most biofilters the autochthonous microflora will adapt to the individual waste gas problem, giving rise to more or less specialised micro-organisms. Usually a filter medium containing a lot of different micro-organisms, like compost, will adapt to the pollutants faster than an almost uncolonised material. However, an extra inoculum added to the filter will reduce the start-up time of the plant. When using a trickling filter or a bioscrubber, it is indispensable to do so, either in adding sewage sludge or an enrichment culture of specialised micro-organisms.

Figure 3. Example for a poorly working biofilter

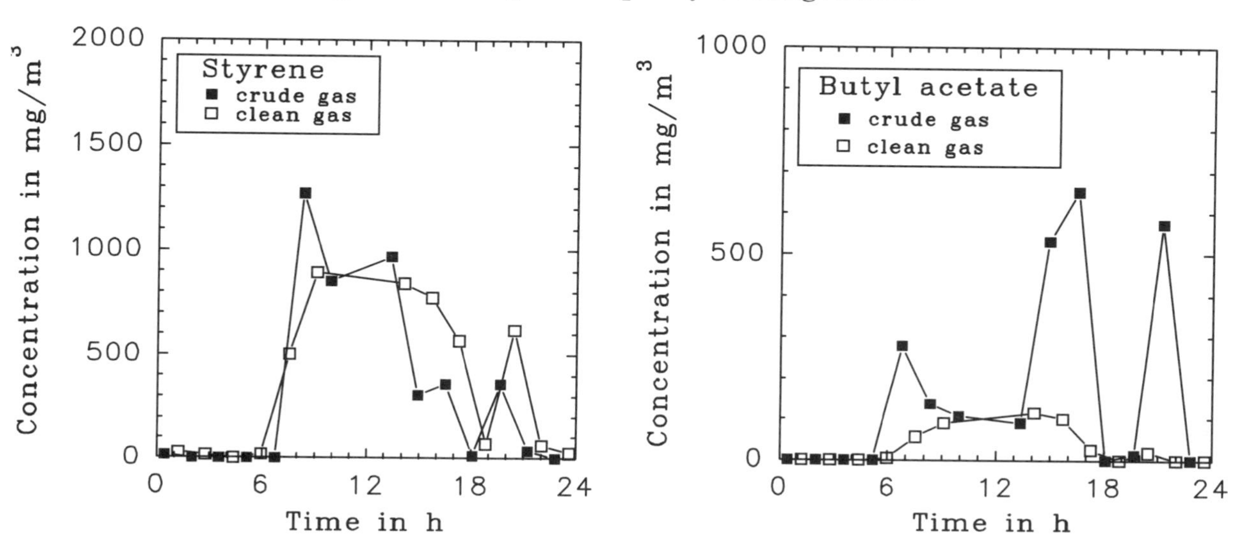

Note:

1. While *n*-butyl acetate is removed from the crude gas quite well, styrene concentration in the clean gas is only affected by sorptional effects. Degradation does not take place, as is shown in desorption during hours 14 to 24.

Source: Reiser, unpublished data

In biofilters, the pollutants may be retarded by sorptional effects. This will not necessarily correlate with microbial degradation of the compound, as is shown by biofilters sorbing when the crude gas concentration is high and desorbing when it decreases (Figure 3). It is up to the legislation to ensure proper air quality directives considering this problem by ordering the continuous monitoring of the emitted gases from bio- as well as of conventional filters.

Long-term stability of biofilters is affected by rotting of the filter material. As a result, the void fraction of the filter decreases, and the specific filter resistance increases. This problem may be solved in using inert filter materials, i.e. building a biotrickling filter. As defined before, micro-organisms attached to the inert filter medium degrade the pollutants while wetted by a separate mobile phase containing nutrients. Structured as well as unstructured packings made of poly-propylene or other materials are currently used in trickling filters, ensuring the realisation of different characteristics according to surface and void fraction. In practice, the specific filter load in biotrickling filters (as well as in bioscrubbers) is higher than in biofilters. This frequently leads to the formation of excess biomass. If not treated, this biomass accumulates in the trickling filter, and clogging occurs. The usually very low specific filter resistance then rises significantly, and it will become difficult to force the waste gas through the trickling filter unit. Approaches to solve this problem mainly focus on lowering the water activity (either by increasing the overall salt concentration of the mobile phase (Van Lith *et al.*, 1994) or by only intermittently wetting the material). This exposes the micro-organisms to external stress, and therefore increases the cell's demand for maintenance energy. As a result, less surplus cell mass is produced.

In fact, it is insufficient to just focus on engineering parameters for designing a working biofilter and seeing biology as a block box. Even correctly designed filters do not necessarily work well, because function depends very much on the microbial flora active in the filter. From an engineer's point of view it is not understandable why a mixture of compounds is often more difficult to degrade than each component for itself. In a simple mixture of two different compounds it is feasible to imagine that one compound is toxic to the other compound's degrading organism. One would find degradation of one compound, while the other compound would not be eliminated.

Outlook: What research has to be performed?

On the basis of the above-mentioned problems, it is indispensable to monitor the filter biology. This may be achieved by developing advanced methods of genetic engineering. Gene probes can serve as a valuable tool. In using these, it will be possible to detect the relevant micro-organisms in real world filters. Highly specialised micro-organisms used as inoculum to enhance the filter performance could be reliably tracked, showing if they prevail against the autochthonous microflora.

This clarifying of the black-box biology will allow an insight into the processes inside the filter. The presence or absence of purification capabilities can clearly be correlated with the presence or absence of micro-organisms and in constructing mRNA probes it will even be possible to detect their relevant metabolic activity. The elucidation of the promising technique of biofiltration will lead to a further acceptance in industrial practice.

Questions, comments, and answers

Q: Long-term stability can be a problem. How do you develop the biofilter in an ecological sense and know when to recolonise the filters?

A: This is a research and technical problem which is still being addressed. Malfunctions are often not detected. Careful monitoring of biofilters is necessary and more studies are required in this area. It may be better to monitor the genes rather than the organisms.

Q: What is the source of the micro-organisms used for styrene degradation. Are pure or mixed cultures used?

A: We have isolated a number of styrene-degrading strains from many sources, from one biofilter to the next. Bark, compost and other humic substances are used for packing the biofilters and the microflora already have this ability.

Q: Are there any problems with restarting the system after it has been stopped or delayed?

A: After a few days of being switched off, the process does take a short while to reach equilibrium again, but this is not a dramatic problem - a question of an hour or two.

Q: How do you monitor the organisms on the biofilter?

A: You could isolate them, but this is often difficult. Genetic probe techniques to identify the strain of choice are better.

REFERENCES

KIRCHNER, K. (1995, "Biologische Abluftreinigung: Anwendungsbeispiele, reaktions- und verfahrenstechnische Grundlagen", in *Abluftreinigung: Theorie und Praxis biologischer und alternativer Technologien*, Umweltbundesamt Österreich, Österreichische Gesellschaft für Biotechnologie Sektion West und Bundesministerium für Umwelt Österreich (eds.), pp. 3-15, Wien.

KOMMISSION REINHALTUNG DER LUFT IM VDI UND DIN (ed.) (1991), "VDI-Richtlinie 3477 Biologische Abgas-/Abluftreinigung Biofilter", in *VDI-Handbuch Reinhaltung der Luft, Band 6*, VDI-Verlag, Düsseldorf.

KOMMISSION REINHALTUNG DER LUFT IM VDI UND DIN (ed.) (1994), "VDI-Richtlinie 3478 Biologische Abgasreinigung Biowäscher und Rieselbettreaktoren", in *VDI-Handbuch Reinhaltung der Luft, Band 6*, , VDI-Verlag, Düsseldorf.

KRILL, H., and H. MENIG (1994), "Vergleich biologischer und konventioneller Abgasreinigungssysteme", in *VDI-Berichte 1104 Biologische Abgasreinigung*, Kommission Reinhaltung der Luft im VDI und DIN (ed.), pp. 429-468, VDI-Verlag, Düsseldorf.

OTTENGRAF, S.P.P. and R.M.M. DIKS (1992), "Process technology of biotechniques", in *Biotechniques for air pollution abatement and odour control policies*, A.J. Dragt and J. van Ham (eds.), pp. 17-31,. Elsevier, Amsterdam.

VAN LITH, C.P.M., S.P.P. OTTENGRAF and R.M.M. DIKS (1994), "The control of a biotrickling filter", in *VDI-Berichte 1104 Biologische Abgasreinigung*, Kommission Reinhaltung der Luft im VDI und DIN (ed.), pp. 169-180, VDI-Verlag, Düsseldorf.

INDUSTRIAL EXPERIENCES WITH BIOFILTRATION: ADVANTAGES AND DISADVANTAGES

by

Chris van Lith
ClairTech BV, Woudenberg, the Netherlands

Introduction

The basis for this presentation is the experience of the ClairTech company, a vendor with only one type of product: biological waste gas treatment systems. The company has completely focused on this business for the last 11 years. ClairTech made many of the mistakes to be mentioned in this presentation, but fortunately, not to the extent that it meant the end of the company. Looking back, however, after having installed in co-operation with our licensees, over 100 turn-key systems with an average value of US $400 000, and having performed over 100 pilot plant tests world-wide, one can also say ClairTech did many things right.

Current situation

Legislation

Legislation almost always determines the development of environmental technologies. Biological waste gas treatment is no exception. The strict legislation on odour nuisance in the Netherlands and Germany caused the early development of biofiltration in those countries. Six of the eight speakers in the Air/Off-gas Session at this workshop come from these two countries.

In the Netherlands, the authorities and consultants focused on trying to do an exact determination of the nuisance level. For this, dispersion models were developed and olfactometric methods were fine-tuned. As a result, the biofilters in the Netherlands were checked more thoroughly than elsewhere, and consequently, biofilters were designed more reliably with enclosed bioreactors and process controls (van Lith, 1990).

In Germany, the authorities focused more on dealing satisfactorily with a problem for the surrounding neighbours. This often resulted in the use of less than optimal designs. Those biofilters were still capable of reducing the nuisance level to some extent, but at much lower cost.

More recently, the two countries have learned from each other's approaches. In the Netherlands, we now know that odour nuisance is difficult to quantify in exact numbers. In Germany, companies now realise that it is better to deal with an odour problem from the beginning -- otherwise the complaints resurface. The end result is that the biofiltration technology, which is developed in the Netherlands, is now used widely in Germany.

In the United States, the interest in biological waste gas treatment has recently increased dramatically. Companies realise that this new technology can be applied to solve many of their environmental problems. Unfortunately, the authorities will also have to go through an educational phase, and some of the mistakes made in Europe will be repeated in the United States (Leson, 1995). In an environment lacking knowledge, low-budget solutions can be very tempting.

Technologies

There are some general rules with respect to the applicability of biological waste gas technologies. The most important factor is the concentration of the pollutants in the offgas stream. At higher levels (over 500-1 000 ppm), biological systems are not normally economically feasible. Lower levels can normally be treated in open or closed reactor biofilters. Open biofilters should only be used with very low concentrations (less than 50 ppm), and when the operational parameters of the biofilter will not influence the production process.

A special version of biological waste gas treatment is the biotrickling filter. Unlike a biofilter, a biotrickling filter has inorganic packing material over which water is continuously re-circulated. Investment costs, and sometimes operational costs, are higher for a biotrickling filter as compared to a biofilter. A biotrickling filter should only be used whenever the pollutants in the offgas stream have poisonous reaction products which should be removed from the reactor (Ottengraf, 1992).

Open systems used to be the only type of biofilter available before the 80s. Industry always was, and still is, very reluctant to use open biofilters because of the lack of control. At this point in time, it is estimated that only 30 per cent of the new biofilters sold by vendors are of the open type. Quite a few open biofilters are built by the companies itself; their numbers are difficult to estimate and are not included in the above percentage. Closed biofilters form approximately 60 per cent of the biological waste gas treatment systems built today world-wide for odour and volatile organics removal.

A strange situation is the case of biotrickling filters. The market for trickling filters is small (10 per cent or less of the total market), and will continue to be so as biotrickling filters will only be economically feasible in niche applications. Nevertheless, most research and development work is carried out on biotrickling filters. An explanation could be that the research is not market-driven, but science-driven. A biotrickling filter is a much more defined system compared to a biofilter, so models can be applied more easily.

Market

The total world market for biological waste gas treatment for odour and volatile organics control is estimated to be US $100 million, but growth will be more than average for two reasons:

- Process-integrated measures to control the emissions from industrial processes will result in more offgas streams, which have relatively low pollution levels, and are thus more suitable for biological systems.

– Many nuisance problems have not yet been solved, and the increased awareness of the population with respect to the quality of life will increase the willingness of industry to deal with odour emissions.

The market for biological waste gas treatment can be divided into two areas: 75 per cent of the systems are applied primarily for odour control; 25 per cent are used for the control of volatile organic compounds, but with some 30 per cent of the latter systems, odour removal is also important.

In the Netherlands and Germany, the concept of odour control is well-established. In the United States, the official target for a system is often volatile organics, while the actual reason for the system is odour control.

Another way of splitting up the market is by areas of industry. About 50 per cent of the biological waste gas systems are applied in foodstuff and related industries. The chemical industry runs approximately 30 per cent of the biofilters, and the remaining 20 per cent can be found in waste water treatment works, composting facilities, and agricultural applications. As mentioned before, the latter percentage may be inaccurate because of relatively high use of home-built systems in these areas.

Bottlenecks

Practice has shown that the market for biological waste gas systems is difficult. Many vendors do not survive, and either go bankrupt or simply terminate their activities. ClairTech, now 11 years in business, may be the oldest vendor active in the field. Tough strategic decisions are necessary on the part of vendors for them to survive.

Controversy purchasing department and plant manager

The barriers to entry in the biological waste gas treatment are high, but unfortunately the barriers are not clear. For this reason many companies, eager to increase their sales, try to penetrate the biofilter market. It is common practice to try to sell a first system at cost to build a reference list. Thus, the controversy emerges. The plant manager has to live with the system, and does not want to take risks; he wants reliable equipment and a reliable vendor. The purchasing department sees two offers: one offer the plant manager prefers, and one offer with the same guarantees, but at a price which is 20-30 per cent lower.

The vendor's choices are:

– He can lower the price.

– He can convince the buyer that he will not solve the problem with the cheap alternative for long.

– He can lower the quality to match the price of the competitor.

Not one solution applies to all situations.

Controversy between consultants and vendors

Industry does not want to be an expert in the field of air pollution control. Industry uses consultants to achieve the optimal solution for its air pollution problems. Consultants specify an optimal solution for the problem based on the knowledge they have. Vendors want their own products specified, because that is the best way to optimise the changes for a sale. Vendors do not want the competition to know the details of their technology, for fear of losing the competitive advantage. The vendor must then decide how much to tell the consultant.

If the vendor tells the consultant too much, the competition receives this knowledge through the specifications. If the vendor does not provide enough information, the consultant cannot judge whether or not certain features are worth including in the specifications.

The worst thing that can happen to an experienced vendor is a sealed bid procedure. Either all self-developed special features are included in the specifications and the competition can simply copy them, or no special features are specified and the vendor cannot optimise the design by balancing investment and operational costs.

Again, there are no rules concerning the relationship between the vendor and consultant. The best situation is when the vendor can trust the consultant and tells him more than will end up in the specifications. A mistake here can be very costly.

Market-driven research investments

The third bottleneck is the level of research and development work done by the vendor. A vendor who does too much will never recover the invested money. A vendor who does too little will lose its competitive edge and market share.

Strangely enough, the mistake of doing too much or conducting inaccurate research can be found more often than not doing enough. The abundance of research on biotrickling filters has already been mentioned.

A partnership between vendor and industry may be the only good way to assure that research money is well spent. The costs can be split between the parties, so in the case of failure, the money lost by the vendor remains within acceptable limits. In the case of success, a reference installation may be installed straight away. This installation recovers the costs, but more importantly, makes the next sale far easier.

Success factors

In the previous paragraph, some reasons for failure were presented. Now it is time to discuss the factors which constitute success. To a large extent, the key to success is dealing with the bottlenecks, but there is more to it.

Technology

A vendor will need a sound technological basis. He needs to have equipment which performs in a reliable way, and needs to have tools to predict the performance of his equipment. Only market-driven research and development should be conducted to build or to extend the technological basis.

It may be true that it is too late for new players to enter the market for biological waste gas treatment. Some key players have accumulated so much know-how on biofiltration that it will be difficult to recover the investments necessary to gain a similar knowledge. The market for biotrickling filters alone still is too small to support any company.

Application know-how

Application know-how is probably at least as important as the technology itself. It takes large amounts of time to investigate new applications; biological processes are slow, and a typical test easily lasts four to six months.

Most applications, within the right concentration range, are suitable for biological waste gas treatment, but tests may be required for various reasons:

- The biodegradability of one or more pollutants is unknown.
- The physical condition of the offgas is unsuitable and the required pre-treatment is unknown.

A vendor will only be effective if he is able to judge projects quickly and make the right kinds of decisions with respect to which projects should be pursued.

Marketing

The market for biological waste gas treatment systems is world-wide. Some geographical areas are more interesting for vendors than others, e.g. north-western Europe and the United States. Keeping in mind the size of the total market, a vendor should have a presence in a large part of the world market to be able to make sufficient sales.

Many branches of industry have a need for biofilters, but not each individual company within such a branch has a need this year. In general, one can say that no company will need a biofilter (or any environmental piece of equipment) unless some of the following conditions coincide:

- The company wants to enlarge its production.
- The company has sufficient financial reserves.
- The environmental officer within the applicable authority is active and has a strong position.
- The company's neighbours are environmentally aware.

A successful vendor does not work its way through each branch of industry looking for this one company in need of a system. The costs for such a sales force would be much too high compared to the systems to be sold. A successful company might present itself to the branches in a general way through conferences and presentations. This way, a company makes sure that a company in need of a biofilter will find him when the time has come to buy a biofilter.

Legislation

The last success factor the vendor has no control over is legislation on air pollution control in the market. Legislation without enforcement remains without effect. A few years ago, when the economic climate was less than favourable, the enforcement of the applicable environmental laws was reduced. The number of projects was reduced as well, and quite a few vendors went through hard times.

On the other hand, legislation makes sure that projects will continue to turn up. The companies are forced to do something. In that respect, legislation can be the vendors' best friend.

Conclusions

Biological waste gas treatment is a well-established technology in north-western Europe and will soon be in the United States. The barriers to entry for new vendors are high, though this is not clear at first sight. The biggest challenge for (potential) vendors will be to balance present investments in new technology with the expected future revenues.

The marketing strategy for the biological air pollution control systems is complex and differs considerably from the traditional business to business marketing. The market is diffuse geographically and within branches of industry. Enforcement of environmental laws determines the market.

Questions, comments, and answers

Q: You say that R&D interest in biotrickling filters is high, but industry is not interested. Why is this?

A: Biofilters are approximately half the price of a trickling filter and can deal with most of the problems. Biotrickling filters are only useful in very specific cases and particular markets.

Q: The market is $100 million per year currently. If the biotechnology could be improved, would the market also grow? Are other technologies, such as bioscrubbers and membrane systems, also dead?

A: The market is still not large enough to receive the appropriate returns. It is, however, predicted to grow at a fast pace and biofiltration will remain the major component. Biotrickling and the others are long-term technologies, with no early profits to support the R&D.

Q: Is there a requirement for R&D to develop specialised strains? How do you cope with very insoluble substrates such as hexane?

A: Specialised strains can solve problem substrates, but the size of the market should be analysed first, i.e. be wise before you invest in R&D. Biotrickling filters and biofilters cannot solve hexane problems but some research is going on, and there is a large market if it is successful.

REFERENCES

LESON, G. and S. DHARMAVARAM (1995), "A status overview of biological air pollution control", Proceedings of the 88th Annual Meeting of the Air and Waste Management Association (AWMA), San Antonio, Texas.

OTTENGRAF, S.P.P. (1992), *Process Technology of Biotechniques: Biotechniques for Air Pollution Abatement and Odour Control Policies*, A.J. Dragt and J. van Ham (eds.), pp. 17-31, Elsevier.

VAN LITH, C. (1990), "Design criteria for biofilters", *TransIChemE* 68 (part B), p. 127.

[illegible]RENCES

LESON, G. and S. DHARKAVAB[illegible] [illegible] "A [illegible] biological air pollution control", Proceeding [illegible] and Waste Management Association (AWMA) [illegible]

[illegible] (1995), [illegible] *biotechniques for Air Pollution Abatement and Odour [illegible]* [illegible] pp. [illegible]

VAN LITH, C. (1996), "Design [illegible]" [illegible] p. 127.

THE STATE OF THE ART OF BIOFILTRATION

by

Ben Koers
Europe CVT Bioway bv, Lunteren, the Netherlands

Biofiltration has been used successfully for the purification of various air flows for many years. Biofiltration has proven to be particularly effective for dealing with certain odour components, and additionally, has been successfully applied to the removal of various types of VOCs.

The technique of biofiltration has developed particularly rapidly over the past 10 years. Many facts which used to be regarded as "black box" aspects have now become clear. The biotrickling filter is one such example. In spite of the fact that the application area has not yet been well-defined, it is already clear that the biotrickling systems may represent an extension of the biofiltration field.

In practice, the application of biofiltration systems is to an increasing extent becoming a precision matter since much has been learned regarding the conditions subject to which biofiltration can be, and can remain, effective.

Biological air purification systems can roughly be subdivided into three types of systems: biofilters; bioscrubbers; and biotrickling filters. Recently, it has been possible to observe a distinct trend towards the application of combinations of systems. The combination of scrubbers with biofilters and the combination of biofilters with active carbon filters are already familiar concepts. More recently, the combination of bioscrubbers with biotrickling filters has become the focus of interest.

Biofiltration is generally regarded as a relatively simple technique. This would appear to be an advantage, although it also has its disadvantages. Since biofiltration is still very much regarded as a relatively simple technique, many simple systems are built and relatively little attention is devoted to the daily maintenance of the biological filter. This is a unfortunate, as it could potentially give this special technique a bad name.

Problems encountered with regard to the management of biofilters in practice are primarily determined by:

1) an imbalance in the distribution of air through the biofilter material;

2) an inadequate or insufficiently even distribution of the moisture content in the biofilter;

3) too high a pressure drop in the system and/or filter material.

The generation of biofilters, which we have created over the past number of years, deals with these problems by, amongst other things, using diagonally slanting pressure chambers and air locks for better distribution of the air, irrigation networks and preliminary humidification for improved water balance, and high-quality filter materials which are tailored to the application.

A relatively high resistance to the application of biological processes still exists in various industrial sectors. The unfamiliarity with this type of system in particular, prevents potential users from applying biological systems. The more expensive conventional purification techniques such as chemical washing, adsorption, and combustion are still used frequently. In many of these cases, a biological system would be at least as effective, and much cheaper.

The resistance to applying new biological systems for the purification of air is further supported by the general purification of air is further supported by the general tendency to assume that this type of system is not very reliable. This assumption is, in fact, partially correct. The technique still has a long way to go before it reaches the stage of development whereby it yields a level of reliability in all situations equal to that from conventional purification techniques such as chemical washing, adsorption or combustion. Each situation differs from the next. The types of components present in the gas outflow, the concentrations thereof, the various combinations of components and the various conditions of gas outflow currents (temperature, dust, humidity) are extremely diverse.

The modelling of these types of biological systems are therefore also essential for predicting the functioning of a biofilter, and hence, for improving the reliability of the system.

However, the extremely irregular structure of the biofilm on the biofilter material and the enormous diversity of filter materials render the modelling of such biological systems very difficult.

At present, and for the foreseeable future, it is important that pilot studies and optimisation tests are carried out in practical situations. The ensuing results of these can prove whether or not a biological system is able to function properly, and whether or not such a system is reliable.

A great number of new applications are being investigated. Substantial development of biofiltration is still under way. Several examples of new applications which are undergoing rapid development include:

- biotrickling;
- the use of systems with fungi;
- the use of systems for the improvement of solubility;
- thermophilic biofilters.

Biotrickling and the application utilising fungi will be discussed in greater depth below.

Biotrickling appears to be a very promising technique

Biotrickling is removing the image which the concept of biofiltration technology has from that of a black box technology. Opportunities have arisen which allow for better control of biological filtration systems. Particularly the ability to control pH has many associated benefits. Since synthetic materials are generally used for biotrickling systems, the durability of the filter materials tends to be longer, and the costs of maintenance throughout the lifetime of the materials tend to be lower. Investment costs will however be higher than those for commonly used biofilters, since the filter materials used for biotrickling are generally more expensive, and because the peripheral equipment used for the control of the system comes associated with extra costs.

Biotrickling is a fine technique, though by no stretch of the imagination should it be regarded as sacred. It expands the field of application for biofiltration technology. But the application of biotrickling filters is limited to very specific situations. Since biotrickling systems are, generally speaking, more expensive than the biofilters currently being used, biotrickling will only be applied in those situations where other biofilters do not function properly.

More specifically, there are two types of situations which are suitable for the application of biotrickling systems. In the first instance, there are those situations in which acidic end-products are formed during the purification of gas outflows. pH monitoring can be used to maintain the pH at an optimal level and hazardous components can be removed. The most well-known and most prevalent acidic components are ammonia (NH_3) and hydrogen sulphide (H_2S).

In the second instance, there are those situations whereby an air flow of mixed composition can be treated in a more effective manner within a phased system. Many air flows contain a mixture of various types of components. Each component has different characteristics in terms of solubility and degradability. By using a combination of biotrickling filters and biofilters, it is possible to achieve a higher removal efficiency of certain components than would be achieved when using only a biofilter or only a biotrickling filter. In a situation where the biofilter alone appeared not to be effective enough, the combination of a biotrickling filter with a biofilter performed excellently. The results below (Figures 1 and 2) have been used to indicate the removal efficiency of various target components. The target components in this particular case were volatile sulphur compounds. These compounds are responsible for the production of a very pungent odour, even at a very low concentration. The concentrations of components were in this case extremely high. Odour removal by the combination of a biotrickling filter and a biofilter in this case proved to be between 98 and 99 per cent. The system was stable for an extended period of time. The volatile sulphur compounds are released during many processes, including composting, water purification, and the processing of manure.

Since most air flows contain a mixture of components, the use of the combination of a biotrickling filter and a biofilter would appear to be promising for many situations.

In addition to providing a more effective form of treatment, such a combined system can save a lot of space, since it can be built so that it is smaller and more compact than the biofilter as it is currently still frequently built.

The use of fungi within the field of air purification technology similarly appears to be a very promising development. As is the case in the field of biological soil purification, there is also an ongoing development with respect to air purification which utilises fungi. The main reason for the use of fungi is to remove restrictions and/or sensitivities which occur when using bacteria.

Figure 1. Biotrickling + biofilter

Volatile sulphur compounds (inlet)

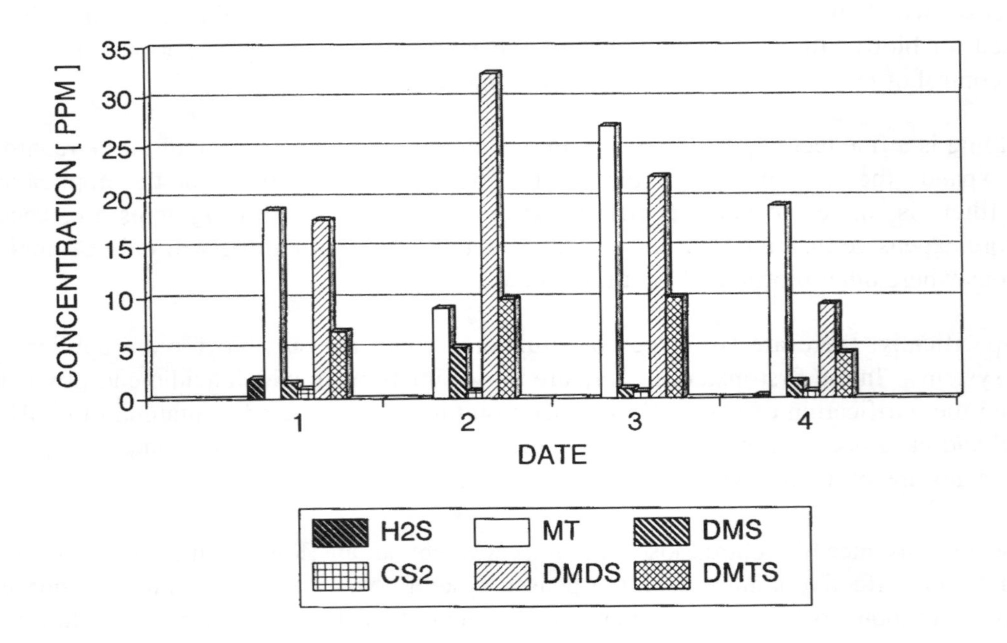

Source: Author.

Figure 2. Biotrickling + biofilter

Volatile sulphur compounds (removal)

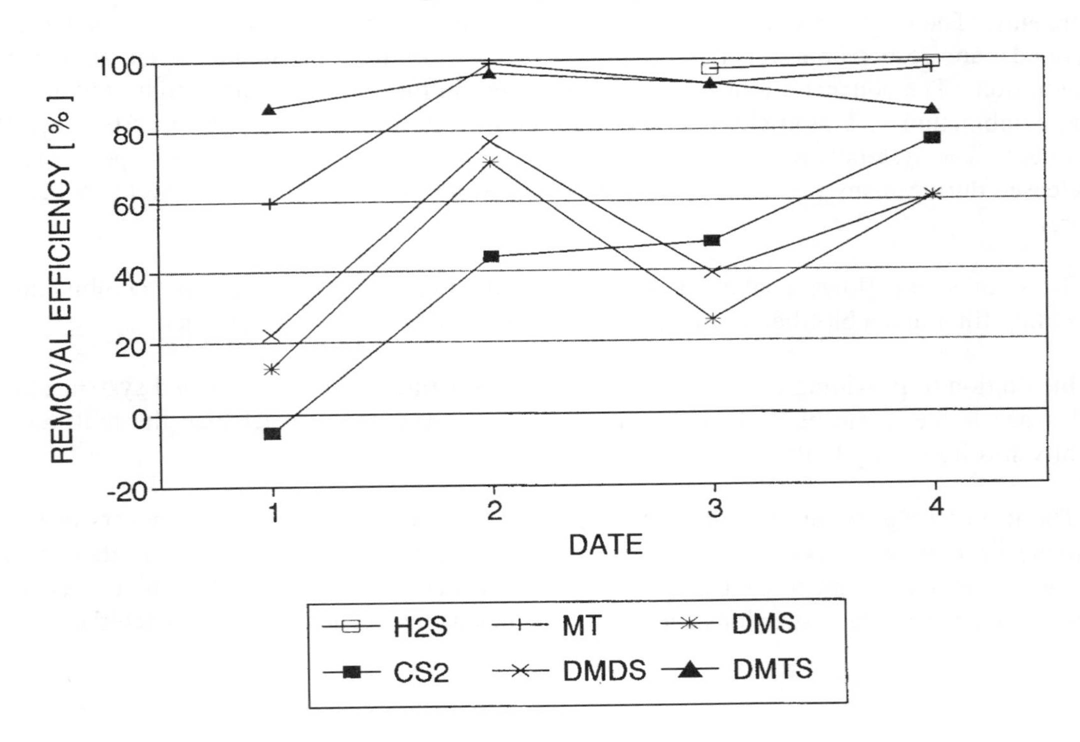

Source: Author.

It should incidentally be noted that there is always a mixed culture present in a biological filter. In addition to bacteria, there are also always fungi, various types of yeast and higher micro-organisms (for example, protozoa) present in the filter. Since fungi grow relatively slowly, they tend to lose the competition with the bacteria. The contribution which is made by fungi in terms of the removal yield in a biofilter is generally very small as compared to the corresponding contribution made by bacteria.

As far as biological soil purification is concerned, fungi appear to exhibit certain advantages in situations where PACs need to be removed. This is due to the fact that the usable fungi (white rot fungi) are capable of reducing the PAC compounds which are difficult to get to outside the cell by means of extracellular enzymes.

As is the case with biological soil purification, it would appear that the use of fungi in biological air purification has a number of advantages for a significant number of applications. Fungi are capable of remaining properly active under conditions, wherein bacteria would become inactive. It is generally known that fungi are able to survive well under relatively arid conditions. It is also a known fact that fungi can be found in relatively acidic surroundings. These characteristics are essential to a number of important areas of application within the field of air purification.

As mentioned earlier, the maintenance of certain degree of humidity is very important in a biofilter. Many problems can be traced to low moisture content or non-homogenous distribution of moisture within the biofilter. When moisture content in the filter material becomes less of a critical condition for the functioning of the biofilter, then the reliability of those same filters can be expected to increase significantly.

It has been demonstrated that the accumulation of toxic intermediary products and/or acidic end-products can severely inhibit the functioning of the filter over time. Again, a biofilter using fungi might offer significant benefits. A third advantage is that fungi are able to survive in a relatively nutrient-poor environment. The probability of less than optimal functioning of the filter as the result of a lack of nutrients is then reduced. This would also be reflected in a substantial increase in the reliability of the system.

It appears that a biofilter with a relatively low graft may have a start-up time of a week and a half. The fungi are thus capable of doing the work they have to do within a relatively short space of time.

Experiments carried out in co-operation with TNO Delft have revealed that stable management of a system is possible for at least one year. In other words, the system is capable of functioning in a stable manner. The focus in this particular case was on the removal of styrene. Preliminary results indicate that removal efficiencies of 98 per cent can be achieved in air flows which contain concentrations of 250-600 mg styrene/m; and with a volumetric load of 50-110 g styrene/m;/h.

Work is currently being carried out to test and optimise the system in practice. Methods of how to deal with peak concentrations are being investigated. The limits of the system are being identified in order to come up with a biofilter system which is, in the first instance, reliable.

Such a system would allow the rubber and plastics processing industry, in particular, to deal properly with their emissions of styrene.

Conclusion

Because biofiltration has passed beyond the stage of being a hasty afterthought to become a complete and systematic product, the implementation time from research development to product has become significantly shorter.

This, in turn, means that the biofiltration technique itself is developing very rapidly, which will open up new applications in new market areas and make it more competitive with existing techniques.

Although this is a very promising area, we must bear in mind that all new technologies require some degree of maintenance, and biological technologies are no exception.

There is still much to be done, therefore, both in terms of design results and maintenance.

Questions, comments, and answers

Q: What is the stability of the fungal community? Do bacteria overtake the fungal colonies? Are you sure it is bioremediation, and not some sort of humification or adsorption to the filters?

A: Competitive growth of bacteria has not been seen so far -- it won't happen because of the pH at which the system is run.

Q: The contribution of fungi to normal biofiltration is minimal. Why is this?

A: Bacteria out-compete the fungi.

Q: Are other pollutants in the styrene gas stream a big problem?

A: In field work, there are lots of other compounds present. We work with several gases together and these do not appear to cause problems.

Q: Is plugging by the mycelial growth of fungal cultures on ceramic supports a problem?

A: If we don't control the system, this certainly happens, but correct supply of nutrients and water can regulate this.

Parallel Sessions/A1 and A2

A1.2. BUSINESS OPPORTUNITIES AND BOTTLENECKS: SOIL

MARKET OPPORTUNITIES FOR *IN SITU* SOIL BIOREMEDIATION[1]

by

H.H.M. Rijnaarts
TNO, Institute of Environmental Sciences, Energy Research, and Process Innovation
Delft, the Netherlands

Introduction

In situ soil bioremediation is more cost-effective compared to other soil remediation methods since it avoids excavation and damage to existing infrastructure. It does not generate waste streams because the contaminants are converted *in situ* to harmless products, and, as a consequence, requires relatively low amounts of energy and raw materials (OECD, 1994). Hence, *in situ* bioremediation is a very competitive technology and can contribute to a significant reduction of the costs of soil and groundwater clean-up. Despite these advantages, a broad implementation of *in situ* bioremediation has not yet been realised, primarily for the two following reasons:

a) *In situ* bioremediation has not yet been demonstrated sufficiently as a proven full-scale technology. This problem has been widely recognised and several national programs, including the Netherlands Research Program on *In situ* Soil Bioremediation (NOBIS) have been started to demonstrate the efficacy of full-scale *in situ* bioremediation.

b) There is insufficient insight into the various potential markets for *in situ* bioremediation. Identification of significant markets is essential for creating a willingness to invest in R&D required to make *in situ* bioremediation a proven and reliable technology.

The study presented here aims to contribute to the elucidation of the second problem, i.e. to identify significant *in situ* bioremediation markets. For this purpose, a new approach was followed for estimating the number of polluted sites that can be remediated with *in situ* biotechnological techniques. In this paper, the application of this method to assess the European *in situ* bioremediation market is discussed. The advantage of this new approach is that from generally well-documented data on population density and soil properties, predictions can be made of the geographic locations and of the size of potential *in situ* bioremediation markets, and the cash flow (actual markets) that will result from these. Moreover, the database and model can be easily recalibrated and used for improved approximations in the near future.

---

[1] Part of this research was funded by the Dutch Ministry of Economic Affairs as a part of the Netherlands Research Programme on *In situ* Bioremediation (NOBIS). J.W.T. Reckman and J. Kooijman of the TNO Institute of Applied Geoscience are acknowledged for performing the calculations with the geographic information software and the interpretation of European soil property data.

The procedure roughly involves an analysis of the situation in the Netherlands from which important parameters were estimated. With these parameters and geographic information on population density and soil properties, the potential markets (number of sites and costs to remediate them with *in situ* biotechnological techniques) in the various European countries were approximated. In a final step, the order of magnitude of the cash flows resulting from these potential markets were assessed.

Analysis of the situation in the Netherlands

In the Netherlands, there are about 110 000 contaminated sites (RIVM, 1993). Approximately 80 000 of them are owned by private companies (Commissie Bodemsanering van in Gebruik Zijnde Bedrijfsterreinen, 1991). We have 300 of these 80 000 sites registered in a database. These 300 sites were screened on type of industry and/or other activities, from which the most probable nature of contamination on these locations was deduced. By comparing these types of contamination with current knowledge on biodegradability (Table 1), it was estimated that 64 per cent of the sites have a type of pollution that can be treated with *in situ* bioremediation. Furthermore, it was assumed that in 50 per cent of the cases, the implementation of an *in situ* remediation is prevented by local circumstances. Examples of such local circumstances are the presence of an industrial installation not permitting the use of essential equipment or the existence of a very high actual risk for ecosystems and/or public health requiring a fast removal and *ex situ* clean-up of the polluted soil. Hence, for the Netherlands, it is estimated that 32 per cent of the contaminated sites can be potentially remediated by a biotechnological technique. These sites are here referred to as potential *in situ* bioremediation (PISB) sites (Figure 1).

Table 1. Occurrence and biodegradability of organic contaminants

Contamination	Degree of occurrence[1]	Biodegradability[2]
Gasoline, oil	+++	++
Aromatic solvents	+++	++
Poly-cyclic aromatics	++	+
Creosote	+	+
Alcohols/ketones/esters/ethers	+	++
Chlorinated aliphatics	+++	+
Chlorinated aromatics	++	+
Polychlorobiphenyls	+	+/-
Cyanides	+	+
Nitroaromatics	+	+
Pesticides	+	+...+/-...-

Notes:

1. Degree of occurrence: +++ = high; ++ = moderate; + = limited.
2. Biodegradable under a broad range of conditions (++), specific conditions (+), very specific conditions (+/-), or under no conditions presently known (-).

Sources: Norris *et al.*, 1994 and the Committee on *In Situ* Bioremediation, 1993.

Figure 1. Number of potential *in situ* bioremediation (PISB) sites per province as a function of the number of inhabitants per province (squares)

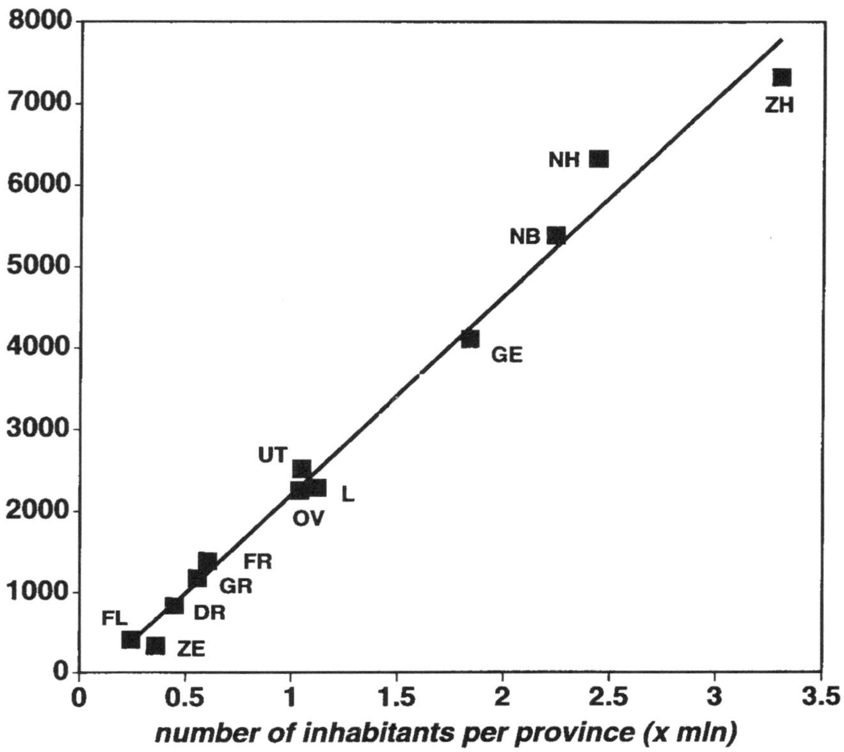

Notes: The line represents the linear regression result.

Provinces: ZH = Zuid Holland
NH = Noord Holland
NB = Noord Brabant
GE = Gelderland
L = Limburg
UT = Utrecht
OV = Overijsel
FR = Friesland
GR = Groningen
DR = Drente
ZE = Zeeland
FL = Flevoland

Source: Author.

The Netherlands consists of 12 provinces for which the number of contaminated sites per province are available (Commissie Bodemsanering van in Gebruik Zijnde Bedrijfsterreinen, 1991). Multiplying these values with the factor 0.32 estimated above, the numbers of PISB sites per province were obtained. In Figure 1 these values are shown as a function of the number of inhabitants per province. Clearly, there is linear relationship between the number of PISB sites and the population density. For the Netherlands, Figure 1 predicts that about 2.2 PISB locations exist per 1 000 inhabitants.

Potential *in situ* bioremediation markets in Europe

The potential markets in Europe were assessed as follows. First, a geographic distribution of the population density for Europe (Figure 2) was created by using a database and geographic information software. The geographic distribution of PISB sites in Europe was directly calculated from the data shown in Figure 2, assuming that the linear relationship between population density and density of PISB sites as derived for the Netherlands also holds for other parts of Europe.

Figure 2. Geographic distribution of the population density for Europe

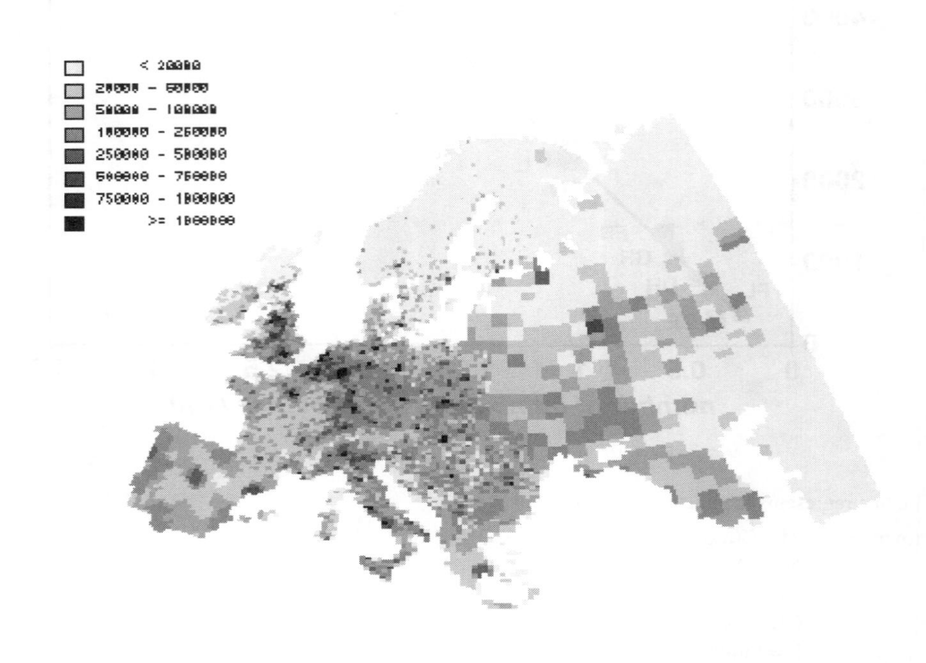

Notes: Grid cells represent approximately 60 km × 60 km.

Population density varies between 1 million people per grid cell (darkest grids) to smaller than 2 000 inhabitants per grid cell (lightest grids).

Source: Author.

In the second step, the geographic distribution of the hydraulic conductivity of the geological top-layer (the first 20-30 m of the sub-surface) was considered using data of the RIVM (Meinardi *et al.*, 1994). Sediments covering aquifers with a high hydraulic conductivity (i.e. greater than 1 metre per day) were identified as sub-surfaces suitable for *in situ* remediation (Figure 3). As can be seen, the primary regions of interest are the larger sedimentary basins, i.e. those of the rivers

Rhine, Schelde, Elbe, Weser, Garrone, Rhône, Po, and Danube, and a number of regions in the Russian Federation.

Figure 3. Geographic distribution of areas with a sub-surface suitable for *in situ* bioremediation (dark regions) for Europe

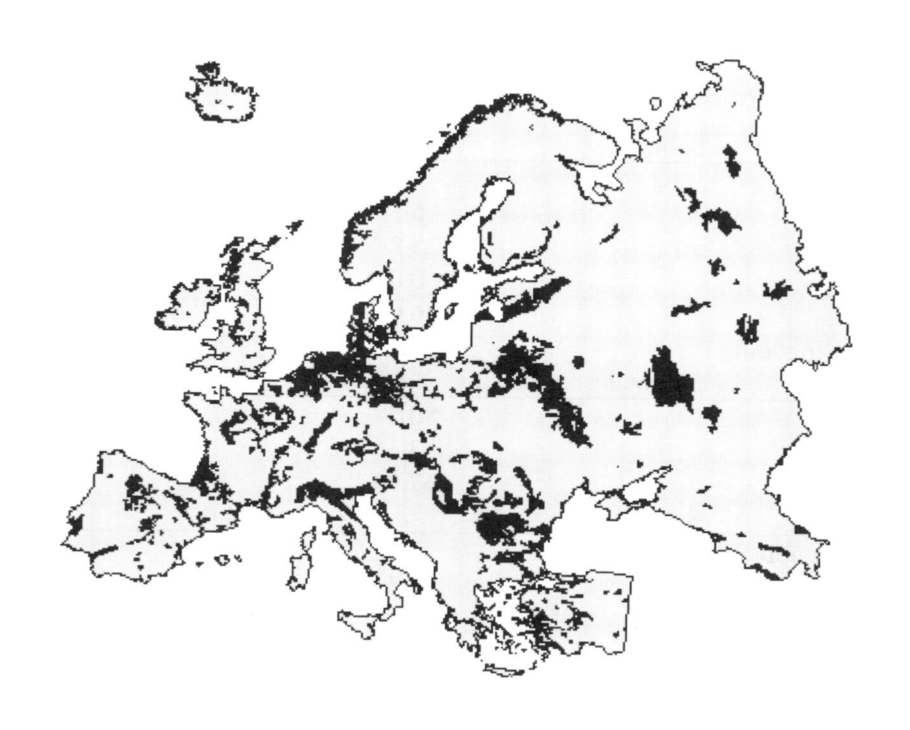

Source: Author.

The number of PISB sites in regions with an *in situ* treatable soil [these sites are real *in situ* bioremediation (RISB) sites], were computed using the data sets represented in Figures 2 and 3 and geographic information software. The potential market for *in situ* bioremediation was calculated for each European country (Table 2), assuming an average size for each RISB site of 7 000 tonnes of contaminated soil and $50 per tonne as the costs for *in situ* bioremediation (OECD, 1994).

The following most predominant potential *in situ* bioremediation markets were estimated (in billion $): Germany (15); Italy (10); Russian Federation (9); Rumania (6); France (6); the Netherlands (5); United Kingdom (2); Spain (2); Poland (2); the Czech Republic (1); and Belgium (1). In total, there are estimated to be 280 000 RISB sites, representing a market of $65 billion. This corresponds to a potential cost reduction of $130 billion, considering that *in situ* bioremediation replaces conventional *ex situ* techniques that cost $150 per tonne (OECD, 1994).

Table 2. Estimated number of real *in situ* bioremediation (RISB) sites and associated potential markets for the European countries grouped in social-economic tiers

Tier[1]	Country	Number of RISB sites (× 1 000)		Potential *in situ* bioremediation market (in billion US$)	
1	Denmark	1.5		0.3	
	Germany	71.4		15.4	
	Netherlands	20.6		4.5	
	Sweden	0.2		0.1	
	Switzerland	0.2		0.1	
	Subtotal		73.2		20.3
2	Austria	5.7		1.3	
	Belgium	6.5		1.4	
	France	25.8		5.6	
	Ireland	0.0		0.0	
	Italy	46.9		10.1	
	United Kingdom	9.9		2.1	
	Subtotal		94.8		20.4
3	Greece	0.2		0.1	
	Portugal	3.3		0.7	
	Spain	7.9		1.7	
	Subtotal		11.4		2.5
4a	Baltic Republic	3.1		0.7	
	Czech/Slovak Republic	5.1		1.1	
	Hungary	6.5		1.4	
	Poland	8.9		1.9	
	Subtotal		23.5		5.1
4b	Albania	0.0		0.0	
	Bulgaria	4.9		1.1	
	Croatia/Bosnia/Yugoslavia	4.2		0.9	
	Romania	26.0		5.6	
	Russian Federation	42.7		9.2	
	Subtotal		77.8		16.8
	TOTAL		281.0		65.1

Notes:

1. Tiers according to Arthur D. Little (1994).

Source: Author.

These estimates are probably conservative, for the following reasons.

i) More and more contaminants, previously considered as completely recalcitrant, are found to be biodegradable. In addition, even non-biodegradable heavy metals can be removed from soils by using biotechnological techniques.

ii) In the estimation procedure, a large number of the sites were identified as not suitable for bioremediation, due to local circumstances (50 per cent of the total number). The real value of this factor is likely to be lower, but there is no realistic estimate available (yet).

iii) The soil property criterium yields a conservative result. Many methods like electro-bioreclamation and activated *in situ* bioscreens (Rijnaarts *et al.*, 1995) are likely to be(come) also applicable to much less permeable silt, clay or peat soils.

For these reasons, the real potential market for *in situ* bioremediation may be much greater than the $65 billion as estimated here.

Actual European *in situ* bioremediation markets

The (soil bio)remediation market is an enforced market (OECD, 1994). Government policies that set environmental standards and demand environmental risk assessments are required for remediation to occur. In other words, this political and social-economic pressure is needed to generate a potential-actual market transfer in terms of cash flow. Policies that stimulate the development of *in situ* bioremediation technologies to proven cost-effective technologies can strongly facilitate the formation of actual *in situ* bioremediation markets.

The environmental policies in Europe, and as a consequence, the actual bioremediation markets (cash flows in $/year), differ greatly among various countries due to differences in economic development and/or in willingness to accept supra-national environmental regulations, i.e. in Europe, European environmental legislation. Arthur D. Little (1994) has presented a classification system of tiers (Table 3) in order to account for this variation in environmental policy and environmental technology status among various countries. The rates for the transfer of potential into actual market (cash flow per year) decreases with increasing tier number. Exact values of these transfer rates are difficult to quantify at present. We estimated the rates on the basis of various sources (Table 4) in order to obtain an order of magnitude for the actual European *in situ* bioremediation market for the next 30 years (Figure 4).

Table 3. Tiers of European countries according to their status of environmental legislation and technology development

Tier[1]	Political-economic pressure to perform remediation	Environmental technological status
1	very high	very high
2	high	high
3	moderate	moderate
4a	low, but growing	low, but growing
4b	low	low

Notes:

1. Tiers according to Arthur D. Little (1994).

Source: Author.

Table 4. Rates of Transfer in percentage of potential market transferred into actual market per year for the next 30 years

tier	transfer rate (in percentage of potential market converted into actual market/year)[1]						
time-period	1995-2000	2000-2005	2005-2010	2010-2015	2015-2020	2020-2025	total conversion (%)
1	1.0	3.0	4.0	4.0	4.0	2.0	90
2	0.4	1.0	3.0	4.0	4.0	4.0	82
3	0.2	0.4	1.0	3.0	4.0	4.0	63
4a	0.1	0.2	0.4	1.0	2.0	3.0	34
4b	0.0	0.1	0.1	0.2	0.5	1.0	10

Notes:

1. Sources on which rate estimates are based: Little, 1994; European Commission, 1994; and OECD, 1994.

Source: Author.

Figure 4. The actual European *in situ* bioremediation market predicted from the potential market values (Table 2) by using the potential-actual market transfer rates given in Table 4

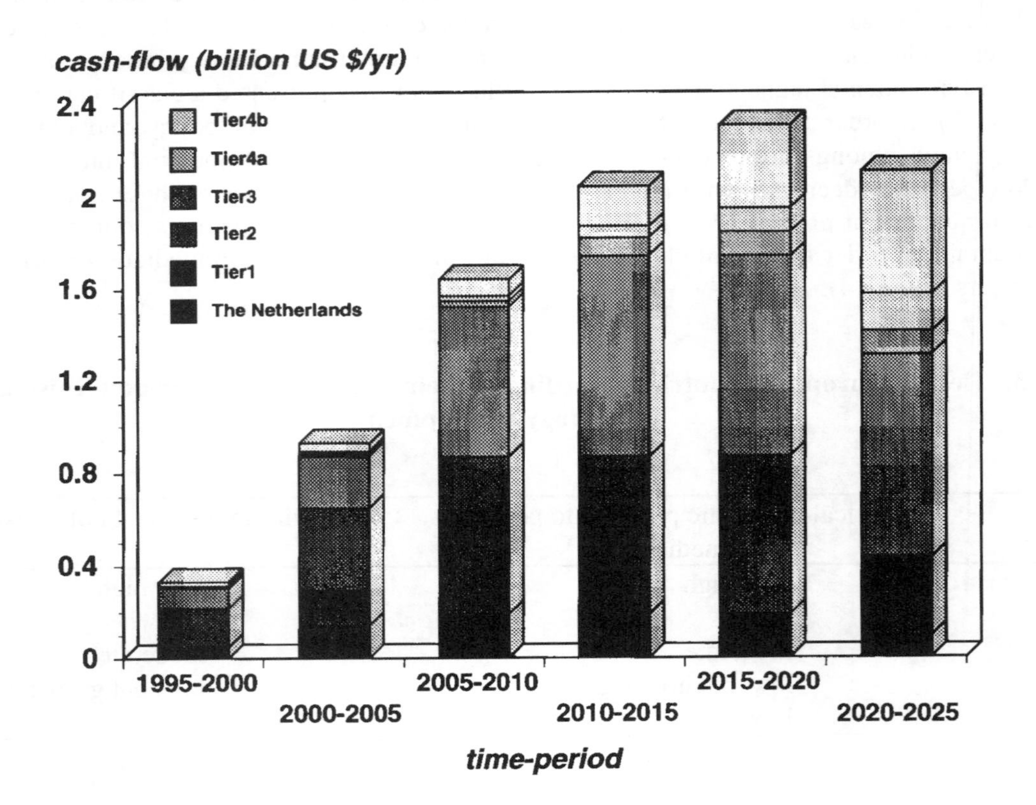

Note: Tiers correspond to Tables 2 and 3.

Source: Author.

The actual *in situ* bioremediation market for the present and near future (1995-2000) is estimated to be about $350 million/year, increasing with about $100 million each year. Although the results shown in Figure 4 must be considered as only indicative, they are in close agreement with the approximations and predictions of Glass *et al.*, 1995 (Table 5).

Table 5. Comparison of various bioremediation market estimates

	Average for 1995-2000 (in million US$/year)	Growth per year
Europe, *in situ* (present study)	350	100
Europe	290	85
Canada	50	7
United States	310	70
World	670	170

Note: Uncertainty in the data is +/- 50 per cent of the presented value.

Source: Glass *et al.*, 1995; Author.

The longer term predictions (Figure 4) indicate that the *in situ* bioremediation market may grow in 20 years from $350 million to about $1-2 billion. The tier 1 countries, Germany and the Netherlands together, form the most significant European *in situ* bioremediation markets for the coming decade. After 2005, the tier 2 countries, France and Italy, probably represent a market as large as tier 1. Tier 3 countries, Greece, Portugal, and Spain, are only of marginal interest while cash flow from *in situ* bioremediation in eastern Europe (tier 4a and b) is predicted to remain relatively small for the next three decades.

The bioremediation market of the world

The bioremediation market and its present and near-future growth are shown in Table 5. As can be seen, the most important markets are those of Europe and the United States. They have a size of approximately $300 million each, and will probably grow with $70-100 million per year.

Conclusions

The new approach used has successfully identified the locations and sizes of the various European *in situ* bioremediation markets. The following estimates were obtained:

- a total potential of these markets of $65 billion; and

- a cash flow in the order of $300 million per year, growing at about $100 million each year.

Germany and the Netherlands are the most important current actual markets; France and Italy will possibly develop into an actual market of equal size after the year 2000.

Recalibration of the potential-actual market transfer rates with observed cash flow values should be performed in the near future which will result in better predictions for the development of the European market. Moreover, the procedure presented here can be extended to other continents which would be of help in getting a better insight into the *in situ* bioremediation world market.

Questions, comments, and answers

Q: Were the trends and forecasts of the markets checked with actual market development during say the past five years, i.e. was the data validated?

A: In market estimates made by other people, the market in 1994 was $100-200 million in Europe, rising to $600 million. Our estimates compare well with these.

Q: How do you determine the biotreatability of soils?

A: Soil permeability at 1 metre depth was analysed, and above a minimum level, the soil is suitable. Our estimate is conservative, however, and *in situ* treatment may eventually be suitable for less permeable sites.

REFERENCES

COMMISSIE BODEMSANERING VAN IN GEBRUIK ZIJNDE BEDRIJFSTERREINEN (Committee on Soil Reclamation of Sites Currently Used by Private Companies) (1991), Eind rapport (Final report).

COMMITTEE ON *IN SITU* BIOREMEDIATION, WATER SCIENCE AND TECHNOLOGY BOARD, COMMISSION ON ENGINEERING AND TECHNICAL SYSTEMS IN NATIONAL RESEARCH COUNCIL (1993), *In situ Bioremediation: When does it work?*, National Academy Press, Washington D.C.

EUROPEAN COMMISSION (1994), "Eco-industries in the EC", in *Panorama of EU Industry 94*, Office for Official Publications of the EC, Luxembourg.

GLASS, J.D., T. RAPHAEL, R. VALO and J. van EYK (1995), "International activities in bioremediation: growing markets and opportunities", in *Applied bioremediation of petroleum hydrocarbons*, R.E. Hinchee *et al.* (eds.), pp. 11-33, Battelle Press, Columbus, Richland.

LITTLE, A.D. (1994), "Seeking market opportunities for Dutch environmental technologies: phase 1", *National market profiles and environmental segment descriptions*, report available at the Dutch Ministry of Economic Affairs.

MEINARDI, C.R., A.H.W. BEUSEN, M.J.S. BOLLEN and O. KLEPPER (1994), "Vulnerability to diffuse pollution of European soils and groundwater", National Institute of Public Health and Environmental Protection (RIVM) report no. 461501002.

NORRIS, R.D., R.E. HINCHEE, R. BROWN, P.L. McCARTHY, J.T. WILSON, D.H. KAMPBELL, M. REINHARD, E.J. BOUWER, J.M. THOMAS and C.H. WARD (1994), *Handbook of Bioremediation*, Lewis Publishers, Boca Raton, Florida.

OECD (1993), "Industrial policy issues and initiatives: section F. Industry-related environmental and energy policies", in *Industrial Policy in the OECD countries: annual review 1993*, OECD, Paris.

OECD (1994), "An economic perspective on environmental biotechnology" in *Biotechnology for a Clean Environment: Prevention, Detection, Remediation*, OECD, Paris.

RIJNAARTS, H.H.M., P.G.M. HESSELINK, and H.J. DODDEMA. (1995), "Activated *in situ* bioscreens", in *Contaminated Soil*, W.J. Brink, R. Bosman, and F. Arendt (eds.), pp. 929-937.

RIVM (National Institute of Public Health and Environmental Protection) (1993), *Nationale Milieuverkenning 3, 1995-2015* (National Environmental Survey 3, 1995-2000), Tjeenk Willlink, Aphen aan de Rijn, the Netherlands.

EVALUATING NATURAL RECOVERY FROM ENVIRONMENTAL PCB CONTAMINATION

by

John F. Brown Jr. and Daniel A. Abramowicz
GE Corporate Research & Development, Schenectady, New York, United States

Introduction

The polychlorinated biphenyls (PCBs) are a group of chemical compounds that were formerly widely used as chemically stable, non-flammable industrial fluids, resins and waxes. Once released into the environment, many of the individual PCB congeners turned out to be persistent, mobile and lipophilic, resulting in widespread environmental contamination and bioaccumulation in aquatic organisms. Concerns over the possible human and environmental health effects posed by such accumulations have repeatedly stimulated interest in the possibilities of remedial actions, such as dredging or encapsulation of contaminated sediments, or, particularly, *in situ* treatment by accelerated biodegradation (Abramowicz and Olson, 1995). Each of these remedial techniques has its limitations, however, and is probably applicable to only a limited number of the world's many PCB-contaminated sites. Thus, in assessing the alternative management options for any individual site, it becomes particularly important to evaluate the progress and prospects of natural recovery in the local environment.

Any such evaluation must address two key issues. First, what is the actual source of the PCBs being found in the biota of concern? Is it a local sedimentary deposit, a distant one, a continuing release into the waterway, or an atmospheric input? Second, are the PCBs in that source already becoming less significant, whether by burial, biodegradation, dechlorination, or some other detoxication process, or does there exist some local obstacle to these natural recovery processes? A partial answer to the latter question may be provided by historical data on fish contamination levels extending over a decade or two. However, if such information is not available, and often even if it is, reliance must be placed instead upon determinations of actual source and detoxication state.

For most environmental contaminants such issues are not easily addressed. For the PCBs, however, we may take advantage of the fact that the chemical composition of an environmental PCB mixture, as determined by capillary gas chromatography, can provide an extensive – albeit coded – record of its source and transformation history. This happens because a) the formerly commercial PCB products were all complex mixtures of the individual PCB congeners, produced in nearly invariant relative proportions by the iron-catalysed chlorination process used, and b) each of the several known types of environmental alteration processes removes its own distinctive group of those congeners. Thus, by characterising the patterns of change in congener distribution, it is often possible to delineate both the original source of the PCBs picked up by the organisms of concern, and also the types of environmental niches subsequently traversed.

Factors affecting environmental PCB composition

Sources

The PCBs found in the general environment are now known to be derived from three sources, one major and two minor. The major source consists of the formerly commercial PCB products, once marketed as Aroclors in North America and the United Kingdom, as Clophens in Germany, and as Phenoclors in France. These products were all prepared by chlorinating biphenyl to a fixed weight gain so as to give either mobile fluids with 21 per cent to 48 per cent chlorine (e.g. Aroclors 1221, 1232, 1016, 1242 or 1248), viscous fluids or soft resins with 54 per cent to 62 per cent chlorine (e.g. Aroclors 1254, 1260 or 1262) or waxes with 68 per cent to 71 per cent chlorine (e.g. Aroclors 1268 or 1270, or the non-Aroclor "Deka"). The most widely used of these products, in terms of per cent of total 1957-1977 North American production, were Aroclors 1242 (51.8 per cent), 1254 (15.7 per cent), 1016 (12.9 per cent), 1260 (10.6 per cent), and 1248 (6.8 per cent). Production of Aroclor 1270 was discontinued before 1957, and those of Aroclors 1221, 1232, 1262 and 1268 each contributed less than 1 per cent to the 1957-1977 total production (Brown, 1994).

One of the minor PCB sources consists of "incidental generation" of PCBs as minor by-products of other chemical processes. These by-product PCBs usually contain only the most lightly chlorinated congeners, and are of little concern as contaminants of the general environment. The other "minor" source, however, consists of emissions from incinerators and other combustion processes. These pyrogenic PCBs contribute only about 1 per cent to the total global environmental PCB load, but an estimated 77 per cent to 99 per cent of the global loading of the toxic and persistent coplanar congeners, PCB Nos. 126 and 169 respectively (Brown *et al.*, 1995). All PCB numbers refer to IUPAC assignments. These congeners are currently being estimated to contribute over 80 per cent of the total dioxin-like toxic risk of PCBs to wildlife in some areas (Alsberg *et al.*, 1995). Thus, if the object of PCB remediation be to reduce toxic risk, it will be important to evaluate the pyrogenic contribution to the local PCB burden.

Interphase partitioning

Once released into the general environment, a mixture of PCB congeners will undergo interphase partitioning between water and air, soil and groundwater, water and biota, etc., with consequent shifts in composition so as to enrich the more mobile phase in the lower congeners, and the more lipophilic phase (e.g., biological tissue) in the more highly chlorinated homologs. The extent of such bioconcentration processes can be correlated with the octanol-water partition coefficients (K_{ow}'s) of the individual congeners (Thomann, 1989) and needs to be considered in the course of any estimate of original Aroclor composition from the homolog distribution observed in fish or of the extent of PCB biodegradation made from a change in congener ratio.

Impact of biodegradation on PCB composition

The various known types of environmental biodegradation can cause dramatic changes in PCB congener distribution, and hence the appearance of the gas chromatogram. Figure 1 shows, first, a DB-1 capillary gas chromatogram of Aroclor 1242 and then that of a 93.5 per cent evaporated specimen. (A less extensively evaporated specimen would give a chromatogram very similar to that of Aroclor 1248, and, in fact, most environmental PCB specimens reported as "Aroclor 1248" are actually Aroclor 1242 at various stages of evaporation or elution.) The next chromatogram on

Figure 1 shows a specimen from a landfill leachate that had been first enriched in lower congeners by extraction into the groundwater and then depleted of most mono-, di-, and trichlorobiphenyls by aerobic microbial action. Below that is the chromatogram of a specimen degraded in the laboratory by the Hudson River aerobic microbe *Alcaligenes eutrophus* H850 (Bedard *et al.*, 1987), and below that two chromatograms of Hudson River sediment PCBs showing how anaerobic microbial dechlorination can remove most of the more heavily chlorinated congeners (represented by peaks on the right side of the chromatogram) and thence generate mono- and dichlorobiphenyls, giving the resultant peaks on the left (Brown *et al.*, 1987).

Figure 1. DB-1 capillary gas chromatograms of Aroclor 1242 before and after exposure to various environmental transformation processes

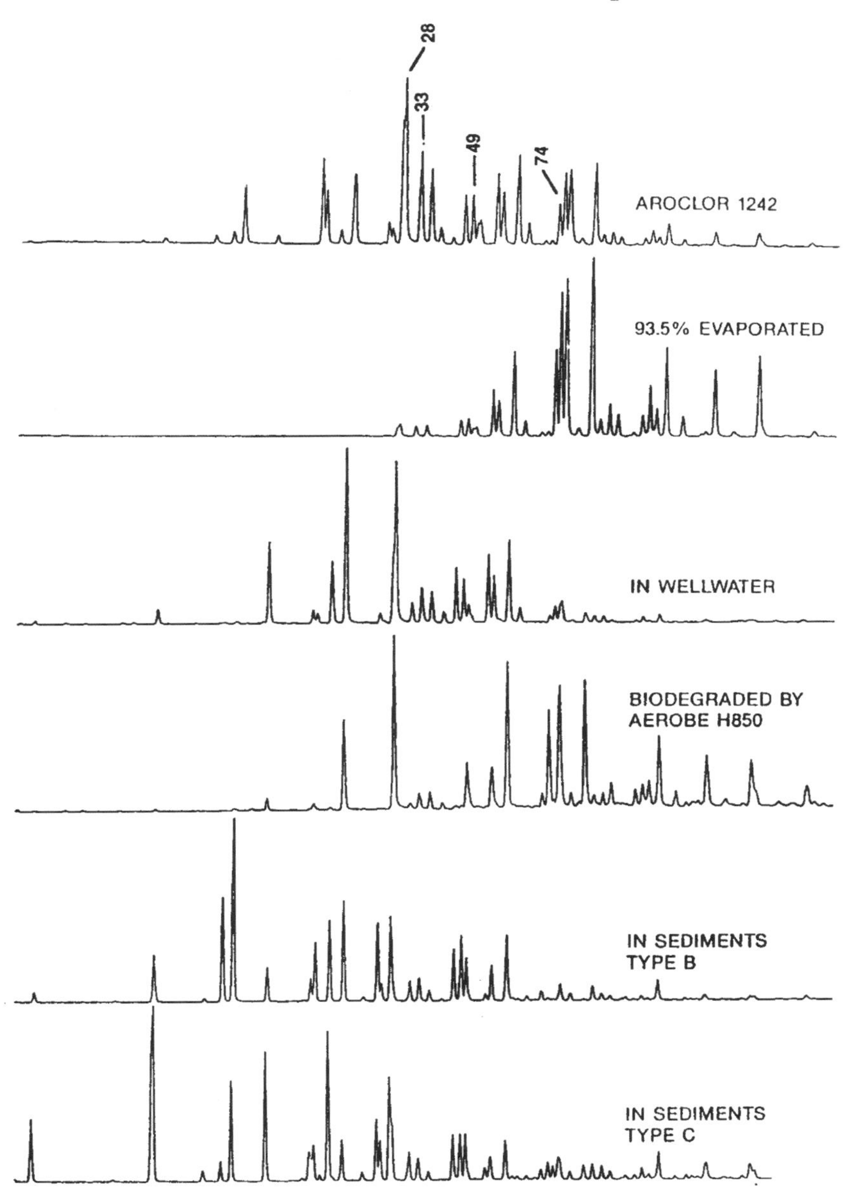

Note:

1. Numbers indicate IUPAC assignments for PCB congeners.

Source: Author.

Aerobic microbial biodegradation

The marked differences in congener removal pattern between different microbial strains has permitted the recognition of several dozen different types of PCB-degrading aerobic bacteria in the environment (Figure 2, taken from Bedard *et al.*, 1986). Some of the individual strains can metabolise most of the mono-, di-, and trichlorobiphenyls, and even some of the tetras, pentas, and hexas, and hence might be attractive for the *ex situ* bioremediation of soils or sediments contaminated with the lower Aroclors (Abramowicz, 1994). The most frequently isolated strains, however, are more selective in their appetites for PCB congeners, and give depletion patterns more like that of the wellwater in Figure 1.

Figure 2. Comparison of the PCB-degrading competence of environmental bacterial isolates

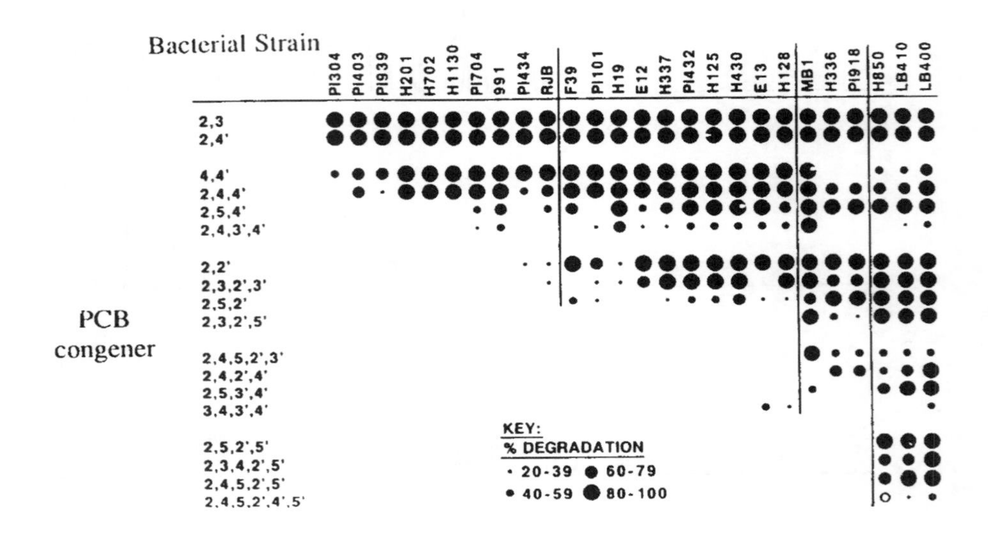

Note:

1. [o] indicates that H850 degraded less than 20 per cent of the congener, but a metabolite was isolated.

Source: Bedard *et al.*, 1986.

Even this extent of depletion can be useful in reducing the total PCB level if applied to an environmentally dechlorinated PCB specimen like those at the bottom of Figure 1. In a large-scale *in situ* field test on the Hudson River, we found that simple oxygenation of the sediments was all that was required to take out 37 per cent to 55 per cent of the PCB present (Harkness *et al.*, 1993). The occurrence of aerobic microbial biodegradation under the accelerated test conditions, as well as in the untreated riverbottom sediments, was demonstrated not only by the depletion of the biodegradation-sensitive lower PCB congeners, but also by the appearance of their oxidised biodegradation products, the hydroxybiphenyls and chlorobenzoic acids (Flanagan and May, 1993). Unfortunately, however, aerobic microbial biodegradation is generally much less effective at reducing levels of the toxic coplanar or "near-coplanar" congeners.

Anaerobic microbial dechlorination

The occurrence of microbial dechlorination in anaerobic aquatic sediments was first discovered in those of the upper Hudson River (Brown *et al.*, 1987) and subsequently found to be occurring at many other PCB-contaminated sites as well, including marine sediments at New Bedford, Massachusetts; Bridgeport, Connecticut; and Escambia Bay, Florida, as well as freshwater sediments at Pittsfield, Massachusetts; Kalamazoo, Michigan; Massena, New York; Waukegan, Illinois; Sheboygan, Wisconsin; and Lake Hartwell, South Carolina. It has also been repeatedly duplicated in anaerobic cultures in the laboratory, although none of the individual organisms involved has yet been isolated. Both the environmental PCB chromatograms (Brown, 1990) and the laboratory cultures (Bedard and Quensen, 1995) show that a variety of anaerobic organisms having widely varying ranges of congener-dechlorinating abilities are present in PCB-contaminated environments. At some sites, the extensive dechlorination to (mainly) mono- and dichlorobiphenyls shown at the bottom of Figure 1 may occur; at others, the dechlorination may be quite selective, attacking only congeners carrying 3,4-, 2,3,4-, 3,4,5-, or 2,3,4,5-chlorophenyl groups. Even such dechlorinations, however, can be toxicologically significant because of their ability to remove the "coplanar" (non-*ortho* substituted) or "near coplanar" (mono-*ortho* substituted) congeners that have been of concern because of their dioxin-like activities in some wildlife species.

A potentially very important recent discovery was that at a site where natural detoxication of a PCB release by anaerobic microbial dechlorination had been proceeding only slowly, the dechlorination process could be greatly stimulated by addition of certain brominated aromatics (Bedard *et al.*, 1995).

Metabolism in higher animals

Another very important form of PCB biodegradation, namely cytochrome P450- (or "CYP") mediated metabolism, occurs in most higher animals. Again, there is some variation in congener selection patterns, the commonest being designated P4501A-like and P4502B-like. A survey of patterns in aquatic fauna (Brown, 1992) showed absence of PCB metabolism in the molluscs, echinoderms, and many of the common teleost fish; weak P4501A-like metabolism in most teleosts; weak to moderate 2B-like activity in most crustaceans and a few teleosts; and strong P4502B-like activity (with or without additional 1A-like activity) in most birds and mammals, including man. These PCB congener selection patterns have been characterised in detail (Brown, 1994), and are useful in assessing the toxic response (if any) of the animal to its xenobiotic burden. In terms of identifying environmental PCB sources and pathways, however, P450-mediated PCB metabolism is often a significant complication since it can attack all but a few of the most highly chlorinated congeners, and thus obscure the record. Accordingly, it is often easier to distinguish PCB sources and pathways in specimens of sediments, molluscs, or metabolically-inactive fish than in those from birds or mammals.

Solar photodegradation

Available data and calculations show that PCBs in the atmosphere (Bunce *et al.*, 1989) or in near-surface waters (Bunce *et al.*, 1978) should be undergoing photolysis at quite significant rates. Thus far, however, we have been unable to identify the congener depletion pattern that is characteristic of solar photolysis in the PCBs of any samples of water, sediments, or biota taken from the environment (Brown *et al.*, 1995). Presumably, this results because only a very small fraction of the total environmental PCB burden resides in niches where exposure to solar ultraviolet radiation may occur.

Evaluation of Environmental PCB Sources

Industrial vs. Pyrogenic TEQ sources

Since most of the dioxin toxic equivalency (TEQ) of an environmental PCB mixture is contributed by the coplanar congeners 126 and 169, and since most of these coplanars could be coming from continuing combustion sources rather than either old or ongoing industrial releases, an evaluation of the former possibility is an important step in remedial response selection. A calculation of the combustion-derived fraction requires data on the levels of PCB congeners 126, 169, and either 156 or (preferably) 153 in a representative biological specimen from the site, and also similar data for the Aroclor (or other commercial mixture) released. In the absence of the latter data, the original Aroclor composition may be assumed to be the average for 1958-1977 production, for which the levels of congeners 126 and 169, relative to either PCB 153 = 1000 or PCB 156 = 127, have been estimated as 0.95 and 0.01, respectively (Brown *et al.*, 1995). That paper also quoted data indicating the mean levels of 126 and 169 in incinerator emissions, relative to the same reference congeners, as 345 and 88, respectively. Thus, any increase in the PCB 169/PCB 153 ratio over 0.01×10^{-3} can be attributed to pyrogenic inputs; the pyrogenic inputs of PCB 126 will be (345/88) times as great, and their contribution to the total original (i.e. pre-degradation) PCB 126 input will be that number divided by the sum of its concentration, plus the calculated PCB 126 input from the Aroclor mix present. (For human and wildlife PCB specimens, this calculated sum is generally greater than the observed PCB 126 level because PCB 126 is not completely immune to the biodegradative processes that usually remove almost the entirety of its next lowest homolog, namely the major coplanar tetrachlorobiphenyl, PCB 77. For dechlorinated sediment specimens, the calculation that assumes the stability of PCB 169 will be invalid; instead, the observed levels of all the coplanars will be very low because their susceptibility to dechlorination is much greater than that of the reference congener, PCB 153.)

Application of the above calculation procedure to the PCB congener ratios reported for both humans and wildlife from various sites (Brown *et al.*, 1995) indicates that the proportionate contribution of pyrogenic PCBs to total PCB TEQ can range from values averaging 60 per cent to 80 per cent in remote sites to near-zero values in the vicinity of Aroclor point sources.

Ongoing vs. Past PCB releases

The fact that the lower PCB congeners can be readily biodegraded in aerobic microbial environments, such as natural waters, lightly contaminated soils, or surficial sediments, provides a means of assessing the "freshness" of an environmental PCB specimen derived from Aroclors 1242, 1016, or 1248. This occurs because these Aroclors contain two congeners having exactly the same volatility, water-solubility, and bioaccumulation potential, but very different susceptibilities to the ordinary, mixed population, aerobic microbial biodegradation. These congeners are the trichlorobiphenyls PCB 28 (24-4 CB) and PCB 33 (34-2 CB), see Figure 1. The relative biodegradative sensitivity is shown by the wellwater gas chromatogram in Figure 1, where PCB 28 is still strong, and the peak containing PCB 33 is gone except for the small (*ca.* 5 per cent) residue provided by a coeluting tetrachlorobiphenyl that is difficult to biodegrade. Thus, the ratio of PCB 33 to PCB 28 in an environmental specimen, relative to that in the original Aroclor, is a sensitive measure of the extent of "weathering". In our experience, we have only rarely seen PCB 33/PCB 28 ratios for temperate zone environmental PCB specimens above 0.5 of the Aroclor values; when such ratios are seen, it can be concluded that the observed PCB is derived from a very recent or continuing source rather than an old environmental deposit.

Alternative Aroclor sources

At heavily industrialised sites there may well be more than one significant source for the PCBs found in the sediments, water, and biota, and the determination of their relative contributions important in remedy selection. Such determinations are possible if the original sources differed in either Aroclor type or co-contaminants present. Procedures for calculating the original Aroclor composition of an environmental PCB mixture need to take into account both the changes in homolog distribution caused by differences in bioaccumulation factors, as well as the changes in isomer distributions resulting from biodegradation, metabolism, or dechlorination, and hence tend to be both site- and specimen type-specific. An example of such a procedure, as applied to sediment PCBs that were both dechlorinated and partially leached, is given by Brown and Wagner, 1990.

PCB dechlorination/detoxication status

An assessment of the progress of dechlorination in the PCBs from a mollusc or fish can be important for determining whether their PCBs came from an old, long-buried, partially dechlorinated bottom deposit, or from more recent releases. It can also indicate the extent of detoxication.

Quantitative assessments of the extent of dechlorination of subsurficial sediment PCBs can often be made from a complete congener analysis by simply summing the total number of *meta* and *para* chlorines per biphenyl present, and comparing it with the original value, as judged by the level of *ortho* chlorines, which are generally not removed by microbial dechlorination. Unfortunately, this procedure is not easily applicable to the PCBs of biological specimens, since bioconcentration and biodegradation processes can also affect the relative levels of *meta* and *para* chlorines. Instead, it is more convenient to focus on ratios of dechlorination-sensitive to dechlorination-insensitive congeners, preferably congeners that are difficult to metabolise with the same molecular weight, so that their bioconcentration tendencies will be similar. We have found the ratio of the readily dechlorinated PCB 74 to the more stable PCB 49 (both tetrachlorobiphenyls) a fairly convenient indicator to use on Hudson River fish, since the original congener ratio is the same for both Aroclor 1242 and 1254. Other indicator ratios, and their use on Acushnet Estuary sediments and water, are described by Brown and Wagner, 1990. Determinations of such ratios have shown that in one stretch of the Hudson River, the PCBs in fish were coming from recent seepage of undechlorinated PCBs from old releases rather than heavily contaminated old sediments, in contrast with long-standing presumptions. In the Acushnet Estuary, PCB sources to fish from undechlorinated, lightly contaminated lower estuary sediments are quite comparable to those of the identified remediation target, dechlorinated heavily contaminated upper estuary "hot spots" (Brown and Wagner, 1990).

It was also found that the toxic coplanar congeners, such as PCB 77, were considerably more rapidly dechlorinated than the more readily analysed mono-*ortho* congeners, such as PCB 74. Thus, in the lightly dechlorinated PCBs of lower Hudson River fish, where the losses of PCB 74 were only about 50 per cent, the losses in PCB 77 were 90 per cent. In upper Hudson River sediments, where losses of PCB 74 might range up to 98 per cent, the coplanars were undetectable. In such areas, natural recovery has significantly mitigated the potential human and environmental health effects via extensive detoxication. The progress of such dechlorination at one repeatedly sampled upper Hudson River site is shown in Figure 3. Even in the lower Hudson River, where loss of the coplanars was only 90 per cent complete (because of the much lower PCB levels and hence slower dechlorination rates), one could effectively argue that natural recovery was making satisfactory progress.

Figure 3. DB-1 capillary gas chromatograms of the PCB residues in the sediments of upper Hudson River site H7 over the period 1982-1990

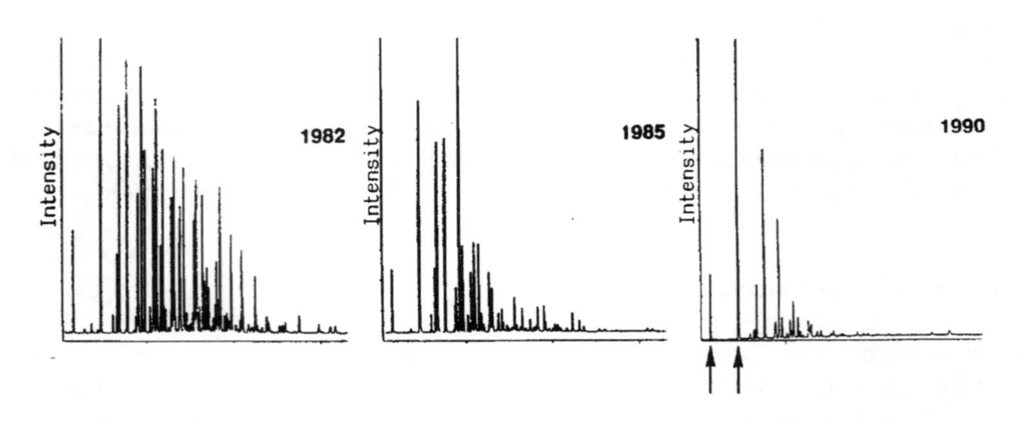

Note:

1. Arrows indicate the peaks given by 2-chlorobiphenyl and a mixture of 2,2´- and 2,6-dichlorobiphenyls.

Source: Author.

Conversely, we are well aware of other PCB-contaminated rivers and lakes where the local microbial ecology is apparently not favorable to dechlorination, and where natural recovery is not yet evident. The gross measures of PCB levels at such sites are often little different from those where detoxication is well advanced. It is only by detailed studies of PCB congener distribution that one can distinguish those sites that are achieving recovery through natural microbiological processes from those that are not.

Questions, comments, and answers

Q: Is weathering occurring in the anoxic zones, and have the PCB's been 'removed' by complexing with the soil particles, etc.?

A: It is happening - we detect dechlorination products entering the aerobic zone. It is, however, relatively insignificant on a mass basis.

Q: How do you collect the sediment data and normalise it since there is fresh deposition each year? How do you translate the fish body burden data to rates of removal?

A: We don't try to normalise it. We do not have enough data to do this statistically. We believe the fish data relates to the rate at which aerobic degradation is going on in the top zone. We were continually challenging the system with more PCB and the degradation was keeping up with this. These are all indirect measurements.

Q: Why does dechlorination not work in anaerobic soils?

A: The only evidence we have for dechlorination is in wet soils, lagoons and wetlands. It may simply be water limited, but we have not looked as hard as we have at sediments. The organisms are there, however.

Q: What type of scouring is going on and does this include catastrophic events?

A: Its physical not biological scouring, e.g. rain, boat movements, etc.

REFERENCES

ABRAMOWICZ, D.A. (1994), "Aerobic PCB biodegradation and anaerobic PCB dechlorination in the environment", *Res. Microbiol.* 145, pp. 42-46.

ABRAMOWICZ, D.A. and D.R. OLSON (1995), "Accelerated biodegradation of PCBs", *CHEMTECH* 24, pp. 36-41.

ALSBERG, T., C. DeWIT, V. ERIKSSON and V. JÄRNBERG (1995), "PCB patterns in herring and pike with special reference to co-planar congeners", Paper presented at Second SETAC World Congress, Vancouver, B.C., Canada, 5-9 November 1995. Abstract No. 238.

BEDARD, D.L. and J.F. QUENSEN III (1995), "Microbial reductive dechlorination of polychlorinated biphenyls", in *Microbial transformations and degradation of toxic organic chemicals*, L.S. Young and C. Cerniglia (eds.), John Wiley & Sons, New York, pp. 121-210.

BEDARD, D.L., L.A. SMULLEN, K.A. DeWEERD, D.K. DIETRICH, G.M. FRAME II, R.J. MAY, J.M. PRINCIPE, T.O. ROUSE, W.A. FESSLER and J.S. NICHOLSON (1995), "Chemical activation of microbially-mediated PCB dechlorination: A field study", *Organohalogen Cmpds.* 24, pp. 23-28.

BEDARD, D.L., R. UNTERMAN, L.H. BOPP, M.J. BRENNAN, M.E. HABERL and C. JOHNSON (1986), "Rapid assay for screening and characterizing micro-organisms for the ability to degrade polychlorinated biphenyls," *Appl. Environ. Microbiol.* 51, pp. 761-768.

BEDARD, D.L., R.E. WAGNER, M.J. BRENNAN, M.E. HABERL and J.F. BROWN Jr. (1987), "Extensive degradation of Aroclors and environmentally transformed polychlorinated biphenyls by *Alcaligenes eutrophus* H850," *Appl. Environ. Microbiol.* 53, pp. 1094-1102.

BROWN, J.F. Jr. (1990), "Identification of environmental PCB transformation processes", *Organohalogen Cmpds.* 2, pp. 21-24.

BROWN, J.F. Jr. (1992), "Metabolic alterations of PCB residues in aquatic fauna: Distributions of cytochrome P4501A- and P4502B-like activities", *Marine Environ. Res.* 34, pp. 261-266.

BROWN, J.F. Jr. (1994), "Determination of PCB metabolic, excretion, and accumulation rates for use as indicators of biological response and relative risk", *Environ. Sci. Technol.* 28, pp 2295-2305.

BROWN, J.F. Jr., D.L. BEDARD, M.J. BRENNAN, J.C. CARNAHAN, H. FENG and R.E. WAGNER (1987), "Polychlorinated biphenyl dechlorination in aquatic sediments", *Science* 236, pp. 709-712.

BROWN, J.F. Jr., G.M. FRAME II, D.R. OLSON and J.L. WEBB (1995), "The sources of the coplanar PCBs", *Organohalogen Cmpds.* 26, pp. 427-430.

BROWN, J.F. Jr. and R.E. WAGNER (1990), "PCB movement, dechlorination and detoxication in the Acushnet Estuary", *Environ. Toxicol. Chem.* 9, pp. 1215-1233.

BUNCE, N.J., Y. KUMAR and B.G. BROWNLEE (1978), "An assessment of the impact of solar degradation of chlorinated biphenyls in the aquatic environment", *Chemosphere* 7, pp. 155-164.

BUNCE, N.J., J.P. LANDERS, J.-A. LANGSHAW and J.S. NAKAL (1989), "An assessment of the importance of direct solar degradation of some simple chlorinated benzenes and biphenyls in the vapor phase", *Environ. Sci. Technol.* 23, pp. 213-218.

FLANAGAN, W.P. and R.J. MAY (1993), "Metabolite detection as evidence for naturally occurring aerobic PCB biodegradation in Hudson River sediments", *Environ. Sci. Technol.* 27, pp. 2207-2212.

HARKNESS, M.R., J.B. McDERMOTT, D.A. ABRAMOWICZ, J.J. SALVO, W.P. FLANAGAN, M.L. STEPHENS, F.J. MONDELLO, R.J. MAY, J.H. LOBOS, K.M. CARROLL, M.J. BRENNAN, A.A. BRACCO, K.M. FISH, G.L. WARNER, P.R. WILSON, D.K. DIETRICH, D.T. LIN, C.B. MORGAN and W.L. GATELY (1993), "*In situ* stimulation of aerobic PCB biodegradation in Hudson River sediments", *Science* 259, pp. 503-507.

THOMANN, R.V. (1989), "Bioaccumulation model of organic distribution in aquatic food chains", *Environ. Sci. Technol.* 23, pp. 699-707.

BIOREMEDIATION OF FORMER GAS MANUFACTURING SITES IN THE UNITED KINGDOM

by

Alwyn Hart
British Gas Plc., Research & Technology, Gas Research Centre
Loughborough, Leicestershire, United Kingdom

Introduction

Natural gas is now a familiar and widely used fuel throughout the world. Its many qualities make it a premium fuel which is more environmentally friendly than alternatives such as coal, oil and nuclear. In the United Kingdom, sources of natural gas under the North Sea and the Irish Sea (especially Morecambe Bay), have been developed by British Gas. However, natural gas has only been widely available since the 1960s; prior to which time a different gas was produced by the reductive gasification of other fuels such as coal and oil. In the United Kingdom, gas manufacture began in the early nineteenth century and remained widespread until the 1960s. At the time of nationalisation in 1948, the newly incorporated Gas Council took control of over 1 000 of the estimated 3 000-5 000 gasworks. Following privatisation, British Gas took over most of these 1 000 sites.

Biological treatment has been used for remediation of many types of hydrocarbon contamination. However, in the United Kingdom, the use of bioremediation on former gasworks sites has been limited. Two full-scale and independently validated projects have been carried out by specialist treatment contractors, but in our view, the results of both have been inconclusive. In particular, it has not been possible to adequately demonstrate that any reductions in contaminant concentration obtained are due to biodegradation. This does not mean that bioremediation is ineffective, rather there remains a pressing need for independent, valid data. Without such validation, future prospects for bioremediation of gasworks waste may be poor.

This paper aims to review the evidence from three bioremediation projects carried out in the United Kingdom: a full scale project carried out in Lancashire in the late 1980s, a similar project in Yorkshire in 1994, and a pilot scale experiment in London by British Gas R&T which is still on-going.

Case studies

Lancashire

This project was carried out by a specialist contractor in 1986 (Bewley *et al.*, 1989) and independent, external validation was done by Lancashire County Analysts Department (UK Department of the Environment, 1988) Microbial treatment of tar contaminated soil was intended to achieve clean-up targets of:

- less than 10 000 mg kg^{-1} for total PAH (sum of 16 US EPA priority pollutants)
- less than 5 mg kg^{-1} for phenols (sum of 9 compounds)

In the event, few of the samples analysed prior to the start of remediation were above the target concentrations. That is, the soil was already cleaner than the remediation targets. The independent validating laboratory in this case reached the following conclusions:

> In those two key areas of phenols and coal tar, direct comparison between the 'before' and 'after' results is almost impossible.
>
> It can certainly be said with confidence that the results... demonstrate compliance with the DOE's post-treatment requirements. It can not be said, though, from our data above that there has been a real reduction in phenols and coal tar levels.
>
> However, examination of the information... indicates a statistically significant reduction in variances after site reclamation. This implies a reduction in the concentrations of the higher levels of contamination which existed on the site prior to its reclamation.
>
> (UK Department of the Environment, 1988, p. 24)

Yorkshire

Background

This project commenced in the spring of 1994. As part of the redevelopment of a former gasworks, a number of treatment methods for contaminated soils from the site were considered. In particular, bioremediation was chosen for some tar-contaminated soil. A number of specialist bioremediation vendors were invited to submit tenders for the project and one was appointed on this basis. Over the course of the redevelopment, approximately 8 000 m^3 of soil were treated by bioremediation.

The issue of validation was also discussed. It was agreed that British Gas R&T, not being directly involved in the remediation, would act as an unofficial validation laboratory. In addition to the vendors' own analyses, two contract laboratories were appointed to monitor progress, carry out chemical analysis related to the site, and validate the treatment. During the course of the bioremediation, sub-contracting parties encountered considerable difficulties in interpreting and validating the results from different laboratories. In particular, adequate validation proved impossible due to lack of data. In this paper, therefore, the results of British Gas R&T's own analyses are presented and discussed. These proved to be broadly similar to the vendors' own data.

Treatment targets

The Interdepartmental Committee for the Redevelopment of Contaminated Land (ICRCL) has set a guideline action value for PAH on land intended for use as landscaped areas, buildings and hardcover (ICRCL, 1986; ICRCL, 1987) of 10 000 mg kg^{-1} in air dried soil.

Treatment: whole landfarm bed

The starting concentrations of PAH reported by the four laboratories are given in Table 1. It was originally intended that all the material for bioremediation would be treated in one landfarm bed. However, due to delays in excavation of the contaminated soil, the bed was set up a series of "bays" or strips with starting times spread over a period of two months.

Table 1. Initial PAH concentrations in soil

Laboratory	Total PAH mg/kg	Pyrene mg/kg	Number of data points
Contractor 1	26	115	22
Contractor 2	251	N/A	15[1]
Vendor	1 352	177	64
R&T	749	119	21

Notes:
1. Includes eight analyses reported as zero.

Source: Author.

The initial soil concentration of total PAH was already considerably lower than anticipated prior to the start of excavation. The delays in establishing the bed also led to problems in the statistical interpretation of the treatment data. The only way to overcome this has been to pool data from different bays on the basis of treatment duration. Thus, for example, samples of soil treated for 60 days may not have all been treated for the *same* calendar 60 days. As a further consequence, sample times are grouped in time bands of approximately 10-day intervals. Table 2 shows the total PAH in samples from the bed taken throughout treatment, plus some statistical data. Since reduction in pyrene was also a condition of successful treatment, pyrene concentration is shown in Table 3.

Table 2. Total PAH (mg kg^{-1} dry soil) vs. time for combined bed data

Time (days)	0.00	9-21	22-34	35-47	48-60	61-73	74-88	>117
Number of samples	14	13	30	11	8	31	20	22
Mean value	955	836	700	556	550	647	596	774
Median	737	690	513	507	494	519	550	699
Standard deviation	634	427	438	136	126	446	170	255
95 % Confidence interval (+/-)	383	258	164	91	105	163	80	114
Upper confidence limit	1 338	1 094	864	647	655	810	676	888
Lower confidence limit	572	578	536	464	445	484	516	660

Source: Author.

Table 3. Pyrene concentration (mg kg^{-1} dry soil) vs. time for combined bed data

Time (days)	0.00	9-21	22-34	35-47	48-60	61-73	74-88	>117
Number of samples	14	13	30	11	8	31	20	22
Mean value	151	125	106	86	78	102	88	106
Median	126	100	86	77	71	85	85	98
Standard deviation	94.	63	62	25	17	72	22	36
95 % Confidence interval (+/-)	57	38	23	17	14	26	10	16
Upper confidence limit	208	163	129	103	92	128	98	122
Lower confidence limit	94	87	82	69	64	75	78	90

Source: Author.

From the data above, it is clear that there has been no large reduction in either total PAH or pyrene concentrations over the course of treatment. This is easily confirmed with simple statistical tests. The Student's t-test and confidence interval tests were applied, and no significant difference in PAH concentrations before and after treatment can be demonstrated. However, one clear effect is a reduction in sample variability as treatment progressed. An obvious mechanism for this is the mixing which occurs as part of the landfarming process. Soil is tilled primarily to increase aeration as an aid to aerobic degradation, but also has the effect of increasing soil homogeneity.

Treatment: individual bay

As treatment progressed and the difficulties caused by the delays in excavation became clear, a decision was made to follow treatment of one area (or bay) more intimately. The aim was to overcome some of the logistical and interpretative problems caused by the piecemeal bed construction. As part of this exercise, the concentrations of lead and zinc throughout the bay were determined as well as PAH, in an attempt to distinguish mixing and biological treatment effects. The results from these analyses are shown below in Tables 4 and 5 .

Table 4. Total PAH (mg kg^{-1} dry soil) in bay vs. time

Time (days)	0.00	31.00	70.00	117.00
Number of samples	10	14	14	14
Mean	1 085	886	873	805
Median	786	844	659	723
Standard deviation	716	491	537	305
95 % confidence limit (+/-)	505	292	320	182
Upper confidence limit	1 589	1 179	1 193	987
Lower confidence limit	580	594	553	623

Source: Author.

Table 5. Pyrene concentration (mg kg^{-1} dry soil) in bay vs. time

Time (days)	0	31	70	117
Number of samples	10	14	14	14
Mean	170	138	127	112
Median	141	127	93	106
Standard deviation	106	70	90	42
95 % confidence limit (+/-)	75	41	54	25
Upper confidence limit	245	179	180	137
Lower confidence limit	95	96	73	87

Source: Author.

Once again, although some trends can be seen in the data suggesting reductions in PAH, and pyrene concentrations have occurred, these are not statistically significant at the 5 per cent level.

Metal concentrations in treated soil

The determination of metal concentrations from the same soil samples collected for PAH analysis revealed different trends for zinc and lead (Tables 6 and 7). Zinc concentrations showed a steady increase through the treatment whilst lead levels decreased markedly then increased again late in the work. The increase in zinc levels was unexpected. Possible explanations include input of zinc as part of the agricultural NPK fertiliser used by the treatment vendor and/or from the machinery used in tilling the soil, and/or from a sand layer laid under the contaminated soil. None of these mechanisms can be confirmed. Since the site is surrounded by major roads on three sites, it was anticipated that lead levels rather than zinc may potentially rise over the course of treatment. Instead, lead levels declined; although the variability once again means that the differences before and after treatment are not statistically significant. The metals data gives no indication of a decrease in the variability of metal content in the samples analysed as treatment progressed. This is in contrast to the PAH data above, and suggests that metals were not affected by homogenisation of the soil in the same way. Zinc in particular was apparently evenly distributed through the soil, presumably in solution. Note that all water leaving the bed as leachate was collected in surrounding purposely built drainage, and recycled as irrigation water.

Table 6. Zinc concentration in bay vs. time

Time (days)	0	31	70	117
Number of samples	9	15	15	14
Mean	130	145	154	178
Median	132	142	145	165
Standard deviation	30	48	45	52
95 % confidence limit (+/-)	26	28	26	31
Upper confidence limit	156	172	180	209
Lower confidence limit	104	117	129	147

Source: Author.

Table 7. Lead concentration in bay vs. time

Time (days)	0	31	70	117
Number of samples	9	15	15	13
Mean value	400	348	192	310
Median	457	253	178	306
Standard deviation	86	212	57	183
95 % confidence limit (+/-)	74	122	33	113
Upper confidence limit	474	470	225	423
Lower confidence limit	327	226	159	197

Source: Author.

Biological analysis

As part of the landfarming treatment, the vendor company applied a suspension of their proprietary bacteria twice weekly via the irrigation system. These organisms were intended to improve PAH degradation rates. In order to assess the microbial population in the bay, soil samples were taken by British Gas R&T for viable count analysis. Soil was vigorously shaken for 10 minutes with one-quarter Ringer's solution, serially diluted and plated onto pre-poured Difco nutrient agar plates or minimal medium plates containing naphthalene as sole carbon and energy source (Cerniglia and Hietkamp, 1990). Naphthalene was chosen as the simplest, most volatile PAH and provided rapid growth for this analysis. Naphthalene is a major component of coal tar. However, it is readily degraded and often little remains in aerated soil on former gasworks. In our experience, organisms which are capable of degrading higher PAH can also degrade naphthalene although the converse does not always apply.

Results are presented in Table 8 and indicate a decrease in the number of heterotrophic and naphthalene-degrading organisms during treatment. Clearly, the representative and reliable assessment of soil organism populations is fraught with difficulties of reliable sampling, organism survival, organism attachment to soil particles, etc. However, since the recovery method used here was identical before and during treatment, it seems reasonable to conclude that this data represents real differences in soil populations.

Table 8. Soil microbial populations in bay before and during treatment

	Cell numbers (cfu/g wet soil)			
Phenotype	Heterotrophs		Naphthalene utilisers	
Time (days)	0	30	0	30
Mean values	7.0E+09	2.2E+06	7.8E+08	9.9E+05

Source: Author.

Anion analysis

Analysis of samples for nitrate, chloride and sulphate was carried out using ion chromatography. Phosphate concentrations could not be determined due to interference from the high concentrations of sulphate present. Constant levels of sulphate (2 400 to 3 000 mg kg^{-1}) and chloride (<500 mg kg^{-1})

were seen but a rapid increase in nitrate concentration was observed early in the treatment (from approximately 100 mg kg^{-1} to 2 000 mg kg^{-1}). This is undoubtedly due to the application of fertiliser by the treatment vendor.

Overall experience

British Gas invested considerable time and resources in the analysis of this bioremediation project. Comparison of PAH and metal concentrations was used in an attempt to assess the effects of mixing, but once again the results were inconclusive. Certainly, mixing cannot be discounted as a possible cause of PAH reduction. It is also unfortunate that the starting concentration of PAH was close to the agreed remediation target.

It appears that the treatment itself and/or the mixed microbial culture applied resulted in a decrease in the soil microbial population. Are there any instances where such inoculation can really be shown to have improved treatment ?

The wide variability in soil contamination levels made confirmation of treatment very difficult to obtain. The sampling density which would have been required to confirm or deny successful treatment would be extremely expensive. Nonetheless, validation is crucial to the identification of effective treatments for contaminated soil.

London

British Gas R&T have carried out in-house investigations into the use of bioremediation for treatment of contamination (in particular polyaromatic hydrocarbons, PAH), on former gasworks sites in the United Kingdom. Following initial laboratory studies, a pilot scale landfarm treatment facility was constructed in early 1994. A site was chosen which was known to contain tar-contaminated soil and where sample material could be excavated without disturbing operational activities.

The treatment bed was built in the winter of 1993 and is approximately 35 m × 5 m in size. A substantial structure was built to ensure that contamination did not affect the surrounding operational site and to protect other workers on the site. The base structure is made of concrete and includes a drain running across the centre of the bed. The base is then surrounded by a breeze-block wall which is coated with a thin layer of concrete. On top of the wall a conventional agricultural polythene tunnel greenhouse was erected to ensure that water entry to the bed could be controlled. Sprinkler pipes along the top of the greenhouse are used to irrigate the bed and supply nutrients. The central drain was connected to an external, underground water storage tank. Run-off water from the bed was collected in the external tank for re-use as irrigation water, and this was supplemented with fresh water from the mains supply as required. Nutrients, in particular nitrate and phosphate, were also supplied through the irrigation system; salts were dissolved in the water tank immediately prior to irrigation.

In this experimental work, somewhat greater reductions in total and individual PAH concentrations have been obtained (Tables 9 and 10). Statistical testing shows these reductions to be significant at the 95 per cent level.

Table 9. PAH in experimental landfarm vs. time

Time (months)	2	12	15
Number of samples	54	54	54
Mean value	1 860	1 293	907
Median	1 570	882	814
Standard deviation	1 292	1 049	558
95 % confidence limit	511	355	249
Upper confidence limit	2 372	1 648	1 156
Lower confidence limit	1 349	937	658

Source: Author.

Table 10. Pyrene concentration in experimental landfarm

Time (months)	2	12	15
Number of samples	18	18	18
Mean value	256	177	122
95 % confidence limit (+/-)	88	84	19
Upper confidence limit	344	261	141
Lower confidence limit	168	93	103

Source: Author.

Clearly, the treatment time scales are much greater than those attempted at either of the two full-scale projects discussed earlier. The longer time was anticipated from the start of the experiment. In our view, it may be extremely unwise to attempt the rapid (one growing season or less) biological remediation of former gasworks waste. Previous laboratory experiments carried out by British Gas have also required similar periods for successful degradation of these wastes.

Summary and conclusions

The use of bioremediation in the United Kingdom for treatment of complex, difficult wastes such as gasworks has been limited to a few sites. Even fewer have been publicly reported. Validation is crucial to the identification of effective treatments and its lack has constituted a major obstacle to the wider use of bioremediation. Many of the treatment claims advertised by vendors are poorly tested, and the need for assessment criteria for bioremediation projects has been identified before.

Some simple steps can be taken:

– Adequate site investigation data must exist before treatment targets are set.

– Chemical analytical methods must be robust and fully validated.

– Untreated control areas should be constructed nearby.

– Any additives (micro-organisms, solvents) should be shown to be at least innocuous.

– Proof of bioremediation is difficult to obtain but other potential processes (e.g. dilution, volatilisation) can be discounted if appropriate measurements are made.

Questions, comments, and answers

Q: Your target levels for PCB remediation are very high. We would be *starting* remediation at these target levels in Germany!

A: The standards in the United Kingdom for gas works derive from a parliamentary committee and they are advisory levels, not enforced. They are currently being revised and are expected to be much lower.

C: We did carry out tests on the site you describe with and without organisms and nutrients on a controlled bed and showed there was reduction in PAHs. It was not possible to demonstrate this at full scale because of the amount of analyses required and the heterogeneity. Here the objectives were to see that targets were met. The UK government targets were set over 10 years ago and were originally based on coal tar.

Q: Can you estimate the number of samples needed? If what was done was not sufficient, what would be? How can we be certain of validation?

A: We know the number of samples we would need to take, but the cost of taking them, using existing techniques, is not sustainable.

REFERENCES

ATLAS, R.M. (1995), "Efficacy of bioremediation: chemical and risk-based determinations", in *Bioremediation: the Tokyo '94 Workshop*, pp. 183-188, OECD, Paris.

BEWLEY, R, B. ELLIS, P. THIELE, I. VINEY and J. REES (1989), "Microbial clean-up of contaminated soil", *Chemistry and Industry*, pp. 778-783.

CERNIGLIA, C.E. and M.A. HIETKAMP (1990), "Polycyclic aromatic hydrocarbon degradation by mycobacterium", *Methods in Enzymology* 188**,** pp. 148-153.

INTERDEPARTMENTAL COMMITTEE FOR THE REDEVELOPMENT OF CONTAMINATED LAND (ICRCL) (1986), "Notes on the Redevelopment of Gasworks Sites", *ICRCL Guidance note* 18/79.

INTERDEPARTMENTAL COMMITTEE FOR THE REDEVELOPMENT OF CONTAMINATED LAND (ICRCL) (1987), "Problems Arising from the Redevelopment of Gasworks and Similar Sites", Revised Edition, Department of the Environment/Environmental Resources Limited.

GRIMES, D.J. (1995), "Validation of reliability and predictability of environmental biotechnologies", in *Bioremediation: the Tokyo '94 Workshop,* pp. 309-315, OECD, Paris.

UK DEPARTMENT OF THE ENVIRONMENT (1988), *Draft report on the investigation of site reclamation at the former Greenbank gasworks*, Blackburn.

NEW PRODUCTS AND STRATEGIES IN SOIL BIOREMEDIATION

by

Beatrix Daei
Lobbe Xenex GmbH & Co, Iserlohn, Germany

Introduction

All human activity leads and has always led to the production of waste, undesired by-products, and pollution. Increasing industrialisation has augmented the spectrum of polluting agents and the intensity of pollution, thereby posing major problems for the environment and human health. Historically, pollution of air and water was noticed first, and protective measures were taken, either through end-of-pipe solutions or changes in production processes, both to avoid the production of harmful substances and to reduce waste. Their success is evident in improved water quality (e.g. rivers) and better urban air, as well as in material flow management (instead of only waste management) and re-use of by-products in the same or other industrial processes.

When soil vulnerability and the overstraining of its self-cleaning capacities were recognised, soil protection and the clean-up of contaminated soil became a fundamental issue for research and political discussion. Conspicuous examples of polluted sites brought this issue to public attention and elicited awareness of the importance and sensitivity of the soil. Land is a finite resource and its re-use is very important to a region's sustainable development and to maintaining greenfield areas. Laws and guidelines developed for soil protection form the political and regulative framework for all activities dealing with soil use, and help avoid differences in criteria for evaluating contamination and setting clean-up requirements for restoring industrial land. However, because of the large number of sites and the costs involved, the initial aim of full restoration of polluted areas soon gave way to partial restoration and the application of effective *in situ* solutions. Moreover, discussions have begun on acceptable levels of contaminants based on the projected use of sites.

Thermal, chemical-physical and biological techniques for soil remediation have proved their practical reliability during the past decade (Weber, 1990). Biological means of remediation are, from both the ecological point of view and often also from the economic one, more advantageous than others, as they make use of natural processes to degrade harmful substances. Yet, despite many successes, many problems remain in the field of bioremediation, along with many restrictions on application of biological techniques. New strategies and further development of existing procedures are necessary to extend their use and to react to changing ecological, economic and political demands for effective and affordable solutions. Today, despite the ecological advantages of bioremediation, other techniques are often preferred, because of the uncertainties, whether justified or not, attendant on the use of biodegradation.

Pointing out some problems

Many aspects of biology, geology, chemistry, technology, legislation, and the economy have to be taken into consideration. Among the problems involved, the following give an idea of the complexity of the issue.

- In large-scale applications, bioremediation is restricted to a narrow field of contaminants, mainly crude oil and refinery products. Although nearly all substances seem to be degradable under laboratory conditions (for a review, see Singleton, 1994), the situation is very different in the field, where their use is severely limited (for a review, see Providenti *et al.*, 1993).

- Some of the limitations are caused by limited availability of the substances.

- Circumstances in the soil may prohibit degradation from occurring, e.g. because of competition from other substances in a complex mixture (Stringfellow and Aitken, 1995) or because of the co-metabolic nature of the process (Bouchez *et al.*, 1995).

- Degradation may start, but metabolites may be either toxic or closely bound to the organic part of the soil, so that remediation is incomplete (due to residual contamination of the original substances or persisting metabolites).

- When fixing the aim of clean-up, the desired use of an area is rarely taken into consideration.

- Validity of monitoring and control mechanisms to evaluate the efficacy of the degradation process in practice (especially for *in situ* treatment but also, for example, in the choice of analytic procedures) need verification and optimisation.

- Because of the quantity of soil to be treated and the many sites to be cleaned low-cost solutions are needed.

- The long time frame of a bioremediation project is a disadvantage in the competition with other techniques.

Searching for solutions: the present state and new strategies

In scientific research as well as in large-scale applications, new strategies to overcome these problems are under investigation in different disciplines and at different levels.

Scientific research

- Basic research to provide a better understanding of interactions between soil, contaminants and micro-organisms. This includes investigation of the influence of tillage on soil characteristics, sorption behaviour of substances on different soil matrices and of parameters such as humidity, pH, temperature, significance of the rhizosphere, structure of communities of soil micro-organisms, including predation mechanisms and changes in this structure due to treatment.

- Basic research to discover presently unknown capabilities of micro-organisms – bacteria and fungi – in order to widen the use of biodegradation. This includes elucidating the degradation processes by the enzymes involved, the genetic basis, expression and regulation of these processes, including possibilities offered by genetic engineering and the production of biosurfactants under varying conditions. In addition to micro-organisms, the potential of plants to overcome metal as well as organic contaminants should be considered.

- Basic research to develop new detection methods to investigate the microbial flora in soil, such as application of PCR or immunological means through use of antisera or monoclonal antibodies.

Practical technical or large-scale approaches

- Application of adjuvants to accelerate the degradation process and solve the problem of non-availability. These adjuvants can either be a surfactant or a combination of nutrients and emulsifiers. At present, various products, which have proved their reliability in practice and are still being improved, are available. The infiltration of micro-organisms that produce biosurfactants in soil is also under intensive investigation.

- Combination of different treatment techniques to solve complex restoration problems, e.g. combining washing techniques and slurry bioreactors or chemical pre-treatment with a subsequent attack by microbes.

- Development of more efficient mechanical treatment techniques to lower costs and assure worker protection. This includes optimised equipment for oxygen and nutrient supply, as well as automatisation of the treatment process and protective measures.

- Extend application of *in situ* techniques to low-cost restoration. It is necessary to develop reliable means of surveying underground, e.g. the use of long-range soil exploration techniques. Activation mechanisms to augment bioremediation in soil and aquifers are needed, with adequate accompanying monitoring mechanisms. While many activation mechanisms are well known, they are handicapped by a lack of validated control mechanisms or by unknown physical subsoil conditions.

Bioavailability and treatment techniques, linked to application of micro-organisms and adjuvants, are discussed below:

Bioavailability

Any effect (toxic or promoting) assumes that substance and organism interact. In general, this interaction depends on contact in a water phase. Most of the relevant contaminants are hydrophobic and are only marginally soluble in water, with the result that they are often not metabolised. At present, the two main approaches to availability in the bioremediation field involve raising availability to improve metabolism/mineralisation and evaluating the significance of a pollutant's availability or non-availability in view of immobilisation.

Efforts to augment availability

Improved availability may be achieved by using tensides or emulsifiers, which may also have negative aspects, such as toxicity or mobilisation of contaminants which are then transported into groundwater. Their effect on bioavailability involves three main mechanisms:

- the dispersion of hydrophobe hydrocarbons;
- increasing the overall solubility of hydrophobic substances;
- facilitating the transport of the pollutant from soil to the aqueous phase.

Studies of their use for accelerated degradation in laboratory or at pilot scale often give opposing results (Fu and Alexander, 1995). Comparison of the overall effects described in 22 publications shows positive effects in 12 cases, negative results in seven, and none at all in three (Liu *et al.*, 1995). Successful application seems to depend on the specific circumstances of the soil, the contaminant and the tenside, and has so far been more empirical than the result of fundamental knowledge.

Induction or support of the production of bioemulsifiers by the degrading organisms might be a more promising approach. However, the intensive search for new forms (Fu and Alexander, 1995) and for circumstances in which the desired substances are excreted takes little account of the actual conditions of a soil clean-up project. These bioemulsifiers are either low molecular weight molecules such as glycolipids and phospholipids that lower interfacial tension or polymers that stabilise emulsions.

Although basic research explains much about the genetic background and the characteristics of different bioemulsifiers, little is presently known about the soil conditions needed for the production and action of these emulsifiers under remediation conditions. In order to be able to use these natural means effectively, research taking remediation and soil conditions into consideration is needed.

Research on the ecological significance of non-availability

The hazardous potential of a pollutant involves its influence on the environment and/or human health. In general, the water path is the main source of influence on living organisms and groundwater or drinking water. Endangering this medium often triggers decontamination activities.

The chemical immobilisation of metal or organic contaminants is a more or less validated means of clean-up (Weber, 1990). Discussion of biological immobilisation and evaluation of this "method" are relatively recent. It is well known that some pollutants, e.g. PAH or pesticides, interfere with the organic matter in soil. They bind to the clay-humus fraction or can be incorporated into the humic matrix. Microbial metabolism increases this tendency.

Experiments with radioactively labelled materials show that part of the contaminants seems to be mineralised (radioactive CO_2) and another part is bound (bound residue fraction) into the humic matrix. The chemical nature of this mechanism is significant for estimating long-term stability, one of the most important criteria for evaluating the reliability of biological immobilisation as a possible remediation method (Mahro, 1994). Furthermore, it is essential to understand the potential for the bound fraction to become a hazard, on the basis of site-specific exposition and uptake rates, taking into account metabolism in different organisms, including humans, as well as possible release of the original substance from humus matrix.

Up to now, the transfer of laboratory results to large scale (in this case as in many others) causes difficulties. It is not possible to simulate the long-term effects, and radioactive markers cannot be used in field studies, whereas most data on laboratory results are obtained this way. New forms of evaluation, perhaps using validated and comparable extraction measures, and research projects on long-term monitoring seem necessary in order to determine the usefulness of biological immobilisation.

Treatment techniques

In addition to the substantive issue of whether a bioremediation technique is applicable, economic efficiency is of great importance. Speed of degradation and optimised technology are decisive in terms of costs.

Acceleration of the process

The degradation process can be accelerated in various way, and there is still much scope for optimisation. Some methods involve the addition of micro-organisms, whether autochthonous or allochthonous, in order to augment the number and distribution of degrading organisms and improve decontamination. This involves techniques for inoculating the organisms, as bacteria cannot migrate in soil over "large" distances; even a few millimetres may be “large”. It is therefore very important to achieve close contact between nutrients, oxygen, degrading organisms, and contaminant through an adequate treatment technology.

Further development includes the idea of constructing the needed organisms, by inserting all the genetic information coding for those enzymes that are necessary for complete decontamination by one organism, without accumulating by-products or dead-end metabolites (Edgington, 1994). Another interesting idea involves the use of micro-organisms that change their surface characteristics. In a dormant state, they are able to survive for a long period and do not adhere to the clay or humus fraction in soil as bacteria generally do. Therefore, it may be possible to insert them in micropores in the contaminated soil over great distances during *in situ* clean-up. Then, with a special trigger, they change to their active form and start to degrade, perhaps using a genetically engineered enzyme repertoire. Micro-organisms have many as yet unknown capabilities, such as the ability to change surface characteristics in different life phases, which could be used in applications. Thus, it would be useful to widen the spectrum of available organisms. Fungi, yeasts, and plants certainly represent a repertoire of untapped capabilities for biotechniques.

Application of enzymes rather than whole organisms is a new approach that needs to be tested. As for micro-organisms as surfactants, the advantages of this method are subject to intensive discussion and seem to be case-dependent. In particular, the use of enzymes raises many questions about the stability and functionality of the enzyme in soil, a topic which must be investigated in order to determine the usefulness of this method. In general, different enzymes – acting in parallel or in sequence – are necessary for metabolisation and especially mineralisation. Although some products are available, further research is necessary to prove their usefulness.

In situ methods

Because of the increasing number of sites to be cleaned, low-cost solutions are needed. *In situ* methods offer some low-cost solutions, such as "pump and treat" techniques, hydroair lift, bioair lift or bioventing techniques, all more or less with microbiological aspects, such as activation of autochthonous flora as well as nutrient and oxygen supply.

One of the problems with these methods is how to monitor and prove the success of the method used. Existing methods require improved monitoring and better specific knowledge of subsoil conditions. Use of geological and geophysical techniques, perhaps transferring knowledge gained in oil fields or ore prospecting, can be very helping for investigating the contaminated area. These methods may include geoseismic, geoelectric or radiomagnetotelluric means. The application of new sampling methods, such as the passive multi-layer sampling system, provides advantages for monitoring the concentration and distribution of the contaminant in the aquifer.

Combination of techniques

Complex clean-up projects often requires various techniques – thermal, chemical-physical and biological – each of which has preferred fields of application. Integration of these processes can help overcome the limitations of a single technique in order to provide an overall treatment. This "technology bundling " represents an effective strategy for stationary soil remediation.

A combination of techniques can also be used for bioreclamation. Up to now, most projects use oxidative processes to metabolise xenobiotics. Sequential promotion of different metabolic pathways – aerobic/anaerobic or co-metabolic - in different phases of the degradation process hold out some promise. Plants can also be used to support degradation, especially to decontaminate large areas with low levels of contamination or reclaimed sites, in order to decrease residual xenobiotics. Transforming fundamental knowledge into a practical process involving adequate technology is very important for development.

Automatisation

While most of the previous discussion concerns the broader applicability of bioremediation, automatisation aims more specifically to reduce costs and help protect workers. Dynamic bioremediation methods in which machines are used to activate soil treatment are much more effective than static methods using stationary drainage systems. Automatisation is a well-tested industrial process that has so far been little used for bioremediation, but it could make it possible to carry out more continuous treatment, regulate the supply of nutrients and oxygen, and control the procedure through the use of computers.

In view of the EG guideline 90/679 EWG, worker protection is of paramount importance when using biological methods. In an automatic process, contact between contaminated material or biological agents is restricted to loading and unloading and perhaps sample taking. Soil movement is carried out entirely in closed machines.

Conclusions

The Council of Experts for Environment (Rat von Sachverständigen für Umweltfragen, Germany) stated in 1995 that "experiences from research and clean-up projects permit a better estimation of the applicability of microbiological techniques without any revolutionary new findings" (Rat von Sachverständigen für Umweltfragen, Altlasten II, February, 1995, p. 261).

Comparison of existing research publications and project descriptions supports this statement. There is a lack of revolutionary findings, but research and experience-based applications are progressing. Progress is evident in the tendency to promote new large-scale strategies or technologies through intensive basic research.

It is extremely important to transfer laboratory results to large-scale applications. Future experiments should take actual conditions more into consideration than they have in the past, for example in the field of emulsifiers or enzymes. Nevertheless, deeper understanding of the failure of micro-organisms to degrade under certain circumstances is necessary.

Extended use of combined methods, in terms both of technologies and of biological strategies, will extend the potential of bioremediation.

In the future, the strategy for soil restoration should take account not only of economic aspects; sustainability and the ecological circumstances of the process also should also be part of the decision-making process.

Questions, comments, and answers

Q: In the sample you describe, is there a limitation to the depth it can be used and how wide are the slices?

A: I believe you can go several metres deep with 10cm slices.

REFERENCES

ANONYMOUS, (1995), "Entsorgung im Magen", *Geo*, October, pp. 197-198.

BOUCHEZ, M., D. BLANCHET, and J.-P. VANDECASTEELE (1995), "Degradation of polycyclic aromatic hydrocarbons by pure strains and by defined strain associations: inhibition phenomena and cometabolism", *Appl. Microbio. Biotechnol.* 43, pp. 156-164.

EDGINGTON, S. (1994), "Environmental Biotechnology: Business and government are looking to biotech for answers about how to clean up the environment", *Bio/Technology* 12, pp. 1338-1342.

FU, M. and M. ALEXANDER (1995), "Use of surfactants and slurrying to enhance the biodegradation in soil of compounds initially dissolved in nonaqueous-phase liquids", *Appl. Microbio. Biotechnol.* 43, pp. 551-558.

LIU, Z., A. JACOBSON, and R. LUTHY (1995), " Biodegradation of naphthalene in aqueous nonionic surfactant systems", *Appl. and Env. Microbio.* 61, pp. 145-151.

MAHRO, B. (1994), "Untersuchung von Möglichkeiten zur gezielten Stimulation der biogenen Mineralisierung und Humifizierung von PAK in Böden", Symposium Umweltbiotechnologie, Bochum, pp. 60-67.

PROVIDENTI, M., H. LEE, and J. TREVORS (1993), "Selected factors limiting the microbial degradation of recalcitrant compounds", *J. of Industrial Microbio.* 12, pp. 379-395.

SINGLETON, I. (1994), "Microbial metabolism of xenobiotics: Fundamental and applied research", *J. Chem. Tech. Biotechnol.* 59, pp. 9-23.

STRINGFELLOW, W. and M. AITKEN (1995), "Competitive metabolism of naphthalene, methylnaphthalenes and fluorene by phenanthrene-degrading pseudomonads", *Appl. and Env. Microbio.* 61, pp. 357-362.

WEBER, H. (ed.) (1990), *Altlasten, Erkennen, Bewerten, Sanieren*, Springer Verlag, Berlin.

SOIL BIOREMEDIATION OPPORTUNITIES IN JAPAN

Osami Yagi and Takeshi Ogawa
Environment Agency of Japan, Tokyo, Japan

The Environment Agency of Japan undertakes risk assessment of bioremediation, including bioaugmentation and biostimulation, and also evaluates field pilot tests for *in situ* bioremediation proposed by enterprises in Japan. In 1993, biotreatability studies on TCE biodegradation using methanotrophs were conducted at a TCE-contaminated site. In 1994, a field pilot test was carried out for *in situ* aerobic degradation of TCE using injection of methane and oxygen. This paper discusses the efficacy evaluation and risk assessment. This field test is the first trial of *in situ* bioremediation in Japan.

Introduction

In recent years, soil and groundwater pollution problems owing to toxic chemicals have become significant issues in Japan. In 1994, 232 cases of soil contamination were reported; the major contaminants were organochlorine compounds such as trichloroethylene (TCE), tetrachloroethylene (PCE), and heavy metals (Environment Agency of Japan, 1995). In February 1994, the Environment Agency of Japan formulated the Environmental Quality Standard for soil, in order to protect human health and to conserve the environment. The levels defined are: 0.03 mg/l for trichloroethylene (TCE); 0.01 mg/l for tetrachloroethylene (PCE); and 1 mg/l for 1,1,1-trichloroethane (TCA). The standards are applicable to all kinds of soil.

Table 1 shows the results of a survey of groundwater pollution by volatile chlorinated compounds. Before 1988, among the wells surveyed for drinking water quality, 2.7 per cent were found to be contaminated by TCE, 4.1 per cent by PCE, and 0.2 per cent by TCA. In 1993, 15 wells were polluted by TCE, and 24 by PCE. More than 2 000 wells were found to be polluted by volatile chlorinated organic compounds.

Various efforts are now being made in Japan on research and development and evaluation of soil clean-up technologies. For heavy metals, the common technologies are solidification, sealing, and digging up. For chlorinated compounds, vapour extraction, pumping up, and air stripping are commonly employed. Because physical and chemical remediation is expensive, less expensive and more effective technologies for destroying pollutants are needed. Attention has primarily focused on *in situ* bioremediation, because most soil and groundwater pollution in Japan involves TCE and PCE. A field pilot test for *in situ* bioremediation of TCE-contaminated soil and groundwater was evaluated.

Table 1. Survey of groundwater pollution by volatile organic compounds

Year	Chemicals	Surveyed wells	Polluted wells	Ratio (%)
1984-88	TCE	26 607	722	2.7
	PCE	26 594	1 078	4.1
	TCA	26 257	46	0.2
1989	TCE	3 388	30	0.9
	PCE	3 388	42	1.2
	TCA	2 569	2	0.1
1990	TCE	5 817	44	0.8
	PCE	5 817	79	1.4
	TCA	4 515	1	0.0
1991	TCE	6 158	27	0.4
	PCE	6 518	44	0.7
	TCA	5 135	0	0.0
1992	TCE	4 762	18	0.4
	PCE	4 762	35	0.7
	TCA	3 952	3	0.1
1993	TCE	4 480	15	0.3
	PCE	4 480	24	0.5
	TCA	3 960	0	0.0

Notes:

1. Drinking standard: TCE 0.03 mg/l, PCE 0.01 mg/l, TCA 0.3 mg/l.
2. TCE: trichloroethylene; PCE: tetrachloroethylene; TCA: 1,1,1-trichloroethane.

Source: Author.

Site characterisation

The site characterisation of the TCE-contaminated area was carried out. TCE contamination of well water was found in 1990. The main contaminant is TCE, and the contaminated soil zone is the upper aquifer, from 14 m to 23 m in depth. The source of contamination is an electrical parts factory. The groundwater pollution is remediated through pumping up and air stripping using active carbon.

To evaluate the effectiveness of bioremediation, the groundwater qualities of the site were analysed (Table 2): pH was 6.28, temperature was 16.2, dissolved oxygen was 4.9. TCE concentration was between 5.3 mg/l and 6.5 mg/l. Total nitrogen and total phosphorus were 7.5 mg/l and 0.01 mg/l, respectively. Methanotrophs were determined at 101 cells /ml.

Table 2. Summary of site characterisation (groundwater at C city)

Depth from surface (m)	14 ~ 23
pH	6.28
Temperature (°C)	16.2
Dissolved oxygen (mg/l)	4.94
TCE (mg/l)	5.26 ~ 6.5
c-DCE (mg/l)	0.042 ~ 0.05
VC (mg/l)	0
Total carbon (mg/l)	-
Total nitrogen (mg/l)	7.5
Total phosphorus (mg/l)	0.01
Copper (mg/l)	0.1
Aerobic heterotrophs (CFU/ml)	2×10^2
Methanotrophs (MPN/ml)	1×10^2

Source: Author.

Figure 1. TCE degradation by methanotrophic enrichments from contaminated sites (Flask test)

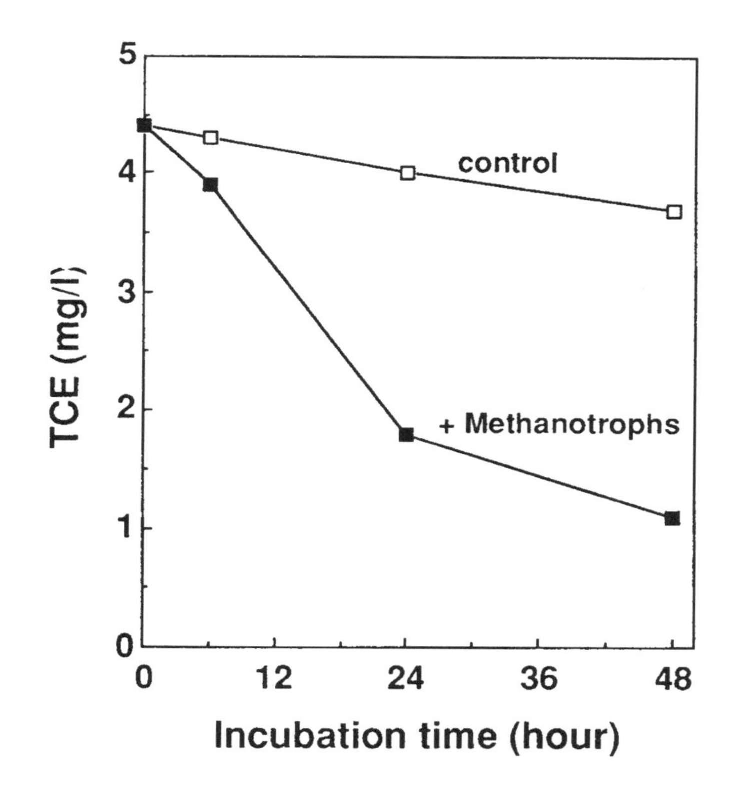

Source: Author.

Biotreatability test

A biotreatability test was conducted. Soil and groundwater samples collected from the contaminated layer were incubated in serum bottles in a mineral salts medium containing methane as sole carbon source. For ten days, the culture was analysed for its ability to degrade TCE; 4.5 mg/l of TCE decreased 1.1 mg/l after 48 hours incubation. Without bacteria, there was a 20 per cent decrease in TCE after 48 hours (Figure 1). From these results, it was concluded that TCE-degrading methanotrophs were present in the contaminated area. Therefore, *in situ* bioremediation was carried out using methane and oxygen injection.

In situ bioremediation

Figure 2 shows the design of the *in situ* bioremediation site. The process involves pumping up 100 t of groundwater a day from the extraction well (EW) and air stripping. Three new injection wells (IW1, IW2, IW3) were constructed, IWl and IW2 for oxygen and IW3 for methane. Two monitoring wells (MW1, MW2) were also constructed. The goal was to stimulate the methanotrophs with methane and oxygen and to reduce the TCE concentration in MW2. A hundred tons of water were pumped up daily, and 50 tons of treated water were injected per day.

Figure 2. *In situ* bioremediation design

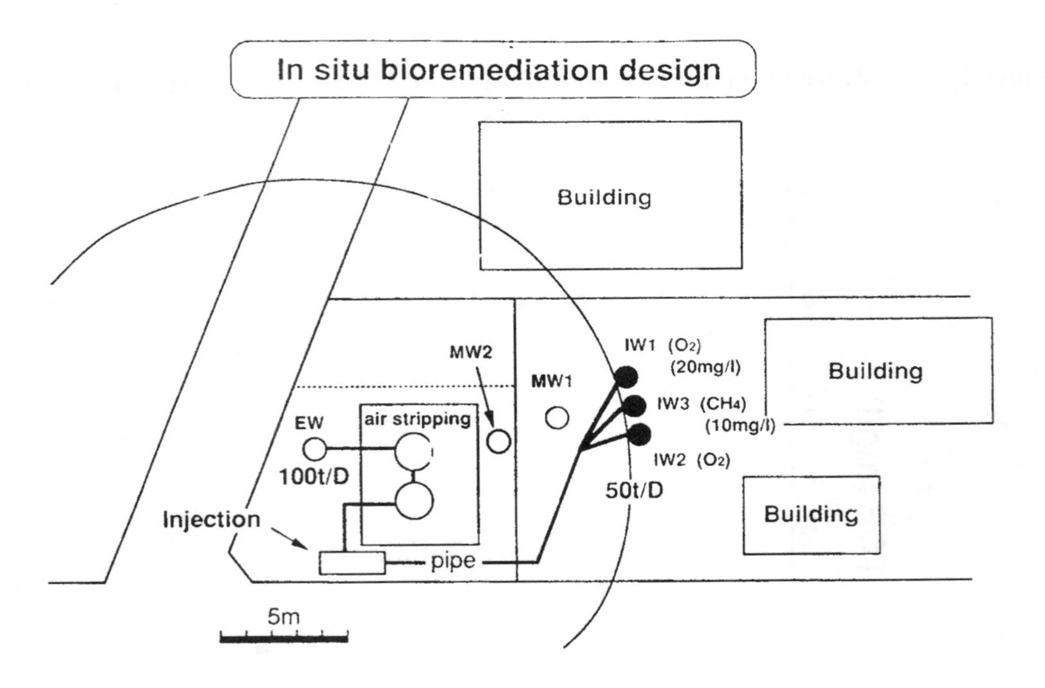

Source: Author.

Figure 3 diagrams the *in situ* bioremediation facilities. The distance from both IW to MW1 and MWl to MW2 is 3.5 m. The distance from MW2 to EW is 5 m. Oxygen, methane, 7 mg/l of nitrogen, and 26 mg/l of phosphorus were added to the injection water. The well was located at a depth of from 14 m to 23 m, where the groundwater and soil layer were contaminated by TCE.

Before bioremediation, the flow rate of the groundwater was normally 10 cm/day. Owing to injection and pumping up, the flow rate increased to 60 cm/day. Therefore, it took ten days for groundwater to run from IW to MW2.

Figure 3. View of *in situ* bioremediation facilities for TCE contaminated groundwater

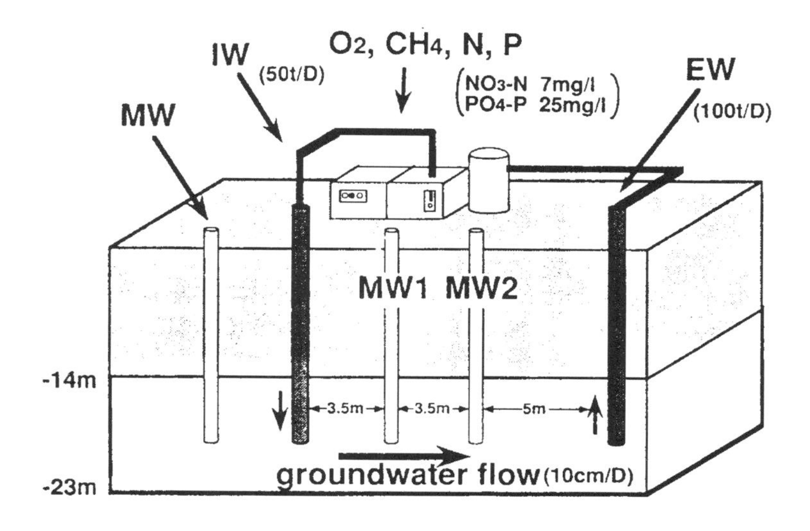

Source: Author.

The monitoring parameters for *in situ* bioremediation using methanotrophs included determining the number of methanotrophs, the activity of soluble methanemonooxygenase in methanotrophs and the number of aerobic heterotrophs. The amounts of trichloroethylene and by-products such as 1,1-, cis- and trans-dichloroethylene, and vinyl chloride were measured. Temperature, pH, dissolved oxygen, methane, nitrogen, and phosphorous were also determined.

Figure 4 shows the change in the TCE concentration in IW3, MWl, and MW2. Before bioremediation, the TCE concentration in IW3 was about 7 mg/l. At first, oxygen-amended water was injected, using IWl and IW2. After ten days, CH_4-amended water was injected, using IW3. After 35 days, oxygen and methane injection were stopped. Therefore, the methanotroph-growing phase lasted from 0 to 35 days, and after 35 days the degradation phase started. Following the oxygen injection, the TCE concentration decreased to less than 0.02 mg/l. Methane emission is considered to play a major role in global warming. However, at the concentrations used in this trial, methane was completely metabolised by the methanotrophs. This low concentration was not due to TCE degradation. The injection of 50 tons of water a day increased the water level of monitoring wells so that contaminated groundwater did not enter MWl and MW2. After oxygen and methane injections were discontinued, contaminated water entered MW1 and MW2, theoretically in ten days. Therefore, if the methanotrophs had not grown, the TCE concentration in MW2 should have begun to increase at 45 days. However, it only began to increase at 90 days. Therefore, it seems that TCE degradation activity continued about 40 days after the injections were stopped.

Figure 4. TCE concentration in injection and monitoring wells

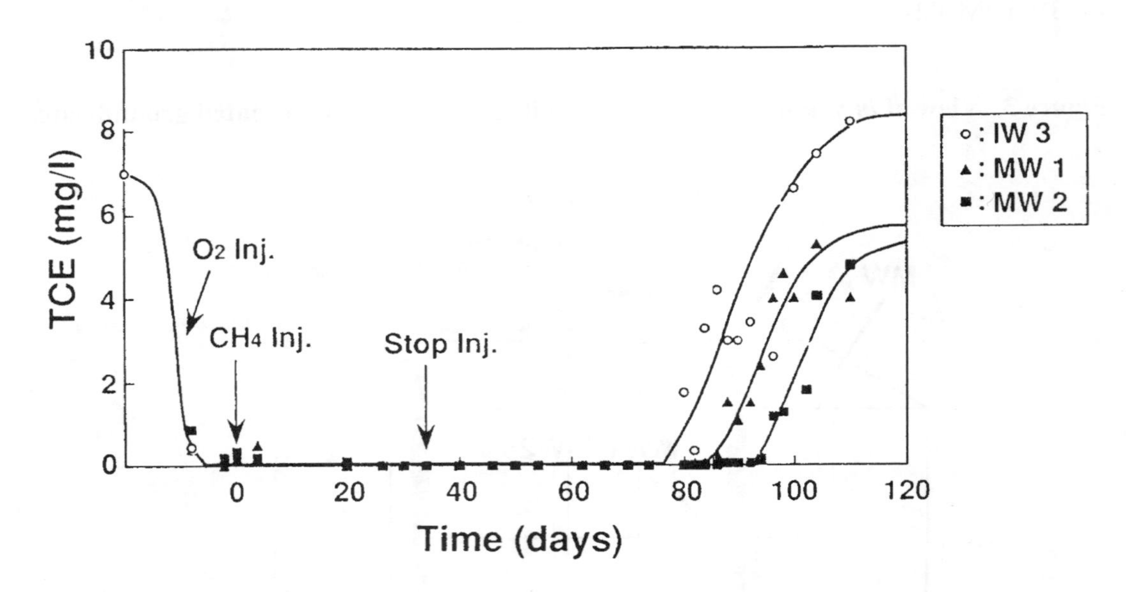

Source: Author.

Figure 5. Change of methanotrophs

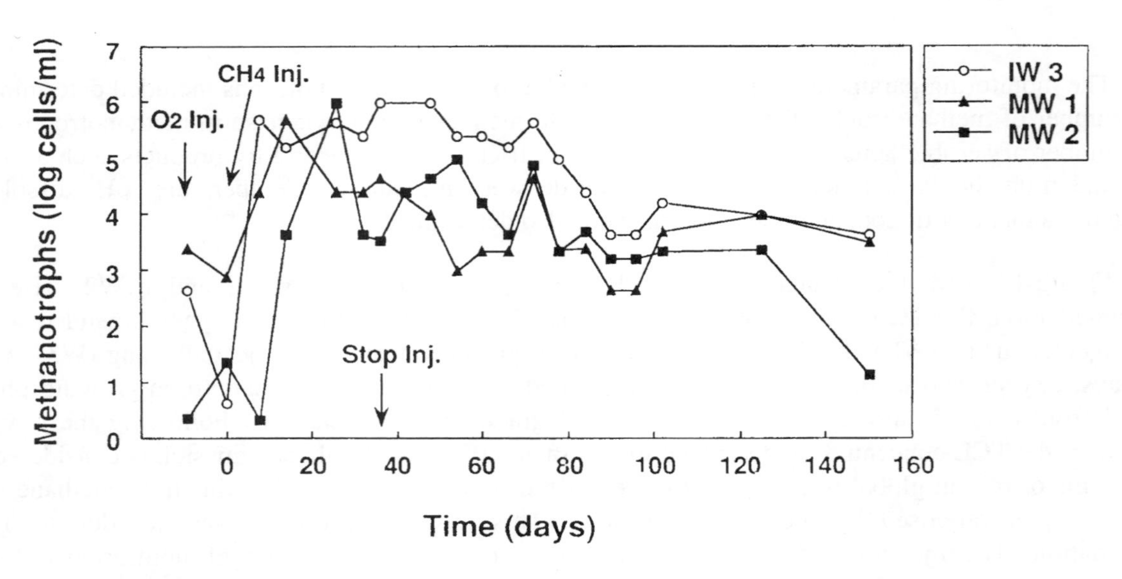

Source: Author.

Numbers of (MPN) methanotrophs in the injection and monitoring wells were determined by the most probable number method using methane as a sole carbon source (Figure 5). In the groundwater, methanotrophs were present to the order of 10 or 10^3. After the oxygen and methane injection, they increased rapidly to 10^5 and 10^6, and did not decrease after the injection were stopped. When the methanotroph number decreased to 10^4, TCE concentration gradually increased in the injection wells. The density of methanotrophs in MW1 and MW2 increased within 20 days after the methane injection. The number of methanotrophs in MWl and MW2 was significantly lower than in the injection wells. After 80 days, the number decreased to 10^3 and TCE degradation activity stopped. Methanotroph density was correlated with TCE degradation activity.

Risk and efficacy evaluation

In bioremediation, risk assessment for metabolite concentration and influence on the ecosystem are very important. In the soil environment, TCE was transformed to 1,1-, cis-, and trans-dichloroethylene, vinyl chloride, and chloroacetic acid (Vogel and McCarty, 1985; Nakajima *et al.*, 1992). Dichloroethylene concentration in the injection and monitoring wells was usually below 0.02 mg/l over 100 days, and vinyl chloride was always below 0.001 mg/l. Dichloroethylene and vinyl chloride concentration were very low. The aquifer seemed to maintain aerobic conditions.

Efficacy measurement was carried out by determining the change in the TCE concentration in the injection, monitoring, and extraction wells during the 40 days of the degradation period. The TCE concentration in MW1, before bioremediation, was 6.7 mg/l. During the degradation period, it was, on average, 0.025 mg/l, and 99.6 per cent of the TCE was removed through the 3.5 m layer of soil.

The amount of TCE degraded was calculated by the change in the TCE concentration in the extraction well (Figure 5). A hundred tons of groundwater were extracted daily, and the average TCE concentration dropped from 0.8 mg/l to 0.5 mg/l. Before bioremediation, TCE concentration was 0.8 (mg/l) x 40 (days) x l00 (ton/l) = 3.2 kg. After bioremediation, TCE concentration was reduced to 0.5 (mg/l) x 40 (days) x l00 (tons) = 2.0 kg. Thus, l.2 kg of TCE was degraded.

Conclusion

This field test indicates that biostimulation technology using methane and oxygen injection remediates TCE-contaminated soil and groundwater effectively (Little *et al.*, 1988; Nelson *et al.*, 1990; Yagi *et al.*, 1994). It is important to establish bioaugmentation technology for cleaning up soil and groundwater contaminated by volatile chlorinated compounds. We are developing bioaugmentation technologies using methanotrophs which were isolated and identified as *Methylocystis* sp. strain M. Strain M can degrade 30 mg/l of TCE (Uchiyama *et al.*, 1989). We are focusing on the development of efficacy evaluation methods, determination of toxicity, movement, and survival of methanotrophs, metabolites such as di- and tri-chloroacetic acids, amended nitogen concentration, fate of TCE and influence on the ecosystem. These data are necessary in order to obtain public acceptance for bioremediation.

Questions, comments, and answers

Q: With both schemes you are using pump and treat. Can you compare the costs of air stripping and biotreatment?

A: Not yet.

REFERENCES

ENVIRONMENT AGENCY OF JAPAN (1995), *Environmental White Paper*, Printing Bureau of the Finance Ministry, Japan.

LITTLE, C.D., A.V. PALUMBO, S.E. HERBES, M.E. LIDSTROM, R.L. TYNDALL and P.J. GILMER (1988), "Trichloroethylene biodegradation by a methane-oxidizing bacterium", *Appl Environ. Microbiol.* 54, pp. 951-956.

NAKAJIMA, T., H. UCHIYAMA, 0. YAGI and T. NAKAHARA (1992), "Novel metabolite of trichloroethylene in a methanotrophic bacterium, *Methylocystis* sp. M and hypothetical degradation pathway", *Biosci. Biotech. Biochem.* 56, pp. 486-489.

NELSON, M.J., J.V. KINSELLA and T. MONTOYA (1990), "*In situ* biodegradation of TCE contaminated groundwater", *Environ. Prog*. 9, pp. 190-196.

UCHIYAMA, H., T. NAKAJIMA, 0. YAGI and T. TABUCHI (1989), "Aerobic degradation of trichloroethylene by a new methane-utilizing bacterium Strain M, Type 2", *Agric. Biol. Chem.* 53, pp. 2903-2907.

VOGEL, T.M. and P.L. McCARTY (1985), "Biotransformation of tetrachloroethylene, dichloroethylene, vinyl chloride and carbon dioxide under methanogenic conditions", *Appl. Environ. Microbiol.* 49, pp. 1080-1083.

YAGI, O., H. UCHIYAMA and K. IWASAKI (1994), "Bioremediation of trichloroethylene contaminated soils by a methane-utilizing bacterium *Methylocystis* sp. M.", *Bioremediation of Chlorinated PHA Compounds*, pp. 28-36, Lewis Publishers.

Parallel Sessions/A1 and A2

A2.1. EFFICACY, RELIABILITY AND PREDICTABILTY: SOIL

ENGINEERING NEEDS AND SOLUTIONS FOR EFFICIENT BIOREMEDIATION

by

E. ten Brummeler, M. Heijnen and J. Bovendeur
Heidemij Realisatie BV, Waalwijk, the Netherlands

Introduction

In strongly industrialised countries all over the world, numerous contaminated sites have been, and will be, identified. The need for soil clean-up depends on government policy resulting in legislative requirements, adopted clean-up standards, and budget. The actual global market for soil treatment and land reclamation in the year 2000 is estimated to be $20-30 billion (OECD, 1995).

During the last decades, Heidemij developed several soil remediation processes to be applied for *in situ* remediation and *ex situ* soil treatment. Typical applications are biotechnological treatment processes for hydrocarbon contaminated soils, methods using photochemical oxidation of pollutants in groundwater, and soil washing for soils and sediments polluted with heavy metals and recalcitrant organics. These techniques are technologically and economically successful for a variety of polluted soils. In recent years the importance of biotechnological remediation processes, both for *ex situ* and *in situ* applications, has increased dramatically.

However, in some cases, full-scale remediation is only partially successful, due to complicated geo(hydro)logical characteristics of the soil, possibly combined with reduced availability of pollutants to degradation processes. To improve the efficiency and the predictability of full scale remediation processes and projects, Heidemij developed several characterisation tests and bioassays. These tests provide accurate information to select optimal remediation processes and strategies. In this paper, the characterisation tests and bioassays are presented as solutions to the engineering needs for efficient bioremediation met in the current practice of *ex situ* soil treatment and *in situ* site remediation.

Microbial and kinetic aspects of bioremediation of contaminated soils

Biodegradability of mineral oil

Large scale utilisation of mineral oil as the main energy resource for energy production as well as for industrial production has its side-effect on the environment. Mineral oil spills of different origin are the major causes of soil contamination in most industrialised countries, since mineral oil is transported, stored and used on enormous scale.

Since mineral oil is the predominant contaminant in several ecosystems all over the world, many case studies are described in literature about the fate of mineral oil after spillage (Atlas, 1984).

Although soil has a rich microbial diversity, the environmental conditions for biodegradation are less favourable than in aquatic ecosystems. Half-life time for mineral oil in soils are much longer than in aquatic ecosystems. Mineral oil as a soil contaminant still poses an intrinsic risk to the environment, since it also can affect groundwater quality.

It has generally been accepted that if conditions for microbial degradation are optimal, mineral oil is relatively easily degradable by aerobic soil bacteria. Anaerobic bacteria are unable to degrade mineral oil, although there are indications for sulphate reducing micro-organisms that are able to degrade oil (Keuning and Janssen, 1987).

Aerobic bacteria and fungi, which are able to degrade mineral oil, are very common in soil. The micro-organisms that are confronted with mineral oil can adapt rapidly to the complex mixture of hydrocarbon and degrade the oil at a significant rate. This is confirmed by microbiological studies which report growth of oil degrading micro-organisms in pure culture at growth rates similar to those of another easily degradable substrate, such as glucose (Persson and Hahn-Hagerdahl, 1993).

Degradation of mineral oil in ecosystems is apparently not limited by the microbial potential. Environmental factors also determine at which rate mineral oil can be biodegraded.

Limiting factors for biodegradation in soil

The key to effective bioremediation engineering is to know how to stimulate the natural degrading capacity in soil. Since the background biodegradation in soil is relatively slow process (half-life time > one year) which is in contradiction to the potential high degradation rate under laboratory conditions. It is clear that biodegradation is severely hampered by the unfavourable environmental conditions in soil. In general, the most important factors which determine the biodegradation of organic soil contaminants such as mineral oil are:

- availability of oxygen;
- temperature range 10-42 °C;
- concentration and availability of the contaminant;
- pH range: 6-8;
- nutrients (nitrogen, phosphorous);
- available water;
- presence of toxic chemicals such as heavy metals or cyanides etc. (Harmsen, 1991).

In general, microbes are not the limiting biodegradation of mineral oil in all types of soils. Seeding with adapted organisms to compensate for eventual limitation by a sub-optimal amount of micro-organisms hardly proved to be efficient in practice (Doelman, 1995). Even in deeper layers (up to 30 metres) soil bacteria can be found, although estimation of the rate of their actual metabolism is complicated (Atlas, 1984).

Optimising bioremediation strategies

Optimising the limiting factors for biodegradation is essential for efficient bioremediation since it is basically an engineering activity. However, since the interactions between bacteria, soil matrix, and contaminant seem to be rather complex, the need for a scientific approach is evident. In addition to an increasing level of engineering knowledge, a growing research effort has been made in the Netherlands as well as in other western countries. Recently, a study of the Amsterdam University has been finished concerning the fundamental aspects of mineral oil mineralization during bioremediation (Freyer, 1994). This study was conducted in close co-operation with Heidemij Realisatie. Some of the results have important implications for efficient bioremediation in practice. It has been shown that bioremediation of oil-contaminated soil shows three distinct stages, instead of two, as is usually accepted (Bosch *et al.*, 1992):

	Stage	Limiting factors
I.	High-rate stage	Oxygen, temperature
II.	Moderate rate stage	Oil concentration, nutrients, water
III.	Low-rate stage	Bioavailability

Depending on the characteristics of the contaminated soil, the best strategy for an efficient bioremediation is to determine the key factor that limits the biodegradation process. Optimising the wrong factor may result in ineffective measures, or may even deteriorate the process.

Predictability of bioremediation

Soil characteristics and effective bioremediation

The limiting factors for bioremediation of contaminated soil are governed by the characteristics of the soil. Determining the soil characteristics shows which corrective measures should be taken to improve the efficiency of the bioremediation process.

Two kinds of aspects can be distinguished in respect to soil characteristics:

- physical aspects;
- chemical aspects.

Physical aspects

The physical texture and structure of the soil at different scales are of great importance. For *in situ* remediation information about subsurface soil structures is more relevant at the scale of sedimentary structures, while information at the scale of grain size distribution is more relevant to *ex situ* remediation.

Geohydrological properties such as groundwater flow (conductivity), humidity (water retention curves) are of direct interest. These properties are usually derived indirectly from geological and geophysical data from boreholes and laboratory measurements.

Other typical physical parameters which are important in respect to bioremediation are temperature and heat capacity.

The fraction of organic carbon is a crucial physical parameter as it influences soil structure and sorption processes of anorganic and organic contaminants.

Chemical aspects

Geohydrochemical parameters like redox potential and acidity/alkalinity have an influence on contaminant behaviour, transport and fate. Besides these effects, pH has a direct influence on microbial activity.

Characterisation

The characterisation of the soil can be done at different levels:

- field studies;
- bench scale studies;
- computer modelling (simulation) as desk top studies.

Figure 1. The *In Situ* Mobile

Source: Heidemij Realisatie BV, 1995.

Field studies

At this scale Heidemij developed its Soil Vapour Extraction Test and Groundwater Pumping Test. Central to these tests is a mobile test rig, called the *In Situ* Mobile (Figure 1). The essence of the tests is to obtain *in situ*, site-specific parameters like feasible extraction rates, the corresponding radius of influence and conductivities. On the other hand, valuable information is collected about contaminant concentration levels at the start of the remediation with which a purification installation can be engineered.

Bench scale studies

Additional to the *in situ* test, soil, groundwater, and soil, vapor can be sampled for detailed laboratory analysis on organic carbon, density, humidity, grain size distribution, distribution of the contaminant over the soil phases and sorption processes.

Modelling studies

Modelling supports the characterisation of spatial and temporal variability of parameters in the soil system. Modelling is a tool for interpretation and analysis of observed data. With modelling the emphasis in the analysis shifts from static observation towards the observation of processes.

Numerous geohydrological models are available, both saturated (MODFLOW) and unsaturated (SWATRE). For simulation of the distribution of contaminant over the soil phases, Heidemij applies ECOSAT (Wageningen Agricultural University).

Biomodels are not yet applicated on a large scale. Generally there is scepticism about the application of biomodels.

The term 'bio' implies that biomodels are specialised models, but usually they include several processes. For instance, on biodegradation in the saturated zone, biomodels include:

- groundwater flow;
- solute transport;
- contaminant decay.

To limit the complexity of the models, the description of the processes is simplified. This simplification again restricts the application of the models.

Another aspect is that the biological degradation generally is limited to:

- one component;
- aerobic degradation;
- one or two dimensional description(s).

As a result of the research at the University of Amsterdam, Heidemij is now implementing a modular biodegradation model. This model integrates several modules that supply the input on soil temperature, unsaturated hydrology, and gas diffusivity.

Bioassays

To optimise the limiting factors for bioremediation, there is a need for biological test methods (bioassays) to predict and support the decontamination process. The most effective treatment scenario can be determined by means of bench scale bioassays. A bioassay should provide relevant information about:

- the presence of an active mineral oil degrading microbial population;
- the need for nutrients and pH correction;
- the actual rate of the oil degradation under field conditions;
- the final residual concentration and bio-availability of the mineral oil.

If such a bioassay can give accurate information about the actual rate and bio-availability, it can be a powerful tool for estimating the course of the biodegradation in oil-polluted soils and the residence time needed for decontamination. Since the results of such bioassays are needed before the start of the actual decontamination, the experimental period of such a bioassay should not exceed one to three weeks.

Figure 2. Degradation of mineral oil during the BAT-test with slurried soil

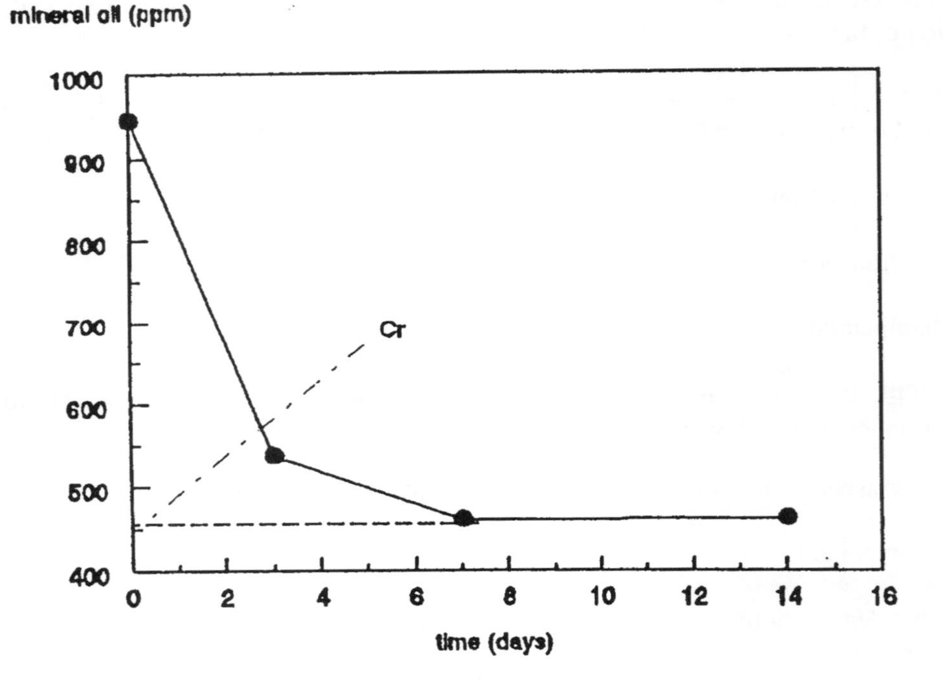

Note: C_r = estimated residual concentration.

Source: ten Brummeler, 1995.

Heidemij Realisatie has developed a test method that meets the criteria mentioned above, the so-called Biological Activity Test (BAT). Recently, the test programme has been terminated successfully, and the bioassay is already in use for estimating the bio-availability in practice. An illustration of the BAT method is given elsewhere (ten Brummeler, 1995).

In Figure 2, a typical plot is shown of a BAT experiment. After plotting the mineral oil concentration against time, the residual oil concentration C_r can be determined from the plot. C_r is defined as the concentration beyond a certain time interval *t*, when the biodegradation rate of mineral oil is lower than the degradation rate before this point.

Figure 3. Confirmation of the BAT-result during indoor bioremediation of an oil contaminated soil

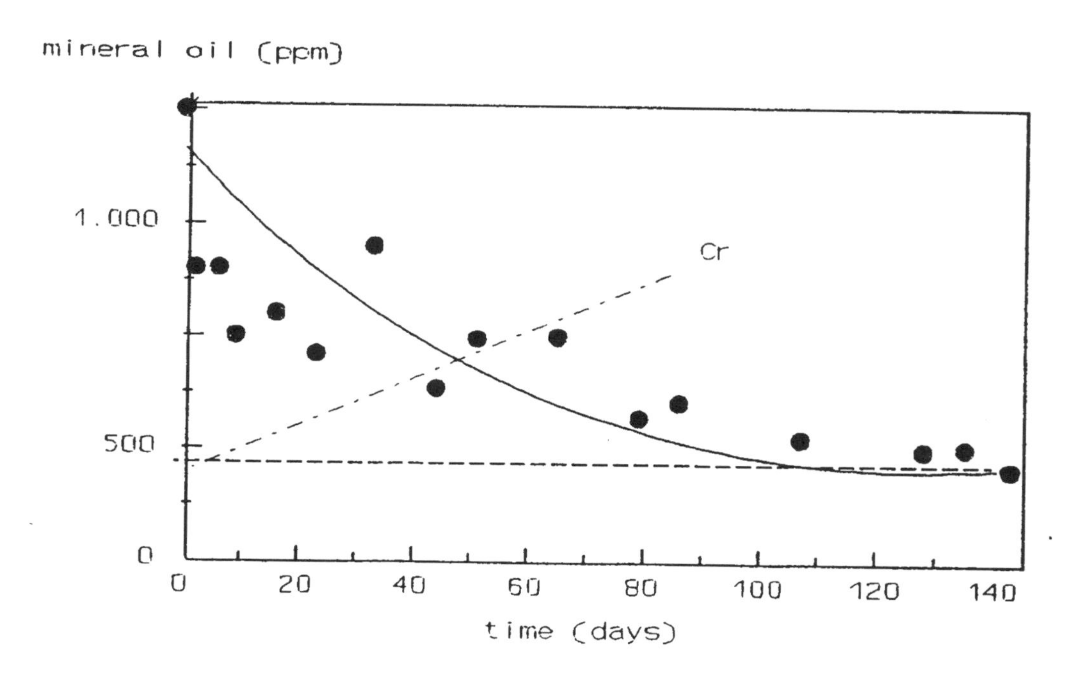

Note: C_r = residual concentration after the treatment.

Source: ten Brummeler, 1995.

Figure 3 shows the plot of the oil degradation versus time. From this plot a residual mineral oil concentration of 500 ppm is determined. At this concentration the biodegradation rate was lower than 5 mg/kg^{-1}/day^{-1} and hence, the treatment was terminated at this stage.

The residual concentrations are related to the very limited bio-availability of the mineral oil, as was estimated with the results of the BAT (Figure 3). The estimation of the bio-availability with BAT method supplies information for supporting biotechnological decontamination in practice. This is valid for *in situ* as well for *ex situ* bioremediation. It gives accurate information about:

- the rate of the oil biodegradation under field conditions (*in situ* and *ex situ*);
- the residence time needed for remediation to a certain oil concentration level (*in situ*, *ex situ*);
- the residual oil concentration that has a limited bio-availability; this can be predicted with the BAT;

- the application of the BAT in biotechnological soil treatment leads to an enhanced process operation. As a consequence, the support of bioremediation with the BAT results in a decreased soil residence time and lower treatment costs.

The 'dry' equivalent of the BAT is a Soil Column flushing test. Soil samples are flushed under controlled conditions in two separate columns. One column is flushed with nitrogen gas. The other is flushed with a nitrogen/oxygen mixture (80/20). The soil column test is normally applied on volatile contaminants and enables differentiation between the distribution of decay and volatilisation of contaminants.

CO_2 evolution

Biodegradation of mineral oil includes aerobic degradation of hydrocarbons to the simple molecules CO_2 and H_2O and biomass. As CO_2 is released to the gas phase at ambient temperatures. The degradation of mineral oil will show a strong correlation with the CO_2 production. For oil degradation in soils, this relation is constant down to a certain level of mineral oil, and depends on the type of mineral oil (Freyer, 1994). At low concentrations (< 1 000 ppm) other mineralization processes are the major source of CO_2. Monitoring the potential and actual rate of bioremediation by measuring the CO_2 production can be a powerful tool for predicting the course of a bioremediation. In addition the equipment that is needed for CO_2, monitoring is relatively simple.

Heidemij is implementing a respirometry test based on CO_2 measurement. From the measurements the actual (natural) and potential degradation rates are derived.

Bioremediation in practice

General aspects

Bioremediation in practice should be effective and efficient as well:

- Effective bioremediation includes optimising the rate-limiting factors to meet the decontamination standards within an acceptable period.

- Efficient bioremediation is doing this at the lowest cost possible.

In the Netherlands, bioremediation of mineral oil polluted soils is a major remediation technology with a total decontamination capacity of 300 000 tons per year (Doelman, 1995). Most of the decontaminated soil is being reused with restrictions to the location (after *ex situ* treatment) and function (after *in situ* and/or *ex situ*).

In general bioremediation is applied to a certain spectrum of soils such as sandy soils and loamy soils. Clays and peat soils are hardly remediated. Although bioremediation in practice is a relatively young technology first applied in the late 70s, several systems have been developed which have proven their effectivity and efficiency. Since bioremediation has found world-wide application, the techniques that are available show an enormous diversity.

As we discussed in the earlier, engineering bioremediation should be effective and efficient. Therefore, it is essential to characterise the contaminated soil with accurate methods such as bioassays, chemical and physical analysis, field tests and modelling. Only when this is done thoroughly, effective and efficient engineering of bioremediation is gained. This implies that a balanced rational evaluation is a necessity before a certain bioremediation technique is selected for a specific soil, valid for both *ex situ* soil treatment and *in situ* site remediation.

Questions, comments, and answers

Q: When treating wet soils and sediments *ex situ*, another approach is needed, for example, suspended treatment. What is your company's experience in this field?

A: We have experimented for five years with slurry reactors; the best way to start is to take out the sand a coarse material and put the fines in a slurry reactor. In large scale reactors, the key is the oxygenation capacity. With oil pollution we have had good results with a residence time of two to three weeks.

Q: Problem-owners want a problem solved on a lump-sum basis and the guarantee of an end-point. You can do this for *ex situ*, but what about *in situ*. Can you make the same guarantees?

A: *In situ* is more difficult but if you want to talk of lump sum, then everything has its price, including taking the risks.

Q: Are there grounds for optimising by pre-treatment before remediation?

A: The last step in the optimisation process has been taken by introducing soil heating. We have the highest degradation rate in untreated soil. Perhaps we could introduce photochemical or other treatments to crack large molecules and then feed them to a bioreactor.

REFERENCES

ATLAS, R.M. (1984), "Microbial degradation of petroleum hydrocarbons: an environmental perspective", *Microbiol. Rev.* 45, p. 180-209.

BOSCH, H., VAN DEN and J. HARMSEN (1992), "Workshop landfarming", Dutch Institute for Health and the Environment (RIVM), Report nr. 736101013, p. 8.

ten BRUMMELER, E. (1995), "Assessment of the bio-activity and bio-availability with a Biological Activity Test", in *Contaminated Soil '95*, W.J. van den Brink, R. Bosman and F. Arendt (eds.), pp. 489-490, Kluwer Academic Publishers, Dordrecht.

DOELMAN, P. (1995), "The practical state-of-the-art of landfarming", Programma Commissie Toegepast Bodemonderzoek (PCTB) report 10.5263.0, IWACO BV, Rotterdam, p. 2.

FREYER, J.I. (1994), "Mineralization of hydrocarbons and gas dynamics in oil-contaminated soils. Experiments and modelling", Ph.D. Thesis, University of Amsterdam, p. 155.

HARMSEN, J. (1991), "Possibilities and limitations of landfarming for cleaning contaminated soils", in *On-Site Bioreclamation*, R.F Hinchee and R.F. Olfenbuttel (eds.), pp. 255-272, Butterworth and Heinemann, Stoneham.

KEUNING, S. and D.B. JANSSEN (1987), "Microbiologische afbraak van zwarte en prioritaire stoffen voor het milieubeleid", Biochemisch Laboratorium, Rijksuniversiteit Groningen, p. 61.

OECD (1995), *Bioremediation: the Tokyo '94 Workshop*, OECD, Paris.

PERSSON, I. and B. HAHN-HAGERDAL (1993), "The influence of the temperature on the regulation of the metabolism in hydrocarbon degrading micro-organisms", in *Soil decontamination using biological processes*, Int. Symposium Karlsruhe, 6-9 December 1993, Dechema, Frankfurt, p. 531.

TRANSFERABILITY OF BIOTREATMENT FROM SITE TO SITE[1]

by

Perry L. McCarty
Western Region Hazardous Substance Research Center, Department of Civil Engineering,
Stanford University, Stanford, California, United States

Historical review

In the early 1980s, society became aware of the widespread contamination of soils and groundwater with organic chemicals that are hazardous to humans and the environment. Prior to that time it was commonly assumed that natural chemical and biological processes in the soil would destroy such contaminants, even though there was good evidence to the contrary. Groundwaters were generally believed to be safe from organic contamination since natural chemical and biological processes would remove chemicals leached from surface soils by rain and surface waters reaching groundwaters through percolation from the surface. Fortunately, this is true for most organic chemicals, especially for those of natural origin. However, for many hazardous xenobiotic chemicals, and indeed for certain natural ones such as aromatic hydrocarbons, destruction by natural processes is exceedingly slow. These then are the organic chemicals of concern and the subject of remediation technologies.

The major organic contaminants most commonly found in soils and groundwater are chlorinated solvents and their degradation products, aromatic hydrocarbons and other petroleum derivatives, and chlorinated aromatic hydrocarbons (Norris *et al.*, 1994). Specific chemicals within these groups and the geological strata where they are generally found are summarised in Table 1. Chemicals with low solubility and/or high octanol-water partition coefficients tend to be surface soil contaminants because of their stronger affinity for soil than for water, and thus their greater tendency to sorb onto and bind tightly with soils. The more soluble chemicals or those with lower octanol-water partition coefficients tend to migrate through soil along with percolating water to become groundwater contaminants.

Bioremediation is one technology frequently used for clean-up of sites contaminated with organic chemicals (Norris *et al.*, 1994). The advantage of bioremediation over other technologies is usually one of cost, but in addition, bioremediation generally results in the destruction of the unwanted contaminants. Thus, with bioremediation, a degree of permanence is obtained in that the contaminants no longer have to be dealt with. This presents an advantage that bioremediation has

[1] This work was supported by the U.S. Environmental Protection Agency sponsored Western Region Hazardous Substance Research Center at Stanford University. This article has not been reviewed by the Agency and no official endorsement should be inferred.

over all other remediation processes, except perhaps incineration in its various forms. Bioremediation may be an *in situ* process or an *ex situ* process. With an *in situ* process the contaminants are not removed from the biogeochemical sphere in which they are found, but are treated without being removed. In an *ex situ* process, the contaminants together with the matrix in which they reside, such as groundwater or soil are removed and taken to a separate treatment system where conditions for treatment can better be optimised. However, *ex situ* treatment may impart additional costs for construction of a treatment system.

Table 1. Common hazardous chemical contaminants, their properties, and where contamination from them is found

Compounds	Log octanol-water partition coefficient	Solubility (mg/l)	Geosphere generally contaminated
Chlorinated solvents	0.6 to 3.0	50 to 20 000	Vadose zone, Groundwater
Chlorinated benzenes	2.8 to 3.4	80 to 500	Vadose zone, Groundwater
BTEX	2 to 3.5	150 to 2 000	Vadose zone, Groundwater
PCBs	5.6	0.03	Surface soil
Pentachlorophenol	5.9	2 000	Surface soil Vadose zone Groundwater
PAHs	4.4 to 5.7	0.001 to 1.1	Surface soil

Source: Sawyer *et al.*, 1994.

Some treatment schemes may be intermediate between *in situ* and *ex situ* (Norris *et al.*, 1994) . For example, bioventing involves air sparging of volatile contaminants out of groundwater and into the vadose or unsaturated zone above, where the contaminants are biodegraded. This is generally considered *in situ* since the treatment occurs within the natural system with perhaps only minor perturbations to it, rather than conveyance to a fabricated treatment system at the surface. Another bioremediation approach is the mixing by ploughing or disking of contaminated surface soil so that oxygen, nutrients, and water can be better mixed with the contaminants and micro-organisms. This also may be considered *in situ*, but there is significant disturbance of the natural environment. As with most attempts to categorise systems, there are grey areas which appear to not fit any category with ease. With this limitation in mind, the generalisation might be made that a much better understanding of the biogeochemical environment is required for *in situ* processes than for *ex situ* ones.

There are various reasons why the contaminants listed in Table 1 are not biodegraded and tend to accumulate in the upper geosphere as listed in Table 2 (Alexander, 1965) . Some of these are related to the properties of the chemicals themselves and the concentrations at which they are found. Other factors are site-specific and dependent upon the biogeochemical environment in which they are found. In order to transfer experiences on contaminant biodegradation from one site to another, an understanding of the biogeochemical setting of the sites as well as the properties of the chemicals themselves must be clearly understood. Some of these considerations are discussed in the following.

Table 2. Factors affecting compound biodegradability

COMPOUND MOLECULAR STRUCTURE SITE BIOGEOCHEMISTRY
Contaminant bioavailability Microbial community structure Enzyme activation Environmental factors
Redox conditions *Nutrient availability* *Inhibitor presence*

Source: Alexander, 1965.

Factors affecting compound recalcitrance

The factors that affect whether or not a compound is likely to be biodegraded at a given site as listed in Table 2 were described by Alexander a few decades ago (Alexander, 1965), and thus are not new. In general, micro-organisms use organic compounds as sources of carbon and energy for growth and reproduction. As such, biodegradation is of benefit to both the organisms and to ourselves. Here, as long as other required inorganic nutrients for microbial growth are present, such as nitrogen, phosphorus, and sulphur, the process is auto-generative and self-sustaining. The chemical process by which organic compound degradation occurs is oxidation-reduction. Here, the organics usually serve as the electron donor or compound being oxidised, while inorganic compounds such as oxygen, nitrate, sulphate, or carbon dioxide serve as the electron acceptors and are in turn reduced. In recent years, important exceptions to these generalisations have been found. Some organic compounds can themselves also serve as electron acceptors. In recent years it has been clearly demonstrated that chlorinated compounds such as chlorobenzoates and chlorinated aliphatic hydrocarbons (CAHs), including tetrachloroethene and trichloroethene, can serve as electron acceptors in energy metabolism (Tiedje *et al.*, 1994). In this process called reductive dehalogenation, a chlorine atom is removed from the compound and replaced with a hydrogen atom. If complete dehalogenation of an organic compound occurs through this process, then it is often beneficial since the less chlorinated structures are generally less toxic than the more chlorinated ones. In other cases, the less halogenated compounds become more biodegradable through oxidative processes than the parent compound. At times, however, dehalogenation is not complete, and the compound formed may be more hazardous than the original one. A common example here is the formation of vinyl chloride, a known human carcinogen, from reductive dehalogenation of tetrachloroethene, a suspected carcinogen to humans (McCarty, 1993; McCarty and Semprini, 1994).

A third and important manner in which organic compounds can be transformed is through cometabolism. Here, the compounds are transformed by enzymes used by organisms for other purposes. Such enzymes may not be very specific and can transform compounds other than those which are beneficial to the organism. The common examples here are oxidation of trichloroethylene by oxygenases used by bacteria to initiate the oxidation of compounds from which they derive energy such as ammonia, methane, and toluene (McCarty, 1993; McCarty and Semprini, 1994). Many reductive dehalogenation reactions that have recently been discovered to occur under anaerobic conditions are also suspected to be cometabolic, although here the case is not yet clear. Examples are the dehalogenation of PCBs and perhaps some of the CAHs. Cometabolism is not autocatalytic and thus requires the presence of other inorganic or organic compounds that serve as electron donors in energy metabolism by the cometabolizing micro-organisms. In some situations, suitable energy sources are available so that cometabolism can occur naturally. In other cases, they may need to be supplied in order to insure that biodegradation will occur.

The above summary suggests that many of the most prevalent, hazardous, and persistent organic contaminants in soils, groundwaters, and sediments, are indeed biodegradable given the proper biogeochemical environment. An understanding of the biogeochemical environment required is of utmost importance if one is to take knowledge of biodegradation gained at one site and transfer it to another. Table 2 indicates many of the biogeochemical factors of importance.

Biogeochemical factors of importance

One of the most significant biogeochemical characteristics of a site is the community of organisms that may be present (McCarty, 1988). Micro-organisms capable of degrading most natural compounds are relatively ubiquitous; one can anticipate their presence in almost any handful of soil, steam sediment, ocean bottom, or subsurface strata almost anywhere. However, for many xenobiotic chemicals, this is not the case. For example, reductive dehalogenation of CAHs often occurs in methane producing anaerobic environments such as in landfills, anaerobic sludge digesters, and heavily contaminated groundwater systems. However, there are many methanogenic environments in which reductive dehalogenation does not occur. Indeed at times, even methanogenic bacteria which are required for methane production to occur are not present in groundwater environments.

In the author's laboratory at Stanford University, several column microcosms containing aquifer solids taken from locations over a 30 km long area of the Santa Clara Valley did not contain bacteria capable of producing methane with relatively common substrates such as glucose, acetone, methanol, or acetate, even when all other environmental conditions were ideal for their growth. The columns were operated for periods of up to two years with no methane production, although in many other ways the micro-organisms present were extremely active, such as in degradation of acetone and xylene. Little transformation of 1,1,1-trichloroethane or TCE occurred in these laboratory microcosms. These laboratory studies confirmed observations made in the field at two Santa Clara Valley major contamination sites where TCA and TCE had leaked from waste storage tanks that also contained high concentrations of acetone and xylene. The acetone and xylene were degraded within the aquifer, but the TCA and TCE persisted. Evidence for lack of a methanogenic microbial population at this location was indicated by the fact that when a small seed of methanogenic micro-organisms was added to the columns, methane fermentation began immediately and continued.

At many other locations, such a combination of chemicals has led not only to methane fermentation, but also to reductive dehalogenation of chlorinated solvents. Examples from Superfund sites are numerous. The St. Joseph, Michigan, site is just one example (McCarty and Wilson, 1992). The experience with Santa Clara Valley aquifer material indicates that a good organic chemical degrading population at one site will not necessarily be present at others. It also illustrates the value of laboratory microcosms to help determine whether appropriate communities of micro-organisms are present. We have also found from laboratory reactor studies that a good methanogenic population with high activity does not necessarily result in reductive dehalogenation of PCE or TCE. Methanogenic conditions are characteristic of the microbial environment in which degrading populations can operate, but do not guarantee that they will be present. Again, we have found that by supplementing the reactor with pure cultures known to dehalogenate, dehalogenation can be caused to occur. This confirms that the problem in some cases is not the environmental conditions *per se*, but the absence of a suitable degrading population. In any situation where bioremediation is to be used, one should operate laboratory soil microcosms under proper conditions to insure that the needed consortia of micro-organisms are present (Dolan and McCarty, 1995).

Even when an active consortia of micro-organisms is present, transformations of a given compound may not occur if the required enzymes are not activated. For example, the cometabolism of TCE, dichloroethylene, or vinyl chloride can be brought about by oxygenases, enzymes that are produced by common bacteria that exist in most aquifer systems. However, to be active, the enzyme must be induced by an appropriate substrate, and an energy producing substrate must also be available to generate the energy for maintaining the activity of the enzyme. Examples of substrates that can both induce the oxygenases for cometabolism of the above CAHs and provide the energy required are methane for methanotrophs, ammonia for nitrifying bacteria, and phenol or toluene for a variety of bacterial species that contain toluene mono- or dioxygenases.

A second most important biogeochemical factor is accessibility to the contaminant by the micro-organisms, that is its bioavailability (Brusseau and Rao, 1989; Farrell and Reinhard, 1994; Pavlostathis and Mathavan, 1992). Biodegradation cannot occur if the organic compound is not accessible to the organism or to extra-cellular enzymes that it might excrete. At a micropore scale, organic contaminants are often strongly sorbed onto the outer surface of soil particles or within small cracks or pores within the soil particles. Diffusion both into and out of small cracks or pores is often a very slow process, sometimes on the order of months to years. It has frequently been observed that if contaminated soil is exposed to an appropriate bacterial community, partial degradation of the contaminant will occur rapidly, perhaps within hours. However, the less accessible organics contained within the microscopic fissures present may not be degraded within months.

The distribution of a given contaminant between readily available and slowly available locations depends upon a complex of factors, including the sorptive properties of the chemical itself (evidenced by the octanol-water partition coefficient), the sorbing properties of the soil particles (often associated with its organic content), the size and quantity of fissures within the soil particle that permit interior access to the particles, particle size, and the time duration over which the soil has been exposed to the contaminant (Brusseau and Rao, 1989; Farrell and Reinhard, 1994; Pavlostathis and Mathavan, 1992. The longer the exposure time to allow diffusion to interior regions, the lower the fast bioavailability. With relatively water-soluble contaminants, such as PCE, TCE, and TCA in very low-organic content clean aquifer sands, rapid bioavailability at the micropore scale may be quite high, on the order of 50-100 per cent. However, with poorly soluble contaminants such as PAHs, PCBs, and many pesticides, in higher organic-containing surface soils, the proportion that may be rapidly bioavailable may be only a few percent. Here, bioremediation for near complete removal may take years. Short-term laboratory experiments with contaminated soils or aquifer materials may be satisfactory for evaluating the contaminants that are rapidly bioavailable, but are insufficient for determining long-term bioavailability of the contaminants. A question often posed is whether the portion of contaminants subject to very slow desorption constitute a significant risk to the public? This is a subject for other discussions.

Another aspect of the bioavailability question lies at the macroscopic scale where heterogeneities play a major role. For example, subsurface systems are often characterised by interpenetrating layers of clay, silt, sand, gravel, and rocks (McCarty and Semprini, 1993). Some of these layers are essentially impermeable by water. Contaminants such as petroleum hydrocarbons or CAHs often penetrate into silts, clays, and fractured rock such that they are inaccessible by groundwater movement or by bacteria. Such materials may slowly diffuse out of these impermeable layers into the more permeable ones, where they then become available. Thus, even though a contaminant may be highly bioavailable when one considers the microscopic pore scale, they may be highly unavailable when viewed at the more macroscopic scale. These factors obviously very greatly from site to site and must be evaluated carefully when considering the potential for bioremediation, or for remediation by any procedure for that matter. These complexities do not impact on bioremediation alone and are limitations to remediation in general.

An additional factor affecting biodegradation is the suitability of the prevailing environment for biological growth and reproduction. Often compounds that are biodegradable may in addition be present at concentrations that are toxic to micro-organisms. For example, phenol is readily degraded as a primary substrate by many micro-organisms, but if it is present at say 500 mg/l, it can be toxic to these same organisms, and thus not be biodegraded. This is the case with many contaminants, such as chlorinated aromatics in general, petroleum hydrocarbons, and CAHs. Frequently, contamination does not include a single compound, it is generally represented by a mixture of compounds that result from leaching from landfills, lagoons, or waste storage tanks containing a variety of organic and inorganic chemicals. While the compounds present that may pose that greatest hazard to human health may not be at a concentration toxic to micro-organisms, the other chemicals may create unsuitable conditions, such as pH that is too high or too low, excessive salt concentration, or simple toxicity by excessively high concentrations of the secondary contaminants, including heavy metal toxicity. An additional factor is the redox environment; this must be suitable for the microbial process of interest. Aerobic biodegradation requires the presence of oxygen, but anoxic processes, such as reductive dehalogenation, require that oxygen be absence. Methane production, if desired, requires the absence of oxygen and nitrates, which are inhibitory to methanogenic bacteria, and sulphate, which provides the electron acceptor for sulfidogenic bacteria. Such potentially inhibiting factors are site-specific and must be considered when attempting to transfer successful bioremediation results from one site to another.

Finally, appropriate nutrients for biological growth must be present and available to the micro-organisms if organic contaminants are to be biodegraded (McCarty, 1988). Of primary significance are the electron acceptors used in oxidative energy-producing processes by micro-organisms. Sufficient oxygen must be present for rapid aerobic biodegradation of PAHs or petroleum hydrocarbons, including complex aliphatic mixtures as well as the aromatic hydrocarbons present in gasoline, such as benzene, ethylbenzene, toluene, and xylene (BTEX). The amount of oxygen available must be sufficient to satisfy the relevant stoichiometry for hydrocarbon biodegration. With highly contaminated surface soils, oxygen may be introduced by frequent turnover or mixing of soil, but in contaminated groundwater, other methods for introducing oxygen must be employed such as air sparging or addition of oxygen-rich compounds such as hydrogen peroxide.

Other required nutrients for growth in macro amounts are nitrogen (ammonia or nitrate) and phosphorus (ortho or polyphosphates) to permit production by micro-organisms of their nucleic acids and proteins. Micro quantities of many other constituents are also required, such as sulphur, magnesium, iron, molybdenum, cobalt, and nickel, as needed for the formation of key proteins and enzymes within the cells. Generally, these materials are sufficiently abundant in soils, but they may not always be. Microcosm studies are often desirable to be assured that both macro and micro nutrients that will satisfy the need of the microbial community of interest are not only present in a given soil but are also biologically available. Soil column microcosms have been used in our own studies to successfully demonstrate that phosphorus and required trace nutrients are present and available in aquifer solids and required nitrogen is available as nitrate in the groundwater to satisfy biological requirements for *in situ* cometabolic biodegradation of TCE. Subsequent field evaluations corroborated the correctness of the results from the laboratory microcosm studies. Again, these conditions are site-specific.

Recommendations

Site-specific conditions as well as the chemical and biological characteristics of a specific contaminant of concern play a major role in determining whether or not bioremediation will be successful (McCarty and Semprini, 1993). For this reason, one cannot assume, because bioremediation for a given contamination is successful at one site, that it will in turn be successful at an alternative site. In any application of *in situ* bioremediation, site-specific characteristics must be determined if a successful operation is to be assured. The generalisations that are applicable from one site to another are concerned both with the basic characteristics of the contaminants themselves and the potential for micro-organisms to bring about their biodegradation. Here, the necessary electron acceptors or oxidation-reduction conditions for biodegradation must be known, as well as the biochemical process involved. The latter includes questions about whether the contaminant can be used as a primary substrate for energy and growth by micro-organisms or whether degradation involves some form of cometabolism. Characteristics of the degrading population must also be known. Additional basic characteristics of the contaminant are its tendency to sorb onto soils as this provides clues as to whether bioavailability at the microscale is likely to be important.

For *in situ* bioremediation, site-specific factors also are very important. Of prime importance is knowledge about the potential of the prevailing population of micro-organisms to bring about bioremediation if other conditions are made favourable. With many natural compounds, it can generally be assumed that suitable populations are ubiquitous and can be counted upon to be present in most situations. With xenobiotic chemicals, and indeed with some natural ones, the presence of suitable populations should always be determined. Examples of bacterial communities that may not be present at given sites are those that can bring about the anoxic biodegradation of the BTEX group of gasoline contaminants that are so prevalent as groundwater contaminants. Others are micro-organisms that bring about the anoxic reductive dehalogenation of a wide range of halogenated organic compounds, including PCBs, pentachlorophenol, chlorinated benzenes, and CAHs. Under aerobic environments, micro-organisms that may not be present are those that are effective at CAH cometabolism and higher ring PAH oxidation, which also may be initiated by a cometabolic processes. Also, some contaminants have been found to be degraded aerobically as primary substrates at some sites, but the degrading micro-organisms are not commonly found. Examples here are organisms found to aerobically biodegrade 1,2-dichloroethane, dichloromethane, and vinyl chloride. In such cases, laboratory microcosms or small-scale field testing should be undertaken with soil or aquifer material obtained aseptically to determine the presence of suitable populations of micro-organisms. Suitable procedures for doing so can be found in the literature.

Bioavailability is another factor of prime importance that varies significantly from site to site. Here, laboratory or limited-scale field studies can be undertaken, generally as part of an overall biodegradability study for the site. Since desorption can be a very slow process, such studies should be long term, on the order of a few months, in order to obtain reliable results. Their are some procedures, such as grinding of the soil or aquifer material to reduce size and increase the rate of desorption, that may help reduce the time scale of such studies.

When dealing with contaminants that have penetrated too far into soil and where groundwater has been significantly contaminated, then heterogeneities of the subsurface become critical for all clean-up schemes. Determining the extent of contamination, its characteristics, and location is a major and often costly undertaking. Highly heterogeneous subsurface environments can be difficult, if not impossible, to remediate in any manner. A knowledge of such heterogeneities and their impact on remediation is essential to a realistic evaluation of alternative treatment schemes, including bioremediation.

Finally, factors such as toxicity and availability of necessary nutrients must be assessed at each site. In addition, an estimate of the quantity of essential electron donors and other chemicals such as primary substrates for cometabolism, as well as nutrients quantities, need to be estimated for a given site. Perhaps of greatest difficulty, especially for bioremediation in the subsurface, is getting the chemicals of need to the micro-organisms and the contaminants. Generally, micro-organisms, contaminants, and nutrients must be brought together to obtain effective bioremediation. The best manner for doing this at a given location is highly dependent upon the biogeochemical setting. Here, experiences from a multitude of other sites are valuable in helping to select the one that may be optimal for a site of interest. Since each site is so different from any other site, application of bioremediation must generally be considered as an experiment in which one applies the best principles and experiences that are available together with whatever knowledge about the site that can be obtained, and then apply good judgement. One generally should then conduct limited field testing to assess the feasibility of the process at that site and to evaluate limitations that might be present. Innovative approaches may then be required to address difficulties that a given site presents. In this way, successful bioremediation can generally be obtained and new or modified approaches that may prove useful at other sites can be developed.

Questions, comments, and answers

Q: Has the Michigan site been anaerobic all the time, or has there been a cycle?

A: Outside the zone, the site is aerobic -- what kept it locally anaerobic is oxygen demand from organics leached into groundwater from a lagoon -- in the plume, it is all methane and no nitrates or oxygen.

Q: Is there a difference in anaerobic microbiology from site to site?

A: In the San Jose (California) area, there were no methanogenic bacteria found over a 20 kilometre zone, even though the conditions were suitable. There was, nevertheless, a very active microflora. On the other hand, at the Michigan site, methanogens were highly active. We must consider that sites have large differences in microbiology.

REFERENCES

ALEXANDER, M. (1965), "Biodegradation: Problems of Molecular Recalcitrance and Microbial Fallibility" in *Advances in Applied Microbiology*, pp. 35-80, Academic Press, New York.

BRUSSEAU, M.L. and P.S.C. RAO (1989), "Sorption Nonideality During Organic Contaminant Transport in Porous Media", *CRC Critical Reviews in Environmental Control* 19(1), pp. 33-99.

DOLAN, M.E. and P.L. MCCARTY (1995), "Small-Column Microcosm for Assessing Methane-Stimulated Vinyl Chloride Transformation in Aquifer Samples", *Environmental Science & Technology* 29(8), pp. 1892-1897.

FARRELL, J. and M. REINHARD (1994), "Desorption of halogenated organics from model solids, sediments, and soil under unsaturated conditions: 2 kinetics", *Environmental Science & Technology* 28(1), pp. 63-72.

McCARTY, P.L. (1988), "Bioengineering Issues Related to *In Situ* Remediation of Contaminated Soils and Groundwater", in *Environmental Biotechnology,* G.S. Omenn (ed.), pp. 143-162, Plenum Publishing Corp., New York.

McCARTY, P.L. (1993), "*In Situ* Bioremediation of Chlorinated Solvents", *Current Opinion in Biotechnology* 4(3), pp. 323-330.

McCARTY, P.L. and L. SEMPRINI (1993), "Engineering and Hydrogeological Problems Associated with *In situ* Treatment", *Hydrological Sciences* 38(4), pp. 261-272.

McCARTY, P.L. and L. SEMPRINI (1994), "Ground-Water Treatment for Chlorinated Solvents", in *Handbook of Bioremediation*, Norris *et al.* (eds.), pp. 87-116, Lewis Publishers, Boca Raton, Florida.

McCARTY, P.L. and J.T. WILSON (1992), "Natural Anaerobic Treatment of a TCE Plume, St Joseph, Michigan, NPL Site", in *Bioremediation of Hazardous Wastes, EPA/600/R-92/126*, pp. 47-50, US EPA Center for Environmental Research Information, Cincinnati.

NORRIS, R.D., *et al.* (1994), *Handbook of Bioremediation*, Lewis Publishers, Boca Raton, Florida.

PAVLOSTATHIS, S.G. and G.N. MATHAVAN (1992), "Desorption Kinetics of Selected Volatile Organic Compounds from Field Contaminated Soils *Environmental Science & Technology* 26(3), pp. 532-538.

SAWYER, C.N., P.L. McCARTY, G.F. PARKIN (1994), *Chemistry for Environmental Engineering*, McGraw-Hill, New York.

TIEDJE, J.M., M. FRIES, J. CHEE-SANFORD and J. COLE (1994), “Anaerobic Degradation of Anaerobic Chlorinated and Gasoline Compounds in Soils and Aquifers”, in *Transactions of the 15th World Congress of Soil Science, Commission III:Symposia, International Society of Soil Science*, pp. 364-374, Acapulco, Mexico.

BIOTREATMENT OF METALS: SITE DEPENDENT?

by

Harry Eccles
Company Research Laboratory, British Nuclear Fuels plc,
Springfields, Preston, United Kingdom

Introduction

During the last decade several techniques have been developed for the remediation of contaminated land, soil and silts. The diversity of these techniques range from excavation of the polluted material, followed by disposal to a controlled disposal site via physical systems such as incineration to biological which employ a specifically selected consortium of micro-organisms. Virtually all of these techniques have been developed for the removal of organic pollutants from soil employing either an *in situ* or *ex situ* operation. Currently *ex situ* applications are more favoured as *in situ* processes generally require some additional and unique considerations.

With the focus now turning to sites contaminated with metals or a combination of metals and organic pollutants (EPA, 1993), the need for *in situ* processes will undoubtedly increase and consequently these additional considerations will have to be addressed.

One major British company, namely British Nuclear Fuels plc, along with its partners, have developed a bioremediation process for the removal of various toxic heavy metals from soil (Eccles, 1995*a*). Principally, the process was intended for *in situ* use, but is equally effective in an *ex situ* application. The process is based on the stimulation of indigenous micro-organisms in the soil to mobilise the toxic heavy metals. Other micro-organisms are employed to recover the toxic heavy metals from the leachate, which has been abstracted from the land.

The process has recently been extended to accommodate organic pollutants (Eccles, 1995*b*).

This paper will, however, concentrate on the *in situ* removal of toxic heavy metals from contaminated land. The major technical considerations will be described and the process developed briefly explained.

Identifying the problem

In situ bioremediation is very site-specific and its successful implementation depends on a thorough understanding of the physio-chemical, hydrogeological and microbiological factors controlling not only the removal of the metal contaminants from the site, but also mass transport of nutrients and oxygen.

Major factors governing the effective removal of metal contaminants from soil are:

- the type and concentration of the metal;
- the physical state of the pollutant;
- the chemical, physical and (micro) biological properties;
- the type of soil, e.g. sand, clay, humic;
- the size and history of the polluted site;
- the presence of co-pollutants;
- the soil permeability;
- the temperature.

The type, physical state and concentration of the metal pollutants can vary strongly as soil pollution has been caused mainly by industrial activities. In addition to the type and concentration of the pollutants, their physical state is a very important factor. The following physical states are observed:

- the metals present in a liquid film around the soil particles;
- the metals adsorbed on the surface of soil particles;
- the metals absorbed in soil particles;
- the metals present as a solid in the pores of the soil particles;
- the metals dissolved in the aqueous phase in the pores of the soil particles.

Regarding the history of the polluted site, three factors have relevance for soil clean-up, in particular for *in situ* remediation of soil: the way the site was polluted, the way the site has been used after it was polluted and the time interval between the moment of pollution and clean-up.

Some of the above considerations will be described in a little more detail, but it cannot be over-emphasised that the success of *in situ* bioremediation is largely dependent on a thorough characterisation of the site, and in particular, the problem.

Soil is probably one of the most complex and complicated matrices on the planet earth. Its composition and characteristics will vary significantly from one site to the next. The fabric of the non-living part of soil is made up of mineral and organic materials, and contains pore space occupied by water and air. The minerals include residues from the parent material, usually mainly quartz and minerals in the clay fraction, which are products of weathering. These clay minerals give the soil important properties, including the capacity to absorb cations and other solutes on negatively charged surfaces. Organic matter has this same property and is also a source of nutrients.

The physical properties of the soil depend largely on the sizes of the soil particles (soil texture) and on their arrangement (soil structure), as in aggregates. To a large extent, texture and structure determine the distribution and movement of water and air in soil. The ability of soil to adsorb/absorb cations will influence the mobilisation of the pollutant metals; the higher the cation exchange capacity (CEC), then a greater quantity of metal sequestration reagent for metal mobilisation is required. Equally, the quantity of metal sequestration reagent will be influenced by the alkalinity of the soil. The CEC and alkalinity, essentially calcium carbonate content, of the soils studied in this investigation, are presented in Table 1.

Table 1. Chemical parameters of three soil types

Soil Type	CEC m.equiv/g	Alkalinity Calcium Carbonate content %/w/w
Silt	0.22	2
Humic	1.13	2
Clay	0.33	25

Source: Author.

The soil structure will influence the rate of water permeability and hence will contribute to the rate of metal mobilisation.

The soil chemistry and microbiology will have a significant influence on the speciation of the metal pollutants. Adsorbed metals will be comparatively easy to mobilise, whilst absorbed metal ions, metal carbonates, metal hydroxides, metal oxides and metal sulphides are increasingly more difficult to remove. Several workers have developed sequential leaching techniques to characterise metal pollutants in soil (Greinert and Poprawska, 1993). The techniques involve the contacting of soil samples with various leach liquors, whose metal complexing ability is successively increased. Such liquors range from water to acid systems.

Although these sequential extraction techniques provides key information with respect to the chemistry and physio-chemical behaviour of the metal pollutants, they do not contribute to the understanding/appreciation of metal-microbial interactions; other separate studies are needed to acquire these data.

It is recognised that micro-organisms require certain heavy metals to sustain their growth. If, however, these metals are present above a threshold value, they can prevent biological processes in ecosystems. Work in our laboratory and elsewhere (Hiroki, 1992) has shown that, in general, certain bacteria can tolerate metal concentrations in soil up to 15mmoles per kg of soil before growth is significantly affected. This lethal concentration of metals is obviously highly dependent on:

- the micro-organisms under investigation;
- the soil characteristics (previously discussed);
- the metal speciation;
- the metal;

– the growth medium.

Other inorganic species can inhibit the growth of micro-organisms, for example, fluorides at low concentrations, i.e. a few hundred ppm, inhibit the growth of *Thiobacillus* species (Ondruschka and Glombitza, 1993).

The removal of metals using an *in situ* bioleaching process based on *Thiobacillus* will be highly dependent on oxygen availability. Therefore, in order to sustain aerobic microbial growth, oxygen must be in plentiful supply. Oxygen, if needed, can be supplied via physical or chemical means. Physical oxygen supply involves forcing air and/or pure oxygen into the contaminated matrix. Chemical oxygen supply involves the addition of substances which can be converted to oxygen, such as hydrogen peroxide, persulphate or substances which can act as terminal electron acceptors directly, such as nitrate.

Invariably, toxic heavy metal pollutants (EPA, 1993) co-exist with organic pollutants and the latter will influence, adversely or even synergistically, the clean-up operation. Recent studies in our laboratory have demonstrated the tolerance of *Thiobacillus* to various organic pollutants.

From the moment that a pollutant comes into contact with a soil, a series of physical and chemical processes are initiated. These processes result in the distribution of the pollutant onto the surfaces and into the pores of the individual soil particles. As the time of contact increases, these and other processes continue. This "ageing" process results in movement of the pollutant to the interior of the soil particles so that less remains on the exterior surfaces. This sequestration of the pollutant over time, that occurs with ageing, has an impact on the availability of the contaminants in soil to living organisms. The bioavailability of organic pollutants with ageing has been extensively studied (Hatzinger and Alexander, 1995), but little or no information is available for metal and/or mixed metal/organic systems. Bioavailability should, under ideal considerations, influence the environmentally acceptable endpoint.

The BNFL process

At the outset, it was realised that a process capable of treating toxic heavy metal contaminated land, and which could be extended to accommodate organic pollutants, would be extremely commercially attractive. Furthermore, if the process was appropriate to both *in situ* and *ex situ* applications, its attractiveness would be further enhanced.

The original concept was therefore of process, comprising discreet stages which, when combined, produced an integrated route generating low volumes of secondary arisings, capable of attaining high clean-up efficiencies. The technologies envisaged could be equally applicable to solving other environmental problems, such as treatment of liquid effluents, acid mine drainage, etc.

The original "block-flow" diagram is illustrated in Figure 1.

Figure 1. Original perceived process

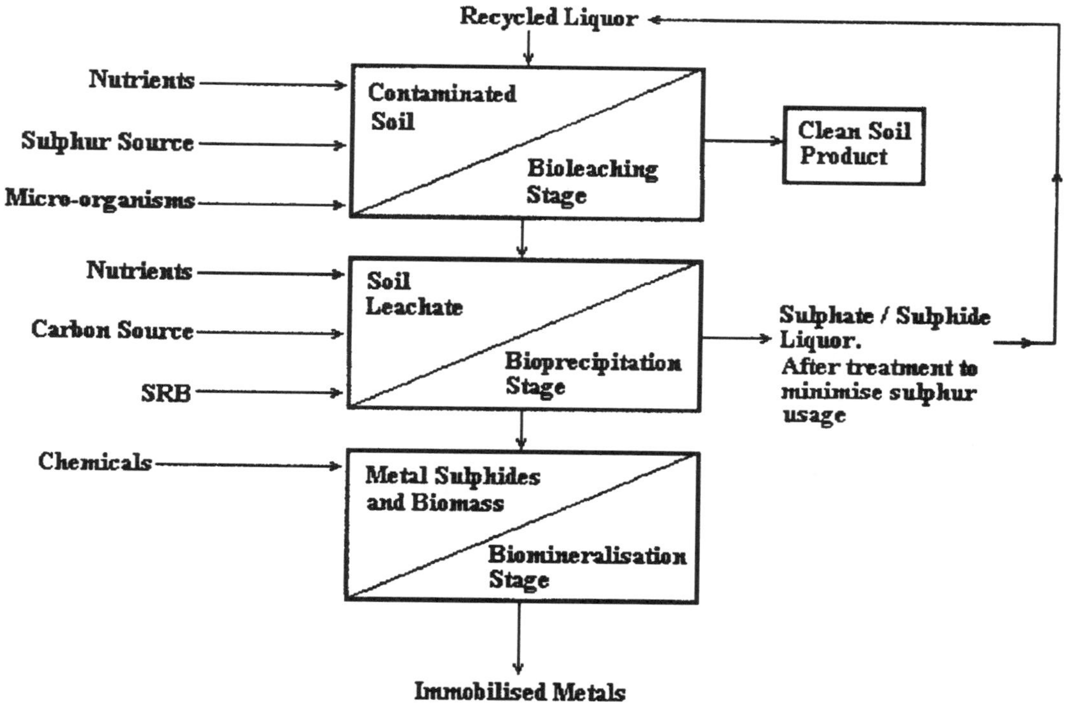

Recycled Liquor
Nutrients
Sulphur Source
Micro-organisms
Contaminated Soil
Bioleaching Stage
Clean Soil Product
Nutrients
Carbon Source
SRB
Soil Leachate
Bioprecipitation Stage
Sulphate / Sulphide Liquor. After treatment to minimise sulphur usage
Chemicals
Metal Sulphides and Biomass
Biomineralisation Stage
Immobilised Metals

Source: Author.

An additional key consideration was the process technology should be based on using indigenous micro-organisms.

Following an extensive literature review, micro-organisms capable of achieving the identified technical objectives of each stage were selected and laboratory studies were initiated. As access to an appropriate metal contaminated site was not possible, micro-organisms from national and international culture banks were initially employed. In the latter stages of the project, indigenous micro-organisms in the selected soil samples proved sufficient.

The studies involved both batch and column studies, single consortia and mixed cultures to verify key process considerations. These included, for example:-

- the growth rate of micro-organisms;
- the tolerance to a variety of inorganic species, in particular pH values and toxic heavy metals;
- the nutrients and nutrient levels;
- the electron acceptors;
- the influence of temperature;

– the rate of metal mobilisation and immobilisation;

– the soil types.

The bioleaching stage mobilises metals from the soil within an acidic leachate. The acidity is generated by *Thiobacilli* bacteria oxidising reduced sulphur compounds, in our case sulphur, and producing sulphuric acid.

The bioprecipitation stage receives the acid, metal loaded leachate and removes the metals through precipitation. This is achieved by sulphate reducing bacteria generating a favourable chemical environment for precipitation. These bacteria are active under reducing conditions and generate sulphide from the sulphate in the leachate. The bioprecipitation stage is a reactor-based process and produces a metal loaded sludge, comprising mainly metal sulphides and microbial biomass.

The biomineralisation stage is optional and will be applicable to bioprecipitation sludges which contain highly toxic metals such as mercury, cadmium or radionuclides; for less toxic heavy metals the bioprecipitation sludge would be treated differently or the metals reclaimed.

Leaching is driven by *Thiobacillus* bacteria in the contaminated soil, oxidising pre-applied sulphur to sulphuric acid. A downward migration of the leachate will be generated by irrigating the site, with abstraction wells intercepting the leachate and pumping it ultimately to the bioprecipitation bioreactors. In these reactors *Desulfobacter* bacteria convert sulphate to sulphide, thus precipitating heavy metals.

One of the key features of the bioleaching stage is the efficiency of sulphuric acid production. The acid available for mobilising metal ions will be in part dependent on the quantity of acid produced by the *Thiobacilli* from the added sulphur source, but also on the cation exchange capacity of the soil.

The acid production as a function of time (days) and temperature for silt soil is illustrated in Table 3 below.

Table 2. Rate of sulphuric acid production versus time and temperature for silt soil

pH value of leachate	Temperature °C		
	25	15	5
2.0	20	40	90
1.0	62	62	220

Source: Author.

Even at a pH value of 2, metal ions are mobilised from the soil, as illustrated in Figures 2 and 3. The release of metals from the soil particles is also influenced by the CEC and alkalinity of the soil, in addition to the acid dissociation constant of the respective metal hydroxide.

Figure 2. Mobilisation of metal ions as a function of pH

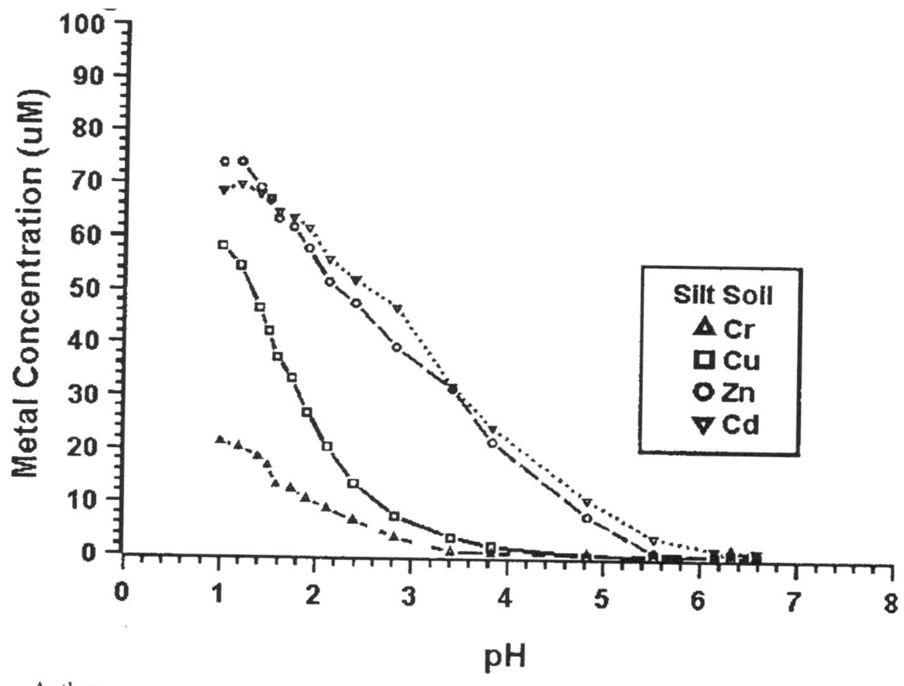

Source: Author.

Figure 3. Mobilisation of metal ions as a function of pH

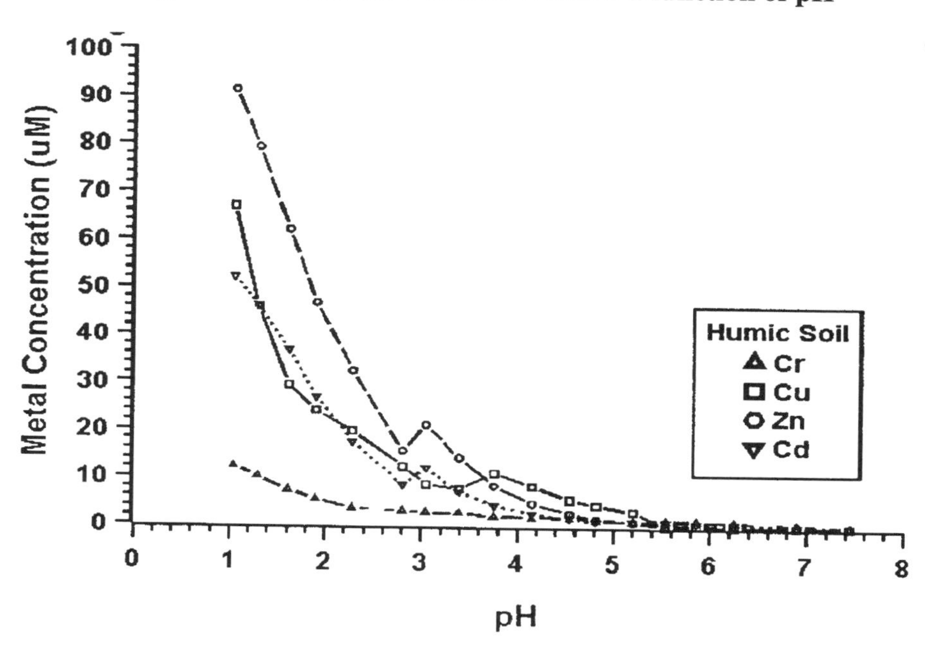

Source: Author.

The final phase of this three year project involved the operation of bioleaching tanks connected to SRB bioreactors to simulate an in-field, albeit *ex situ*, demonstration. Approximately 7kg of metal-contaminated soil was leached, and the acid metal sulphate leachate treated in a one litre SRB bioreactor. The efficiency of metal removal for most of the toxic metals studied was in excess of 90 per cent.

Conclusions

Remediation of metal and co-contaminated sites will in the near future receive increasing attention. The biological removal of toxic heavy metals can be relatively easily accomplished in an *ex situ* mode, as the technique is largely based on the technology currently employed by the copper, uranium and gold mining industries.

Extending the technique to an *in situ* operation requires some additional considerations, many of which will be specific to the site under investigation.

The imposition of organic pollutants on metal removal has not been studied sufficiently to determine if these additional pollutants will cause further insurmountable problems.

Future consideration

With the extension of land clean-up to include inorganic pollutants such as toxic heavy metals, the need for more fundamental data, whilst appreciating the strengths of alternative technologies for metal removal from dilute acid liquors will be paramount.

The need for a multi-disciplinary approach will be important if environmental biotechnology is to achieve its full potential.

Equally, a greater understanding of environmentally acceptable end points (EAE's) and their relevance to bioavailability of the pollutants is crucial. More environmentally realistic test procedures are needed to measure EAE's and not the current over-pessimistic chemical techniques.

Questions, comments, and answers

Q: When you have added S compounds to the system, how do you get them out again? Once the pH is down to 2, what are the costs of getting it back again with neutralising agents?

A: Conversion of S to sulphate is not 100 per cent and there may be an excess. We looked at the cost of using lime (the cheapest if not the most effective) to bring back the pH and factored this into the *in situ* remediation. Provided the pH of the soil returns to about 7-8, then excess S may not be a problem. This is a technology that no-one has explored and therefore we are on a steep learning curve. We are not developing a process but rather a set of protocols for engineers.

Q: Is there a small sub-set of site characteristics which will give you the confidence to predict clean-up or do you have to do a full site characterisation each time?

A: We are developing a set of models which will allow us to tell the customer how long we will be on site. A robust, flexible process is the aim and engineers ask do we have to go through all the tests and if so how can it be robust. Until we have a large database we have to rely on good characterisation.

Q: How do chelating agents compete?

A: Prior to developing this process we developed a chelation process. Using chelates, you have to break the metal chelate bond efficiently to recover the metal and you also need to be able to recycle the chelating agent. This is all expensive.

Q: At what stage of the learning curve are you? Is it desk design or are clean-ups taking place?

A: We have not yet demonstrated the process. A pilot plant is being designed and will be demonstrated in the United States early next year.

REFERENCES

ECCLES, H., (1995*a*), "Process for the treatment of contaminated material", WO 95/22374, 24 Aug.

ECCLES, H., (1995*b*), "Process for the treatment of contaminated material", WO 95/22375, 24 Aug.

EPA (1993), "Cleaning up the nation's waste sites: Markets and technology trends", 542-R-92-012, April.

GREINERT, H., and B. POPRAWSKA (1993), "Heavy metal removal from mineral soils with various extractants", in *Contaminated soil '93*, F. Arendt, G.J. Annokkee, R. Bosman and W.J. van den Brink (eds.), pp. 1389-1391, Kluwer Academic Publishers, Netherlands.

HATZINGER, P.B., and M. ALEXANDER (1995), "Effect of ageing on chemicals in soil on their biodegradability and extractability", Environ. Sci. Technol. 29 (2), pp. 537-545.

HIROKI, M. (1992), "Effects of heavy metal contamination on soil microbial population", Soil Sci. Plant Nutr. 38 (1), pp. 141-147.

ONDRUSCKHA, J., and F. GLOMBITZA (1993), "Inhibition of natural microbiological leaching processes", in *Contaminated soil '93*, F. Arendt, G.J. Annokkee, R. Bosman and W.J. van den Brink (eds.), pp. 1195-1196, Kluwer Academic Publishers, Netherlands.

ROLE OF CONSORTIA IN OPEN SYSTEM BIOREMEDIATION

by

I. Janssens and W. Verstraete
Laboratory of Microbial Ecology, University of Gent, Gent, Belgium

Introduction

The central problem in environmental biotechnology is that it generally relies on mixed cultures, microbial associations or communities. Unfortunately, the composition and functioning of these communities is, as yet, poorly understood. Environmental biotechnology will only attain complete acceptance at the moment one is able to re-assemble, re-construct the bioprocesses with defined, pure cultures. Only at that moment will environmental biotechnologists be at the level of other technologists in general, and of other biotechnologists in particular. Indeed, the latter are, for example, capable of producing products of predictable composition and quality such as lactic acid, beer, wine, monosodium glutamate, etc., using well-defined biochemical processes. In contrast, environmental biotechnologists start with one or more poorly defined inocula. They subsequently wait until the desired phenomena occur, and until they acquire a more or less steady-state level. The latter cannot, in most instances, be predicted and must simply be awaited (Verstraete and Top, 1992).

One of the major problems the industrialised world is facing today is the contamination of soils, groundwater, sediments, surface waters and air with hazardous and toxic chemicals. The need to remediate these contaminated environments has led to the development of a number of technologies that emphasize the detoxification and destruction of the contaminants rather than the conventional approach of disposal. Bioremediation, the use of micro-organisms or microbial processes to detoxify and degrade environmental contaminants, is among these new technologies. However, despite the rapid growth in the use of this technology, bioremediation is not yet universally understood or trusted by those who must approve of its use, and its success is still an intensively debated issue (National Research Council, 1993). Therefore, further research and better education of those involved in bioremediation is necessary to understand the fundamentals behind bioremediation and to improve the ability to apply a logical strategy for bioremediation. From this perspective, this paper presents a general evaluation and some critical thoughts on the applicability of bioremediation based on past experiences and developments in the field of soil microbiology, and stresses the role education can play in overcoming the inherent complexity involved in performing bioremediation projects.

Background information

Throughout this century the evolution of soil microbiology reveals a clear shift from agricultural to environmental applications. Until the seventies, the focus of soil microbiology was mainly oriented towards improving soil fertility and increasing crop yield. Only during the past decades the attention in the field of soil microbiology switched to the use of micro-organisms for environmental clean-up.

The following examples illustrate this evolution:

- In the first half of the century various types of soil micro-organisms such as nitrifiers, N-fixers, S-oxidisers, etc., were discovered. The focus was on the minerals and the relation to plant nutrition. For two nutrients, i.e. N and P, micro-organisms were considered to play an essential role and hence, attempts were made to inoculate specific strains.

  Actually, in the seventies, the introduction of N-fixing strains into the soil was expected to solve the world's hunger problem in the developing countries. The extensive work on Rhizobium and the tenacity of people like Professor Döbereiner in Brazil, who tried to inoculate soils with *Azospirillum*, were fuelled by this type of inspiration (Day and Döbereiner, 1976). Recently, considerable interest has been devoted to the inoculation of various P-mobilising bacteria and fungi. Yet, both with respect to N and P, the overall successes have been rather equivocal.

- Soil bioremediation, i.e. the use of micro-organisms to clean up polluted soils, only became significant about 10-15 years ago. Although the first studies in hydrocarbon metabolism (naphthalene, anthracene, etc.) date from the late thirties (Tausson, 1927), a first full emphasis on the soil microbiology came in the late sixties, with Martin Alexander's as yet unsurpassed book *Soil Microbiology* (Alexander, 1967). Only in the early eighties, major advances were made in our fundamental knowledge of the biodegradation of chlorinated organics (Bouwer *et al.*, 1981; Vogel and McCarty, 1985; Bedard *et al.*, 1987; Quensen *et al.*, 1988; and many more). The first lab-scale bioremediation experiments with inoculation of degradative strains have been reported with varying success (Brunner *et al.*, 1985; Chatterjee *et al.*, 1982; Focht and Brunner, 1985; Golovleva *et al.*, 1988).

  In recent years, some reports have been published which illustrate that strains and genes can successfully be introduced into microbial communities. Knackmuss (1981) reported the introduction, via a laboratory strain, *Pseudomonas sp.* B13, of the capacity to degrade chlorinated aromatics in an activated sludge wastewater treatment plant. McClure *et al.* (1989) noticed that a strain with plasmid-borne catabolic genes, introduced into an activated sludge unit, did not enhance the degradation of 3-chlorobenzoate (3CB), while a total breakdown could be achieved in batch cultures. Autochthonous bacteria, on the contrary, previously isolated from the activated sludge on 3CB-containing medium, could enhance the biodegradation after inoculation in the activated sludge unit (McClure *et al.*, 1991). Oldenhuis *et al.* (1989) reported the introduction of appropriate pure bacterial cultures into soil suspensions which enabled chlorinated solvents such as chloro- and *o*-dichlorobenzene to be degraded. A constructed *Ps. aeruginosa* strain, carrying a degradative plasmid, has been shown to be useful for cleaning up soil contaminated with kelthane residues (Golovleva *et al.*, 1988). Apajalahti and Salkinoja-Salonen (1986) isolated a *Rhodococcus chlorophenolicus* strain capable of degrading polychlorinated phenols. These strains continued to degrade polychlorinated phenols in natural soil when immobilised on polyurethane foam (Briglia *et al.*, 1990). Shu-Yen *et al.* (1990) noticed that a *Streptomyces* strain, introduced in a sterile soil containing the acylanilide herbicide metolachlor, transformed the herbicide, while introduction in a native soil only gave transformation upon increasing the pH of the soil.

The catabolic plasmid RP4::Tn4371, containing the genes for biphenyl and 4-chlorobiphenyl catabolism, was transferred from *Enterobacter agglomerans* DMK_3 to indigenous bacteria in biphenyl amended sandy soil (De Rore *et al.*, 1994). Proliferation of transconjugants above a detectable level required presence of the concomitant pollutant biphenyl.

– Successful soil inoculation of micro-organisms requires survival and/or growth of the introduced strain in a highly competitive environment. Micro-organisms introduced in the soil without exogenous C sources are not expected to remain active, because the soil environment is already occupied by a diverse and well-adapted community competing for limited substrates (Devliegher *et al.*, 1995). The problem of competitiveness might be overcome by creating a temporary niche for the guest organism through the addition of a substrate only utilisable by this strain, in other words biostimulation. Devliegher *et al.* (1995) tested four detergents as selective C sources for the plant-growth-promoting rhizobacteria *Pseudomonas aeruginosa* 7NSK2 and *Pseudomonas fluorescens* ANP15. Co-720 (Igepal CO-720) or DOS (dioctyl sulfosuccinate), dosed at 0.2 per cent in the soil, increased the number of detergent-adapted, inoculated strains by almost 1.5 log units after 25 days, accounting for virtually the entire increase in total bacteria. The same dose of Tween 80 or N-laurylsarcosine, on the other hand, increased the indigenous populations by almost 2.5 log units, with only increases in the number of detergent-adapted inoculated strains. When CO-720 or DOS was initially supplied, the number of detergent-adapted 7NSK2 organisms was about 2 log units higher after 3 months of incubation than for the detergent-unadapted strain. This better survival resulted in a significantly higher root colonisation of maize in a pot experiment with soil inoculation, with a significantly ($P < 0.05$) higher shoot dry weight (18 per cent to 33 per cent).

Yet, how do we at present visualise a microbial community in the soil. Figure 1 summarises our current understanding quite well. Moreover, we now generally accept the following :

– One kg of normal soil contains some 50 mg microbial biomass carbon; this corresponds with some 2 000 kg live biomass per ha.

– This biomass consists for 2/3 out of fungi and 1/3 out of bacteria; their normal maintenance "diet" amounts to some 5 000 kg of organic matter (dry weight) per ha per year.

– The microbial community is not a constant assemblage of populations of particular species. Some species increase and then decline due to biological phenomena such as parasitism and predation. Yet, the functions they achieve are, in a stable microbial community, taken over by other species. In this respect, the concept of "climax pattern" is used. Microbial associations in which all functions (niches) are fulfilled to the extent that additions of other genomes cannot alter or improve them, are considered to be in a "climax pattern" state (Verstraete and Top, 1992). Consequently, the introduction of new species will normally result in a rapid elimination by the "homeostasis" of the system.

– The overall behaviour of populations in soils can be described by the Gompertz model (Vandepitte *et al.*, 1995). In particular, the concept of the plateau value is interesting. The Gompertz equation is $dX/dt = -A.X (\log X - \log M)$. The symbol A represent the exponential growth or decay rate (d^{-1}), X is the population size at time t (CFU/g soil) and M is the asymptotic number of bacteria reached in time per gram of soil (same units as X). The time t is given in days. Survival kinetics of some *Bradyrhizobium japonicum* strains in a sandy, loam and clay soil are represented in Figure 2 (Corman *et al.*, 1987). Note the large difference between strains and the faster evolution in sandy soil relative to heavier textured soils. These aspects are too often only qualitatively interpreted in bioremediation studies.

Figure 1a. Model of aggregate organisation, showing relative size of the components and the major binding agents

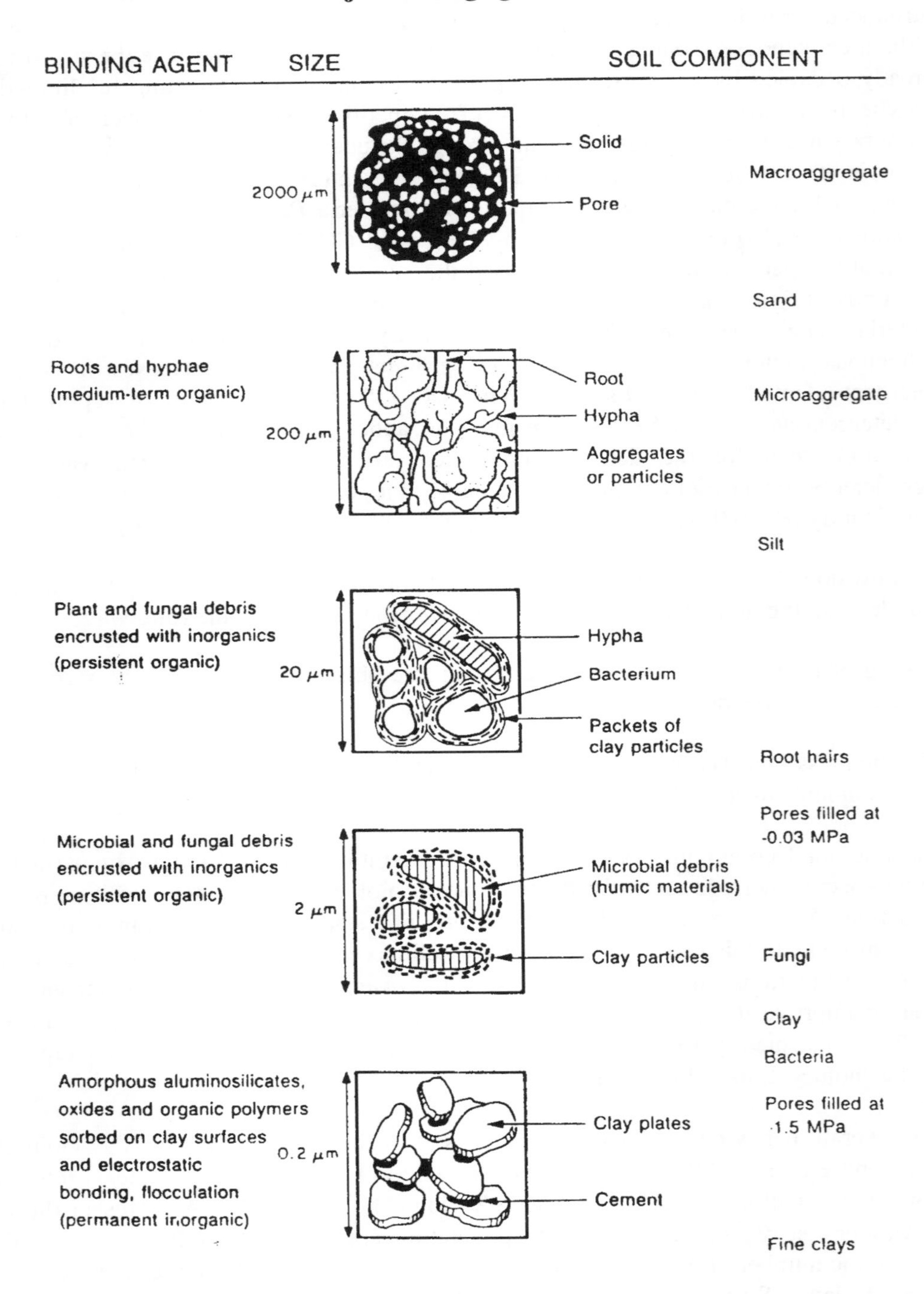

Source: Paul & Clark, 1989.

Figure 1b. Model of soil aggregate, showing organic matter protected from attack by crypt formation

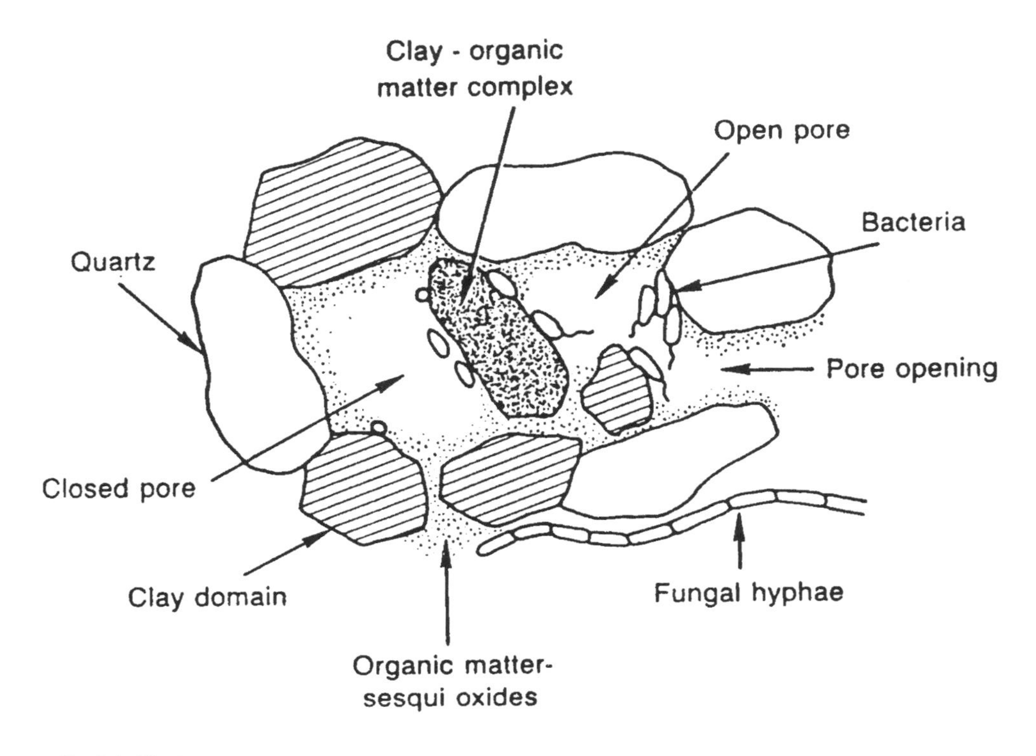

Source: Paul & Clark, 1989.

Lessons to remember

Throughout this evolution from "soil" to "environmental" microbiology, it has become clear that i) the soil and its inhabitants are intrinsically heterogeneous and therefore generalised information is of limited value, and ii) there is a massive lack of quantitative approaches, including mass balances and mathematical models. A few examples to highlight these aspects:

- It is well known that soil systems are spatially variable with respect to all their properties. Boekhold *et al.* (1990) have shown that chemical properties which play a role in the process of adsorption, such as pH and organic matter content, show highly variable distributions. Transport of reactive contaminants, which are adsorbed by clay or soil organic matter, is affected by the heterogeneity of both physical and chemical soil properties. Prediction of pesticide leaching to groundwater is more realistic when variability of soil hydraulic functions and chemical parameters are taken into account. The spatial variability of soil properties undoubtedly plays a role in the exposure of organisms to contaminants.

- Kinetic modelling of microbially mediated soil processes and the mass balances of such transformations have been cried out for. Yet, when it comes to the control of processes of direct agronomic and at the same time environmental importance, such as nitrification, we must admit that nothing thus far has succeeded. Notwithstanding all the efforts made, one has no single chemical yet which in practice has proven to be useful to modulate the nitrification process to the tune of the crop needs or the environmental quality standards.

Figure 2. Survival kinetics of *Bradyrhizobium japonicum* GMB1 ka, G49 and G2 5p in a sandy, loam and clay soil.

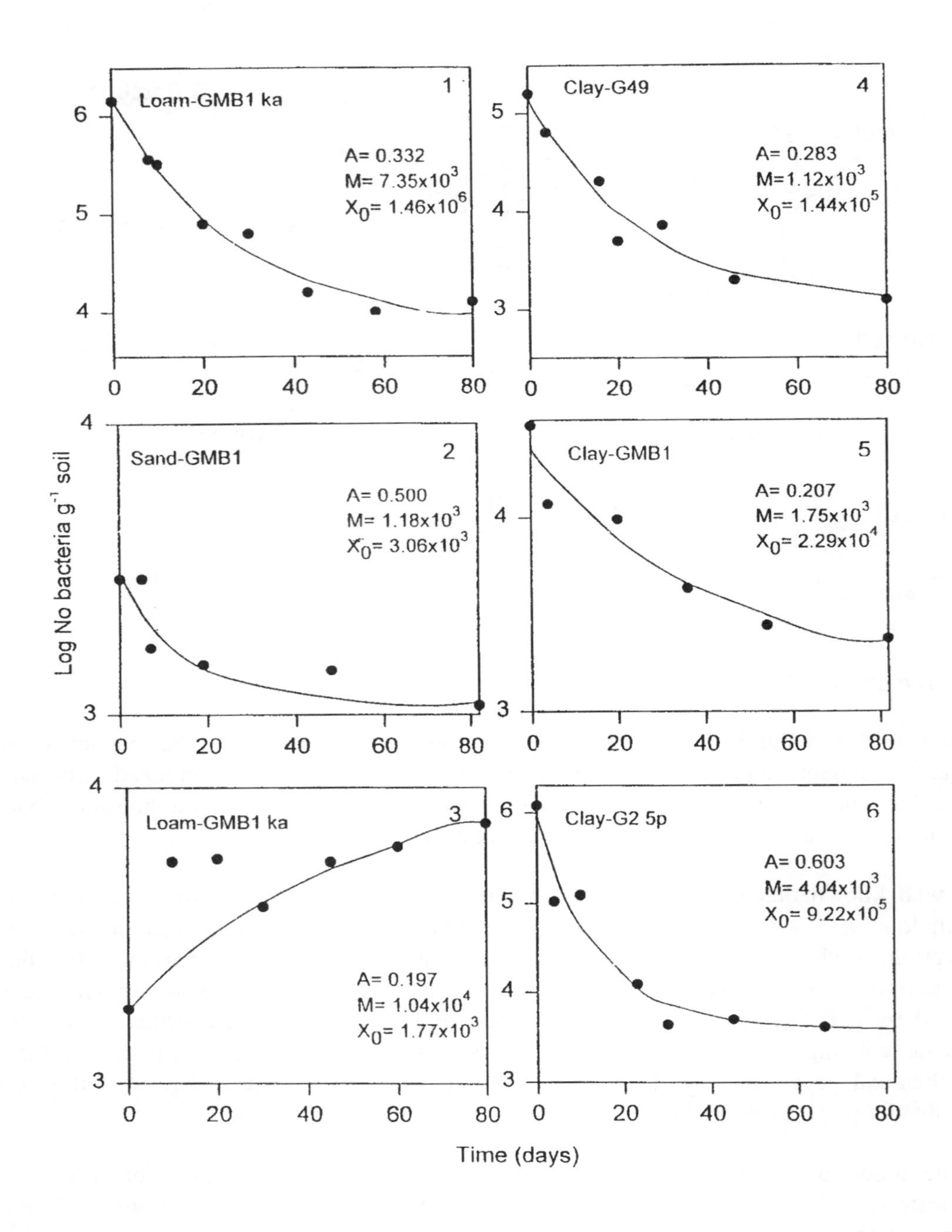

Note:

1. A = exponential decay rate (d-1); M = asymptotic number of bacteria reached in time per gram of soil (CFU/g soil); Xo = initial population size at time 0 (CFU/g soil).

Source: Corman *et al.*, 1987.

- The current hit-list of pollutants should be considered carefully. Indeed, the PAHs, the PCBs, the Heavy Metals, etc., are out there in so many hectares and constitute a threat to our biosphere. However, to some extent they belong to the past because they are by now stigmatised, and by the imposed life cycle assessment legislation they will gradually be banned. A group of chemicals which is constantly needed, and constantly spread in the environment at a rate of 2×10^9 kg per annum are the pesticides. Yet, they are not longer in scientific fashion. The reason for this is certainly not the fact that they do no longer constitute a problem. A very extended legislation has been developed in recent years (e.g. the EU United principles, EC Directive 91/414/EEC). A close look shows us that a majority of the active ingredients now applied are large molecules which have two things in common: they have a rapid DT50 because some moiety of the molecule is programmed to release relatively easily, but at the same time the remainder of the molecule is highly recalcitrant and disappears due to its "non-extractable" character. In other words, the residue is hardly or not metabolised to CO_2 (a few per cent per year) and is integrated in the soil humus complex, as has already been shown for some residues 20 years ago (Hsu and Bartha, 1974). Clearly, this organo-chemistry is extremely complex. Yet, it is essential that these "non-extractables" receive at least a fraction of the attention the superfund-chemicals receive.

The evolution in the field of microbiology has made it clear that the answer to the central question of bioremediation is not evident. This question is not whether bioremediation processes are successful, but how to improve these natural processes in the few cases where the natural processes are too slow to meet our needs or desires. Indeed, until now men's impact on the soil biota has been quite restricted. Only a few forms of imposed colonisation or attempts to artificially change the ecology of the microbial community have been successful so far. At the end of this millennium, it appears wise to accept the lesson from the practise: the microbial community of the soil is not different from any other society of animals or even people; it is highly resistant to changes.

The tasks ahead of us

Bioremediation is neither understood nor trusted (National Research Council, 1993). An additional reason for this mistrust is the lack of predictability of the bioremediation process, partially caused by missing information on items such as the complex actions of micro-organisms, environmental technology, appropriate cost estimations, etc. Until today, the use of bioremediation to clean up pollutants other than petroleum hydrocarbons has been limited due to three main gaps of knowledge, i.e. inadequate understanding of:

- how introduced microbes behave in the field (bioaugmentation);
- how to supply the microbes with stimulating agents (biostimulation);
- how to obtain good contact between the pollutant and the microbial cell (bioavailability).

Advances in fundamental knowledge of microbial ecology, microbial physiology, and the link of the latter with engineering are needed in order to obtain public acceptability. It should not be forgotten that the first axiom in microbial ecology is "the environment selects" (Alexander, 1971). One central aspect in this respect is the design of the circumstances which activate the microbial community to bring about the processes desired in terms of the overall soil quality. Another approach should concentrate on what the change of free energy is for the specific compound under the given environmental conditions. As a function of such a thermodynamic niche, existing or created around a particular compound, one can then construct communities, respectively organisms and genes to bring about the processes. Only through a better understanding of these aspects can the full potential of this technology be realised.

The future

There is an increasing amount of evidence that, in microbial associations, horizontal gene flow takes place (Stotzky and Babich, 1986). Although the transfer of DNA by conjugation, transduction or transformation does not directly alter the taxonomic identity of the organism involved (Slater, 1984), it certainly makes identification of the organisms much more difficult. Furthermore, by means of plasmid borne genes, the members of a microbial community maintain a gene pool which is probably far greater than the capacity of its individual member (Day, 1987). The horizontal gene transfer is of special interest in the communities of polluted soils because specific adaptations to pollutants are often plasmid-bound. Correlation between selection pressure exerted by pollutants and plasmid emergency is in fact suggested in many cases. Plasmids seem to play a major role in the adaptation of bacteria to xenobiotics and in the acquisition of new genetic traits due to pollution (Mergeay *et al.*, 1990). Plasmid-encoded pathways are ecologically advantageous because they provide genetically flexible systems and can be transferred between bacterial species (Sayler *et al.*, 1990). A particularly important aspect is the occurrence of some broad host range plasmids specialised in the degradation of synthetic chemicals (Burlage *et al.*, 1990). Due to their broad transfer and replication range, their introduction into a microbial community could provide the latter with enhanced degradative capacities.

The increased knowledge of enzyme structure, function and corresponding DNA sequence opens exciting new areas in environmental biotechnology. In the future, enzymes for the degradation of compounds, which could, until now, not be degraded microbially, will be constructed by protein engineering and inserted into desired micro-organisms. Because of this development, and the increased knowledge of controlled gene transfer, it appears necessary to make a step forward in microbial ecology, i.e. to study the ecology of genes in the microbial community rather than the organisms they are packaged in (Verstraete and Top, 1992).

Many recalcitrant pollutants are present in heterogenous mixtures and consequently are difficult to remove through high rate bioremediation technology. Therefore, it seems interesting to examine the possibility of adding a plain sorptive agent to the soil, such as activated carbon. For several pollutants, 1 kg of powered activated carbon (PAC) can sorb up to 100 g of pollutant and still guarantee an equilibrium concentration C of the order of 1 mg/l. Examination of the effect of such a treatment on the physiology of the soil microbiota appears interesting. Indeed, the work of Bally *et al.*, (1995) demonstrates that for surface waters, when organic carbon is limiting at the level of 20 μg/l, the organisms induce enzymes which can use all organics which are present as a source of carbon. In other words, under such harsh conditions, it is possible that enzyme systems are induced which otherwise might be repressed. Clearly, if high rate bioremediation is not applicable, it seems reasonable to openly opt for deliberately making the pollutants less bio-available and thus decrease their eco-toxicology (Bilbao *et al.*, 1995).

Questions, comments, and answers

Q: How is the introduction of plasmids into the environment regulated?

A: Under EU regulations this is not a particular restriction. We declare and test under well controlled conditions. We have not yet introduced it into the field but I do not see a major problem. Bioremediation is one of the best places to demonstrate that genetic manipulation can have a future.

Q: What do you mean by controls for gene flow?

A: I want to know what is happening - a plasmid with a specific catabolic gene is marked so we know where it is going. We need good experiments and cases to demonstrate what can be done and the biphenyl example is a good one.

Q: Which consortia or micro-organisms are involved in increasing sorption in the soil?

A: I am not thinking of micro-organisms but rather chemical and physical methods to bind to achieve non-extractability. This is well known in pesticide chemistry. We need to know more about microbial response to very low pollutant concentrations.

Q: How do you trace and identify each organism in consortia - in open systems?

A: This is a task for the next century. For 20 years there has been this need but there have been no major breakthroughs. Plating gives you a pattern of biochemical reactions which allows you to differentiate communities and see drift if you introduce new species or change the environmental conditions.

REFERENCES

ALEXANDER, M. (1967), "Introduction to soil microbiology", J. Wiley & Sons Inc., New York.

ALEXANDER, M. (1971), "Microbial ecology", J. Wiley & Sons Inc., New York.

APAJALAHTI, J.H.A. and M. SALKINOJA-SALONEN (1986), "Degradation of polychlorinated phenols by *Rhodococcus chlorophenolicus*", *Applied Microbiology and Biotechnology* 25, pp. 62-67.

BALLY, M., E. WILBERG, M. KUNHNI and T. EGLI (1995), "Growth and regulation of enzyme synthesis in the nitriloacetic NTA degrading bacterium *Chelatobacter heintzii* ATCC 2 9600", *Microbiology* 140, pp. 1927-1936.

BEDARD, D.L., M.L. HABERL, R.J. MAY and M.J. BRENNAN (1987), "Evidence for novel mechanisms of polychlorinated biphenyl metabolism in *Alcaligenes eutrophus* H850", *Applied and Environmental Microbiology* 53, pp. 1103-1112.

BILBAO, V., H. DE RORE, J. DRIES, W. DEVLIEGHER, E. TOP and W. VERSTRAETE (1995), "Soil remediation through binding of pollutants to the soil matrix", Proceedings of a Conference on Technology of Remediation of Contaminated Soils, Zamudio, Spain, 5 October 1995, pp. 1-30.

BOEKHOLD, A.E., S.A.T.M. VAN DER ZEE and F.A.M. DE HAAN (1990), "Prediction of cadmium accumulation in a heterogeneous soil using a scaled sorption model", in *Model Care 90: Calibration and reliability in groundwater modelling*, IAHS Publication No. 195, pp. 211-221.

BOUWER, E.J., B.E. RITTMANN and P.L. McCARTY (1981),"Anaerobic degradation of halogenated 1 carbon and 2 carbon organic compounds", *Environmental Science Technology* 15, pp. 596-599.

BRIGLIA, M., E.-L. NURMIAHO-LASSIA, G. VALLINI and M. SALKINOJA-SALONEN (1990), "The survival of the pentachlorophenol-degrading *Rhodococcus chlorophenolicus* PCP-1 and *Flavobacterium sp.* in natural soil", *Biodegradation* 1, pp. 273-281.

BRUNNER, W., F.H. SUTHERLAND and D.D. FOCHT (1985), "Enhanced biodegradation of polychlorinated biphenyls in soil by analogue enrichment and bacterial inoculation", J*ournal of Environmental Quality* 14, pp. 324-328.

BURLAGE, R.S., A.B. LYNN, A.C. LAYTON, G.S. SAYLER and F. LARIMER (1990), "Comparative genetic organization of incompatibility group P degradative plasmids", *Journal of Bacteriology* 172, pp. 6818-6825.

CHATTERJEE, D.K., J.J. KILBANE and A.M. CHAKRABARTY (1982), "Biodegradation of 2,4,5-trichlorophenoxyacetic acid in soil by a pure culture of *Pseudomonas cepacia*", *Applied and Environmental Microbiology* 44, pp. 514-516.

CORMAN, A., Y. CROZAT and J.C. CLEYET-MAREL (1987), "Modelling of survival kinetics of some *Bradyrhizobium japonicum* strains in soils", *Fertility and Soils* 4, pp. 79-84.

DAY, J.M. and J. DÖBEREINER (1976), "Physiological aspects of N_2-fixation by a *Spirillum* from *Digitaria* roots", *Soil Biology and Biochemistry* 8, pp. 45-50.

DAY, M. (1987), "The biology of plasmids", *Scientific Program Oxford* 71, pp. 203-220.

DE RORE, H., K. DE MOLDER, K. DE WILDE, E. TOP, F. HOUWEN and W. VERSTRAETE (1994), "Transfer of catabolic plasmid RP4::Tn*4731* to indigenous bacteria and its effect on respiration and biphenyl breakdown, *FEMS Microbiology Ecology* 15, pp. 71-78.

DEVLIEGHER, W., M. ACH SYAMSUL ARIF and W. VERSTRAETE (1995), "Survival of plant growth of detergent-adapted *Pseudomonas fluorescens* ANP15 and *Pseudomonas aeruginosa* 7NSK2", *Applied and Environmental Microbiology*, Nov. 1995, in press.

FOCHT, D.D. and W. BRUNNER (1985), "Kinetics of biphenyl and polychlorinated biphenyl metabolism in soil", *Applied and Environmental Microbiology* 50, pp. 1058-1063.

GOLOVLEVA, L.A., R.N. PERTSOVA, A.M. BORONIN, V.M. TRAVKIN and S.A. KOZLOVSKY (1988), "Kelthane degradation by genetically engineered *Pseudomonas aeruginosa* BS827 in a soil ecosystem", *Applied and Environmental Microbiology* 54, pp. 1587-1590.

HSU, T.S. and R. BARTHA (1974), "Interactions of pesticide-derived chloroaniline residues with soil organic matter", *Soil Science* 116, pp. 444-452.

KNACKMUSS, H.J. (1981), "Degradation of halogenated and sulfonated hydrocarbons", in *Microbial degradation of xenobiotics and recalcitrant compounds*, T. Leisinger, A.M. Cook, R. Hutter and J. Nuesch (eds.), pp. 189-212, Academic Press, London.

McCLURE, N.C., A.J. WEIGHTMANN and J.C. FRY (1989), "The survival of growth of *Pseudomonas putida* UWC1 containing cloned catabolic genes in a model activated sludge unit", *Applied and Environmental Microbiology* 55, pp. 2627-2634.

McCLURE, N.C., J.C. FRY and A.J. WEIGHTMANN (1991), "Survival and catabolic activity of natural and genetically engineered bacteria in a laboratory-scale activated sludge unit", *Applied and Environmental Microbiology* 57, pp. 366-373.

MERGEAY, M., D. SPRINGAEL and E. TOP (1990), "Gene transport in polluted soils", in *Bacterial genetics in natural environments*, J.C. Fry and M.J. Day (eds.), pp. 152-171, Chapman & Hall, London, New York.

NATIONAL RESEARCH COUNCIL (1993), *In situ bioremediation: When does it work?*, p. 207, National Academy Press, Washington, D.C.

OLDENHUIS, R., L. KUIJK, A. LAMMENS, D.B. JANSSEN and B. WITHOLT (1989), "Degradation of chlorinated and non-chlorinated aromatic solvents in soil suspensions by pure bacterial culture", *Applied Microbiology and Biotechnology* 30, pp. 211-217.

PAUL, E.A. and F.E CLARK (1989), *Soil microbiology and biochemistry*, Academic Press, London.

QUENSEN, J.F., J.M. TIEDJE and S.A. BOYD (1988), "Reductive dechlorination of polychlorinated biphenyls by anaerobic micro-organisms from sediments", *Science* 242, pp. 752-754.

SAYLER, G.S., S.W. HOOPER, A.C. LAYTON and J.M.H. KING (1990), "Catabolic plasmids of environmental significance", *Microbial Ecology* 19, pp. 1-20.

SHU-YEN, L., L. MIN-HUA and J.M. BOLLAG (1990), "Transformation of metolachlor in soil incubated with a *Stroptomyces sp.*", *Biodegradation* 1, pp. 9-17.

SLATER, J.H. (1984), "Genetic interactions in microbial communities", in *Current perspectives in microbial ecology*, M.J. Klug and C.A. Reddy (eds.), pp. 87-93, American Society of Microbiology, Washington, D.C.

STOTZKY, G. and H. BABICH (1986), "Survival of, and genetic transfer by, genetically engineered bacteria in natural environments", *Advances in Applied Microbiology* 31, pp. 93-138.

TAUSSON, W.C. (1927), "Napthalin als Kohlenstoffquelle für Bakterien", *Planta* 4, pp. 214-256.

VANDEPITTE, V., P. QUARAERT, H. DE RORE and W. VERSTRAETE (1995), "Evaluation of the Gompertz function to model survival of bacteria introduced into soil", *Soil Biology and Biochemistry* 27, pp. 365-372.

VERSTRAETE, W. and E. TOP (1992), "Holistic environmental Biotechnology", in *Microbial control of pollution*, J.C. Fry, G.M. Gadd, R.A. Herbert, C.W. Jones and I.A. Watson-Craik (eds.), pp. 1-17, Cambridge University Press, Cambridge.

VOGEL, T.M. and P.L. McCARTY (1985), "Biotransformation of tetrachlorethylene, dichloroethylene, vinyl chloride, and carbon dioxide under methanogenic conditions", *Applied and Environmental Microbiology* 49, pp. 1080-1083.

Parallel Sessions/A1 and A2

A2.2. EFFICACY, RELIABILITY AND PREDICTABILITY: AIR/OFF-GAS

DEVELOPMENT OF BIOFILTRATION PROCESSES AND STRATEGIES FOR THEIR APPLICATION[1]

by

Jan Páca
Department of Fermentation Chemistry and Bioengineering
University of Chemical Technology, Prague, Czech Republic

Introduction

Intensive research on the biological treatment of waste gases from volatile organic compounds (VOCs) and other odours began in Western Europe and in the United States in the mid-1980s (Ottengraf, 1986, 1992; Fouhy, 1992). In the Czech Republic, waste gases on an industrial scale have so far been treated using chemical or physico-chemical procedures. The only exception is an installation of biological gas scrubbing systems in four veterinary sanitation plants focused on animal utilisation. The equipment and technology were supplied by ÖSKO Environmental Technology, Austria, between 1985 and 1989.

Over the last five years, several private companies have been founded in the field of remediation (Drobník, 1995). To clean contaminated soil and groundwater, some of them use stripping of VOCs followed by adsorption. Some experiments performed during these remediation procedures replaced the adsorber by a biofilter. However, the experimental results were not published, probably because of the rather empirical design of the biofilter, the composition of pollutants in the gas phase, and the microflora present in the packing material.

Biological treatment of waste gases can also be used in cases of intensive animal farming (van Langenhove *et al.*, 1988; Rodhe *et al.*, 1988) and intensive composting. In addition to the odour nuisance to people living nearby, these are considered to be the most important sources of ammonia emissions. There are not yet any waste gas treatment plants in the Czech Republic for these sources of air contamination for two reasons: the incomplete transformation of the former communist co-operative farms and the new farm owners' serious lack of finance.

Despite all of this, we started systematic R&D on biological waste gas treatment four years ago. The work was initiated under new ownership of enterprises, a group of new small private companies, and the pressure of new laws on environmental protection, specifically the Czechoslovak Clean Air Act of 1991, variously amended up to 1994. In addition, international collaboration, notably in CORINAIR 90, has contributed positively to R&D in this field.

---

[1] This work was supported by the Grant Agency of the Czech Republic (Grant No. 204/93/2439).

Biodegradation of pollutants from gaseous waste

A biofiltration process can be successfully used to clean waste gases from volatile pollutants from biologically degradable compounds. Among these pollutants are both organic and inorganic compounds such as aliphatic and aromatic hydrocarbons, alcohols, aldehydes, ketones, esters, mercaptans, amines, sulphides, and ammonia.

In the past, interest focused on the elimination of odorous compounds using a biological treatment of waste gas (Rodhe *et al.*, 1988). More recently, the degradation of toxic chemicals has also become important (Zilli *et al.*, 1993; Singleton and Kant, 1995).

We focused our efforts on developing biofiltration processes for two significant sources of air pollution: solvents used in painting shops (with xylene and toluene as basic compounds) and ammonia. The development of a biofiltration process for degrading xylene and toluene degradation in the gas phase is described below.

Isolation, selection, and adaptation of mixed culture

Many soil samples were collected from various places with long-term pollution by aromatic hydrocarbons. The microflora present in the samples were eluted and cultivated in a mineral salt medium. The cell suspension was used to inoculate packed bed reactors in which selection and adaptation to solvents had been carried out over several months.

Biofilter set-up

We used a packed bed reactor of cylindrical shape with an inner diameter of 50 mm. The experimental biofilter set-up consisted of blower, air humidification stage, vessel for xylene and toluene evaporation, measurement and control of volumetric gas flow rate, measurement of the pressure drop, and sampling ports in the inlet and outlet gas piping of the biofilter.

Packing materials and their properties

We used a mixture of peat or compost, bark and wood as packing material and tested the effects of changes in the following parameters:

- changes in the ratio of peat to bark and to wood;
- changes in the size of particles of each component, use of various kinds of bark, replacement of peat by compost;
- changes in pH value and humidity of the packed bed;
- addition of nutrients to the immobilised cells;
- shrinkage of the bed during long-term operations.

Adaptation period of the biofilter

In order to achieve high elimination capacity and degradation efficiency in the biofilter, a period of adaptation is necessary. During this period, the organic load has to be gradually increased. We investigated the effect of the manner of loading (starting with only one pollutant or with both pollutants) and of the rate of increase of the organic load. Figures 1 and 2 show typical curves for the elimination capacity and the degradation efficiency in the adaptation period for toluene and xylene respectively.

Figure 1a. The increase of elimination capacity (EC) of toluene during the adaptation period at pH 6.6-5.9

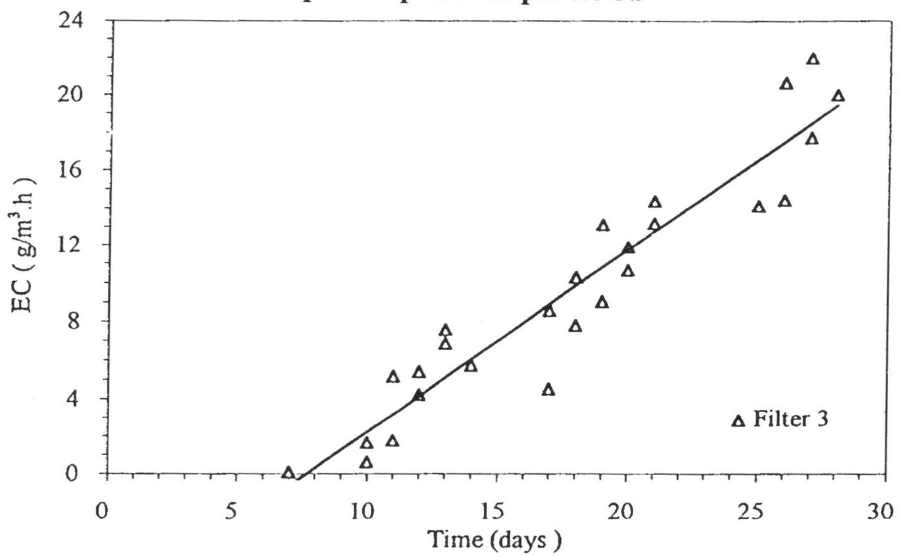

Source: Author.

Figure 1b. The increase of degradation efficiency (DE) of toluene during the adaptation period at pH 6.6-5.9

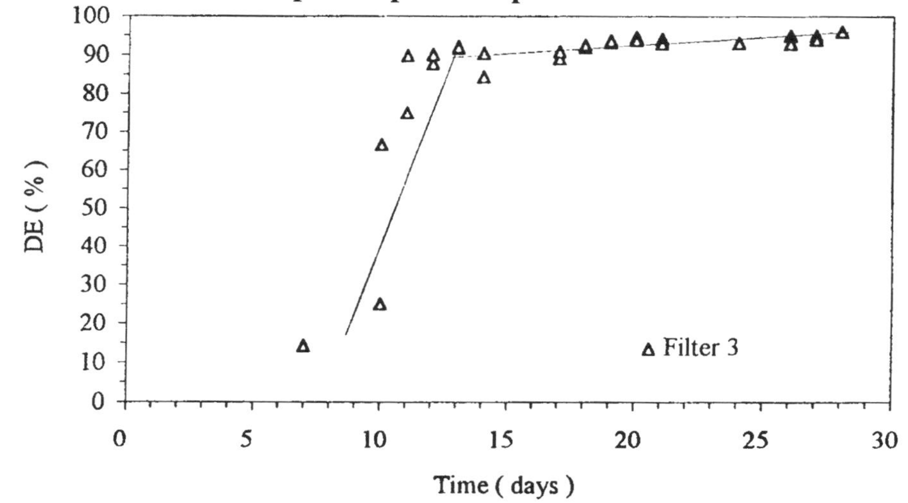

Source: Author.

Figure 2a. The increase of elimination capacity (EC) of xylene during the adaptation period at pH 6.6-5.9.

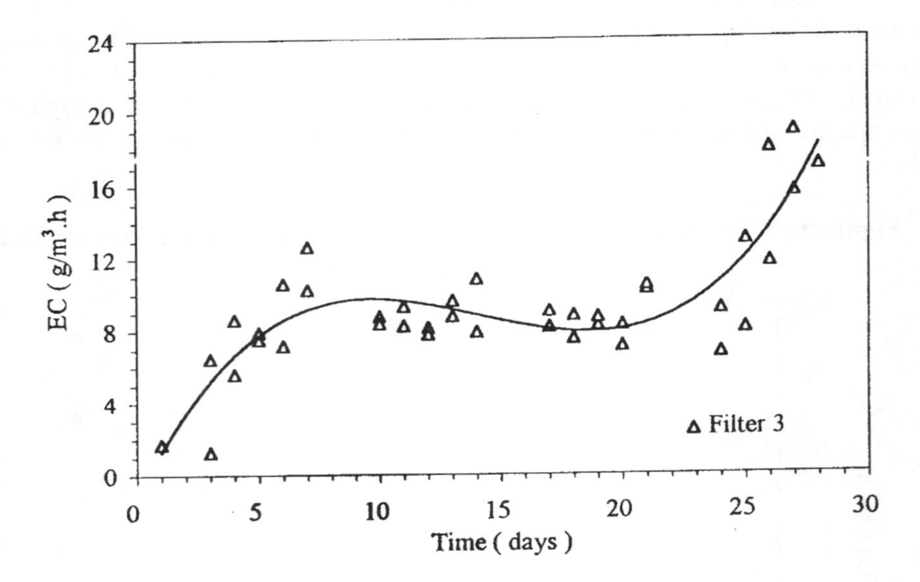

Source: Author.

Figure 2b. The increase of degradation efficiency (DE) of xylene during the adaptation period at pH 6.6-5.9.

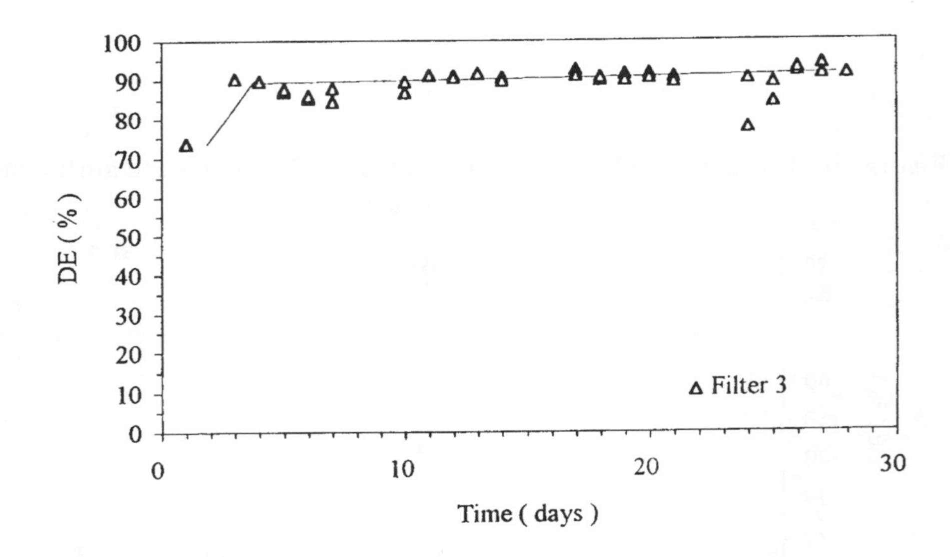

Source: Author.

In addition, we studied the effect of the pH value of the packed bed on changes in the ratio of prokaryotic and eukaryotic cells in mixed culture during the adaptation period (Figure 3).

Figure 3. Effect of pH value in the packed bed on changes in mixed microbial culture during the adaptation period

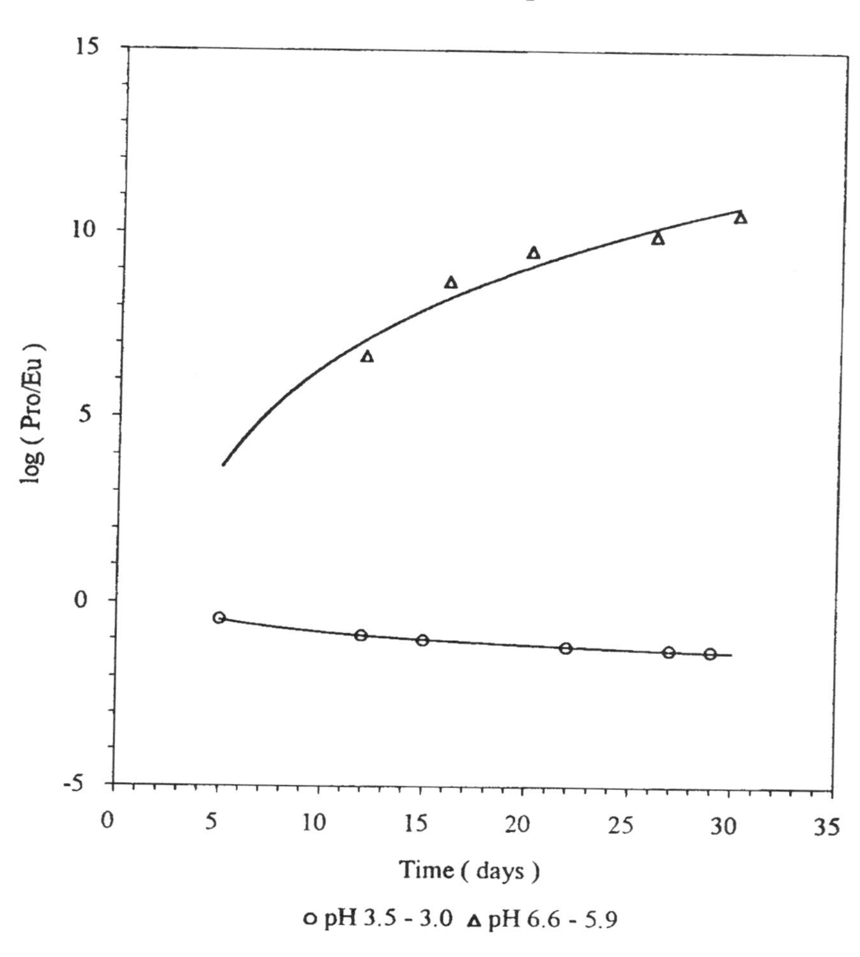

Notes:

1. Pro - Number of prokaryotic cells per unit weight of packing material.
2. Eu - Number of ekaryotic cells per unit weight of packing material.

Source: Author.

Performance characteristics

We studied the effects of the following on the elimination capacity and the degradation efficiency of the biofilter:

– increased concentrations of both pollutants in the inlet gas at the constant gas flow rate;

– increased gas flow rate at constant inlet gas concentration of the pollutants;

– parameters of the packed bed (particle size and composition of packing materials, humidity, pH, nutrient supply);

– stability during long-term experiments.

The results revealed a very important relationship between the rate of loading and of pollutant degradation, as substrate inhibition or limitation or too short a contact time among molecules of pollutants and cells can affect performance (Páca and Koutský, 1994).

Dynamic behaviour and reliable performance

For practical application of the biofilter, it is necessary to know the dynamic stability of the waste gas cleaning process. This was tested under conditions of repeated loading shocks, which were simulated by shifting up and down the inlet gas concentration of both pollutants. The frequency of changes in the inlet gas concentration and the number of repetitions were taken into account. The results showed that the biological system was very stable and did not tend to vary (Figure 4).

Figure 4. Dynamic stability of the biological systems. Effect of shift-up and shift-down in inlet gas concentration (Cin) on degradation efficiency (DE)

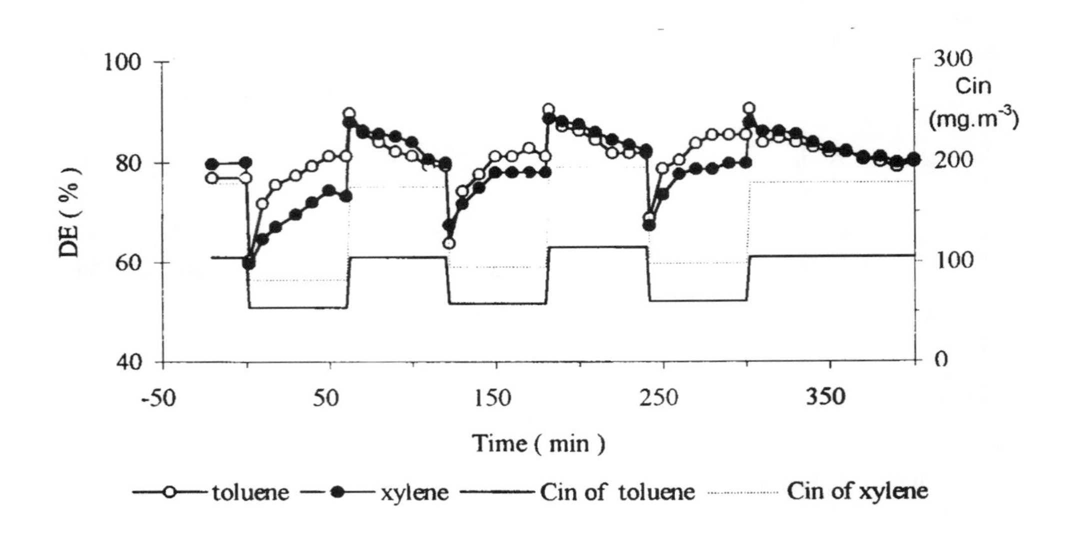

Source: Author.

Interruptions of biofilter performance

When work is not being done in a painting shop, the biofilter stops working. Work can be interrupted for different intervals: about 15 minutes (a coffee break), overnight, or a weekend. Therefore, we investigated the elimination capacity and the degradation efficiency following restarting of the biofilter after interruptions lasting from 15 minutes to 72 hours. Following an interruption of up to 48 hours, the degradation efficiency after a new start-up remained as high as it was before the interruption. After longer interruptions, degradation efficiency first dropped strongly, by about 25 per cent, but achieved the previous level after a few hours (Figure 5). The length of time of decreased degradation efficiency can be significantly shortened by using a lower volumetric gas flow rate after the start-up of the biofilter.

Figure 5. Transition states of toluene degradation efficiency (DE) induced by interruption of biofilter performance

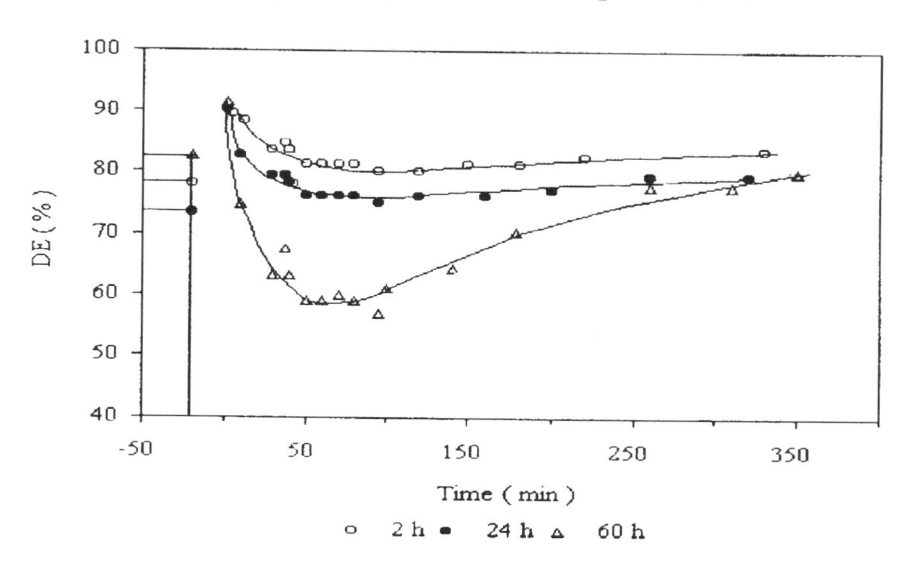

Notes:

1. Toluene Cin = 86-94 $mg.m^{-3}$.
2. The interruption period: 2 h, 24 h, 60h.

Source: Author.

In addition, the long-term performance (six months) and the reproducibility of important performance parameters (elimination capacity and degradation efficiency) were repeatedly tested under the same loading rates.

Current activities

We have developed technology using the biofilter design for waste gas cleaning for solvents containing xylene and toluene as the main compounds. At present, we are intensively studying a biofilter for ammonia degradation. This year, we started an investigation into styrene degradation by a biofilter, and we now plan to start a study of acetone and alcohol degradation. The rate, efficiency, and progress of our research will depend on what financial support is available.

Strategies for practical applications

We are able to offer technology and biofilter design for small painting shops using solvents containing xylene and toluene. Because our students are involved in this research programme, several of our graduates are very well informed about this field. Through lectures at congresses, conferences, and workshops and seminars at other inland universities and abroad, we disseminate information to the public. We also engage in consulting activities in this field.

Practical applications of our results will strongly depend not only on processors' efforts to satisfy the requirements of the Czech Clean Air Act, but on their financial possibilities as well.

Questions, comments, and answers

Q: Why is there a sharp decrease in efficiency at high organic load?

A: Possibly because of substrate concentration. The slope depends on the quality of the mixed cultures and the time of contact. The use of a single substrate does not give as steep a slope.

Q: Why do you use that particular packing material and what was your start-up culture?

A: Peat, bark and wood were used because the materials are cheap and available and show good results. Inorganic materials are also used to increase porosity. We prepared special mixed cultures isolated from highly polluted areas.

REFERENCES

DROBNÍK, J. (1995), "The challenge for bioremediation: A Central European perspective", in OECD (1995), *Bioremediation: The TOKYO '94 Workshop*, pp. 117-124.

FOUHY, K. (1992), "Cleaning waste gas, naturally", *Chem. Eng.*, December, pp. 41-46.

OTTENGRAF, S.P.P. (1986), "Exhaust gas purification", in *Biotechnology,* Vol. 8, H.J. Rehm and G. Reed (eds.), pp. 425-452, VCH Verlagsgesellschaft, Weinheim.

OTTENGRAF, S.P.P. (1992), "Biological systems for waste gas elimination", *Trends in Biotechnol.* 5, May, pp. 132-136.

PÁCA, J. and B. KOUTSKÝ (1994), "Performance characteristics of a biofilter during xylene and toluene degradation", *Med. Fac. Landbouww. Univ. Gent* 59 (4b), pp. 2175-2184.

RODHE, L., L. THYSELIUS and U. BERGLUND (1988), "Biofilters for odour reduction: Installation and evaluation", *AFRC Engineering*, Bedford, United Kingdom.

SINGLETON, B. and W. KANT. (1994), "Application of biofiltration for the treatment of aromatic hydrocarbons, H_2S, Methyl Mercaptan, Dimethyl Sulfide and Dimethyl Disulfide", paper presented at the AIChE Spring National Meeting, 29 March-2 April, Houston, Texas.

VAN LANGENHOVE, H., A. LOOTENS and N. SHAMP (1988), "Elimination of ammonia from pigsty ventilation air by wood bark biofiltration", *Med. Fac. Landbouww. Rijksuniv. Gent* 53 (4b), pp. 1963-1969.

ZILLI, M., A. CONVERTI, A. LODI, M. DEL BORGHI and G. FERRAIOLO (1993), "Phenol removal from waste gases with a biological filter by *Pseudomonas putida*", *Biotechnol. Bioeng.* 41, pp. 693-699.

SCALING UP BIOFILTRATION FOR RELIABLE PROCESSES IN PRACTICE[1]

by

S.P.P. Ottengraf
Eindhoven University of Technology, Department of Chemical Engineering, Eindhoven;
University of Amsterdam, Department of Chemical Engineering, Amsterdam, the Netherlands

M.C.J. Smits
University of Amsterdam, Department of Chemical Engineering, Amsterdam, the Netherlands

R.M.M. Diks
Eindhoven University of Technology, Department of Chemical Engineering, Eindhoven, the Netherlands

Introduction

As a result of the increasing concern about environmental pollution in many countries nowadays, a statutory control takes place on the emission of toxic compounds into the environment. Stronger regulations have been put into action, and hence, in many different branches of industry, much interest exists in reliable, simple and cheap purification techniques for the elimination of undesirable contaminations present in waste gases. One promising technique is the biological treatment of contaminations. In the last few decades, the number of applications of biological treatment systems for off-gas purification has strongly increased. Although this development is quite recent, the principle of waste gas purification by contacting a contaminated gas flow with a suitable microbial population is much older. Already in 1923, the biological elimination of H_2S emissions from waste water treatment plants was discussed (Bach, 1923). Afterwards, in 1934, probably one of the earliest patents in this field was applied for, claiming a biological purification system concerning 'die Reinigung von luft- oder sauerstoffhaltigen Gasgemischen die biologisch zerstörbare Riech und/oder Feststoffe enthalten...' (the purification of air or oxygen containing gas mixtures, which contain biologically degradable odorous compounds or particles...) (Prüss and Blunk, 1941). Reports on the actual application of the technique on a larger scale date back to the early fifties, when soil bed filters were mostly applied to purify odorous waste gases from municipal sewage treatment plants. Ever since, a lot of microbiological, as well as process engineering, research has been carried out on the development of biological elimination systems for the removal of volatile organic and inorganic compounds, as will be discussed below. Although at the outset biofiltration was mainly applied for odour abatement, it has nowadays become an important alternative to many physical and chemical methods of waste gas purification, as the application of biofilters generally appears to be quite

[1] This article is based on material published in the Proceedings of ISEB-1 (International Symposium on Environmental Biotechnology), April 1991, Oostende.

reliable and effective at relatively low cost. Table 1 gives a rough indication of the relative capital and operational costs involved in the application of different waste gas purification techniques.

Table 1. Relative capital and operational costs for off-gas purification

Reference	Maurer (1979)		Jäger & Jäger (1978)	Lith (1990)
Process	Investment costs $DM/(m^3/h)$ (Application not specified)	Operational costs $DM/1\ 000m^3$	Total costs $DM/1\ 000m^3$ (Composting works) Price level 1974	Total costs $Dfl/1\ 000m^3$ (VOC: 100-2 000 mg/m^3) Gas flow 10 000 m^3/h
Thermal incineration	12-14	1.4-1.7	9.10 (fuel costs only)	7-9 (50% energy recovery)
Catalytic incineration	14-16	1.3-1.5	*	6-8 (50% energy recovery)
Adsorption	5-20	0.5-1.0	1.5 (incl. regeneration incineration)	14-18 (incl. steam by regeneration)
Absorption	8-10	0.8-1.0	4.20 (chlorine)	*
Ozone oxidation	6-8	0.4-0.6	4.2	*
Biofilter open	3-10	0.3-0.5	0.6	*
Biofilter closed				0.5-3

Notes:

Capital costs are given per ($m^3_{waste\ gas}$/h) (related to the gas flow).
Operational costs are given per 1 000m^3 of waste gas treated.
* = Not available from the specified literature source.

Sources: Maurer, 1979; Jäger and Jäger, 1978; Lith, 1990.

Microbial substrate degradation

The application of a biological treatment system is primarily based on the microbial degradability of the compounds present in the waste gas. However, the performance of a continuously operating bioreactor is the ultimate result of a complex interaction between the microbiological and physical phenomena, often denoted as the macro-kinetics of the process. The physical phenomena include the mass transfer between gas and liquid phase, the mass transfer to the micro-organisms, the average residence time of the mobile phases, etc. Some of the microbial phenomena are, for example, the reaction rate of the degradation, the substrate or product inhibition, and diauxic phenomena.

The micro-kinetics of the degradation process are generally investigated and modelled for pure cultures of suspended micro-organisms. However, in bioreactor systems for environmental purposes, mainly heterogeneous mixed cultures of micro-organisms are present rather than mono-cultures, which means that the application of the micro-kinetic results may be limited for bioreactor design purposes.

In many bioreactor systems, the degradation also takes place within fixed biofilms, which means that additional mass transfer phenomena (i.e. mass transfer to the biofilm and internal diffusion) should be taken into account.

In the following presentation, some aspects concerning the micro-kinetics will be reviewed. Thereafter, essential differences in substrate degradations within heterogeneous biofilms will be discussed.

Micro-kinetics of suspended micro-organisms

The elimination of organic substrates by micro-organisms results from the fact that those organisms can use the organic compounds as their sole energy (catabolism) and carbon source (anabolism) (Figure 1). Approximately 50 per cent of the carbon of the organic substrate is involved in each reaction.

Figure 1. Substrate elimination due to microbial oxidation

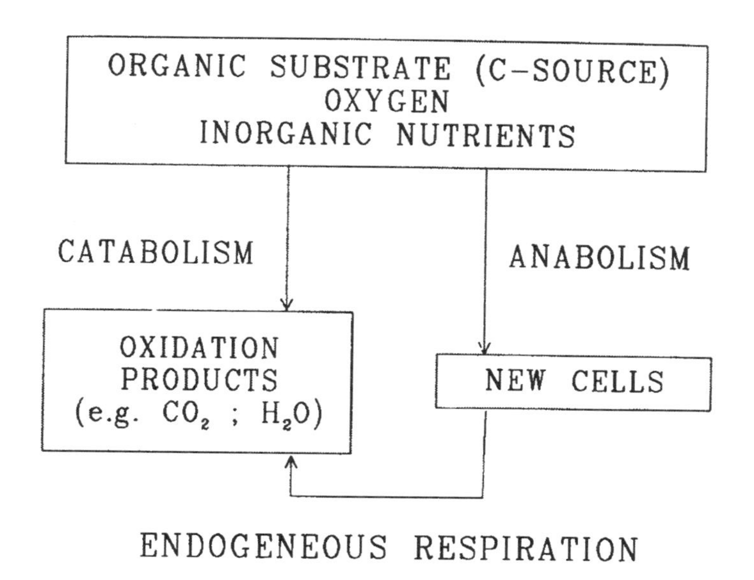

Source: Author.

The degradability of a compound often reflects its origin: biogenic compounds are easily biodegradable, while anthropogenic (i.e. man-made) compounds sometimes possess such unnatural structures (xenobiotics) that biological degradation is difficult (recalcitrant compounds), or even impossible (persistent compounds). However, due to intensive microbiological research in recent years, much progress has been made in isolating, selecting or constructing strains or mixed cultures of micro-organisms (mainly bacteria) which can degrade some recalcitrant compounds to such an extent that the application of a biological waste gas treatment system, even in such cases, offers real prospects. This is illustrated by Table 2, which lists the growth rate of some bacteria on xenobiotic chlorinated hydrocarbons that have been isolated from, among others, activated sludge, contaminated soils. It is surprising to conclude from the data presented in Table 2, that the growth rate of suited strains on many xenobiotic compounds is of the same order of magnitude as encountered in the degradation of many biogenic substrates. This stresses the importance of thorough isolation and adaptation procedures.

Table 2. The aerobic degradation of some xenobiotic chlorinated hydrocarbons by pure cultures of micro-organisms

Micro-organism	Substrate	Growth rate [h^{-1}]	Source
Hyphomicrobium	methylchloride	0.09	Hartmans *et al.*, 1986
Pseudomonas DM1	dichloromethane	0.11	Brunner *et al.*, 1980
Methylobact. DM11	dichloromethane	0.17	Scholtz *et al.*, 1988
Xanthobacter GJ10	1,2-dichlorethane	0.12	Janssen *et al.*, 1985
Mycobacterium L1	vinylchloride	0.05	Hartmans *et al.*, 1985
Pseudomonas AD1	epichlorohydrine	0.20	Wijngaard *et al.*, 1989
Pseudom. WR1306	chlorobenzene	0.79	Reineke *et al.*, 1984
Pseudom. GJ60	1,2-dichlorobenz.	0.33	Oldenhuis *et al.*, 1989
Pseudomonas	1,3-dichlorobenz.	0.07	Bont *et al.*, 1986
Alcaligenes A175	1,4-dichlorobenz.	0.13	Schraa *et al.*, 1986

Sources: As above.

The degradation of substrates and the subsequent growth of micro-organisms is generally described by the Monod equation. The value of the Monod-constant K_s for organic substrates generally amounts to 1-10 g/m^3, while for oxygen it is about 0.1 g/m^3 (Cooney, 1981). However, the concentrations of substrate and oxygen in the liquid phase in bioreactor systems often exceed the value of K_s. The degradation can hence be described as a zeroth order process.

Figure 2. The influence of the temperature on the maximum growth rate of *Hyphomicrobium* GJ21

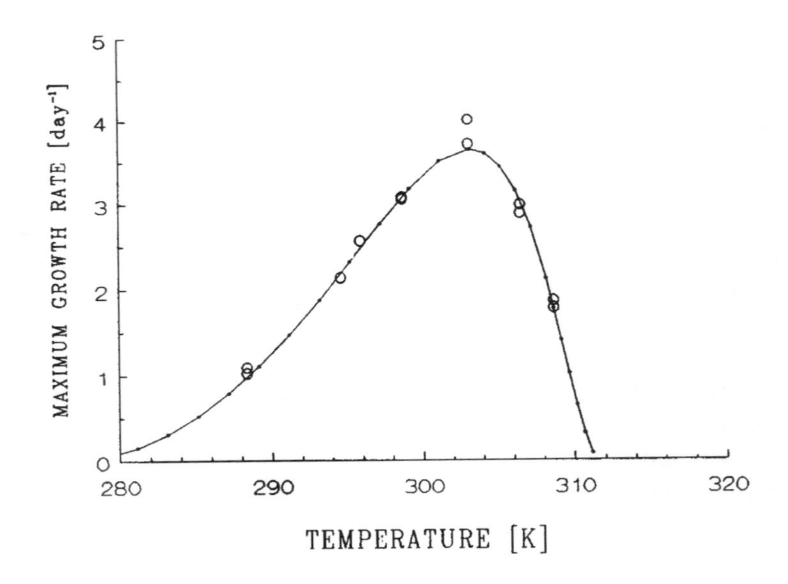

Notes: For the range of 15-25 °C, an activation energy of 76 kJ/mol was calculated.

Source: Diks, 1992

Most micro-organisms are able to grow over a pH range of about 4 pH units, but the growth rate generally has a pH optimum around 6.5-7.5. Above 8.5 and below 5, the growth rate may become very low.

Apart from the substrate availability, the growth rate also depends upon the physiological conditions, e.g. the temperature, pH and inorganic nutrients. Figure 2 shows the influence of the temperature on the maximum growth rate of *Hyphomicrobium* GJ21 growing on dichloromethane (Diks, 1992). An optimum of around 30 °C can be observed, whereas at higher temperatures the growth rate quickly decreases. Within the temperature range of practical interest (15 °C-25 °C) this influence can be described by an activation energy of 76 kJ/mol, according to the Arrhenius equation. In general, the temperature influence on the growth rate of many mesophylic bacteria is similar to the one shown in Figure 2.

Substrate degradation in biofilms

Opposed to most micro-kinetic experiments performed on microbial suspensions in laboratory, biological purification systems are often based upon fixed film degradation processes. The micro-organisms are immobilised inside the pores or on the surface of a carrier material. The spontaneous formation of aggregates sometimes occurs as well. The immobilisation of micro-organisms has the advantage that biomass concentrations, hence the volumetric reaction rates, can be considerably increased. However, the kinetics of immobilised cells may substantially differ from that of freely suspended cells.

The most important feature of a biofilm is the existence of concentration gradients of substrates and products. These gradients result from the internal mass transport by diffusion and the substrate depletion by reaction. The reaction rate therefore may vary throughout the biofilm, and also, serious diffusion limitations can occur. This may result in starvation and decay of cells in deeper parts of the biofilm, which may eventually result in detachment.

Figure 3. The effectiveness of a flat biofilm plotted versus the zeroth order Thiele number for different values of α

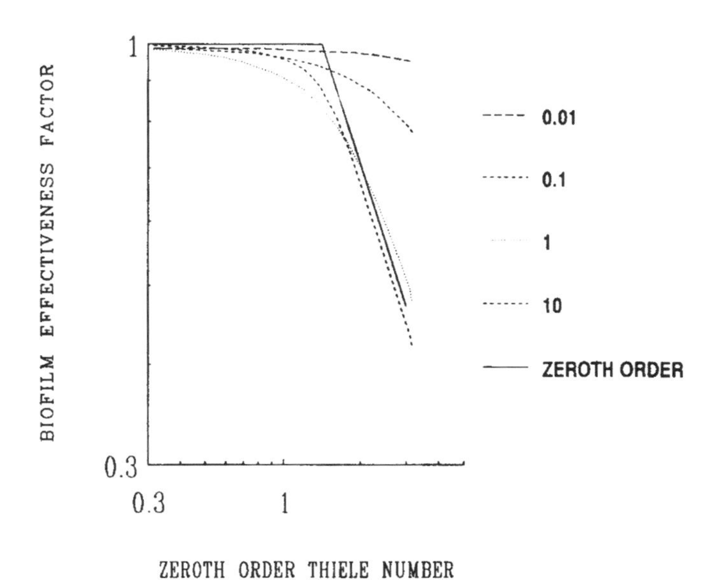

Source: Author.

If Monod kinetics are applied to describe the biological reaction rate, it is obvious that the substrate gradient will also considerably affect the overall reaction rate. This is illustrated in Figure 3, in which the effectiveness of a homogeneous, flat biofilm is plotted versus the zeroth order Thiele number for different ratios of the biofilm surface concentration and the Monod-constant. The Thiele number reflects the ratio of the maximum rate of degradation and the maximum rate of diffusion in the biofilm. For low values of the Thiele number, the biofilm efficiency approaches unity, whereas at higher Thiele numbers, the efficiency decreases. This effect increases at lower values of α. From Figure 3 it can be concluded that already at $\alpha=1$ the biofilm effectiveness factor can be approximated by a zeroth order behaviour.

The diffusion of oxygen in aerobic biofilms is known to have even more pronounced effects upon the overall activity of a biofilm. Typical oxygen penetration depths amount to 100-200 mm, while the whole biofilm may be as thick as several millimetres (Figure 4) (Cooney, 1981; de Beer, 1990; Harris et al., 1976).

Figure 4. Concentration gradients of ammonium (□), oxygen (o) and nitrate (Δ) in a nitrifying biofilm

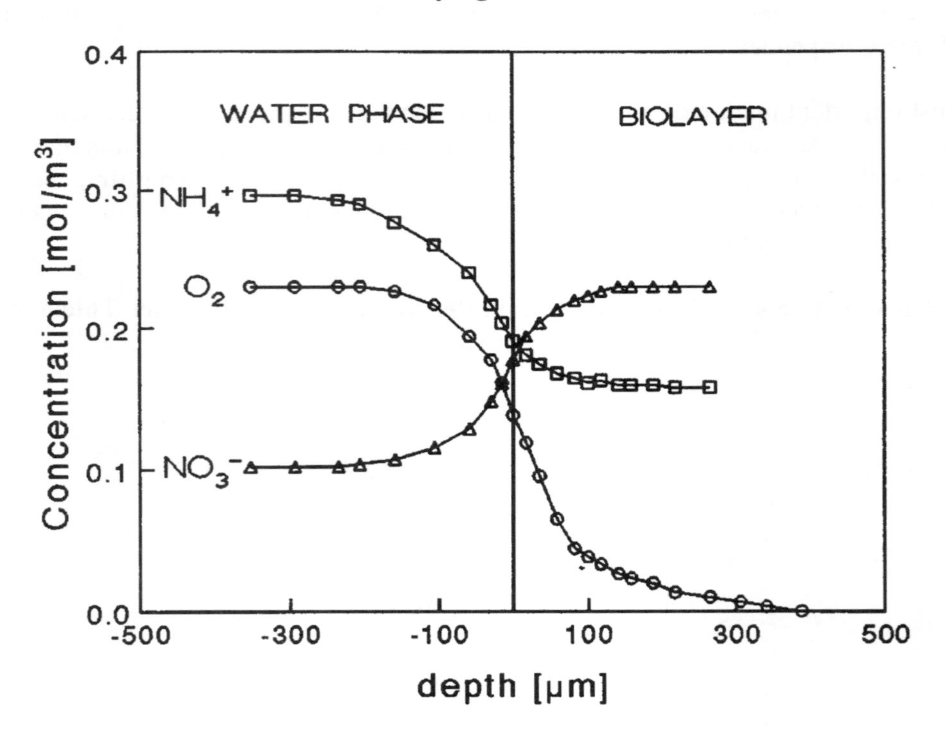

Source: Adapted from de Beer, 1990.

Diffusion limitation aspects also hold for inorganic nutrients. The minimal nutrient requirement during the growth of micro-organisms may follow from the stoichiometry of the elemental composition of biomass. However, this stoichiometrical composition is not sufficient to calculate the minimum concentrations required. In a fixed film process, diffusion limitations may further increase the minimal requirements.

Inhibiting products which must be transported outwards will also influence the local activity in the biofilm. Figure 5 shows the existence of a pH gradient which has been determined in a dichloromethane degrading biofilm, in which HCl is produced. Due to the influence of the pH on the reaction rate, the overall biofilm effectivity will decrease.

So far only gradients of diffusing compounds in biofilms of mono-cultures have been discussed. As bioreactors for environmental purposes generally are continuously operated under non-septic conditions, a heterogeneous population of micro-organisms may develop in the system.

Heterogeneous population may develop for two reasons. First, in the field of waste gas purification, the carbon source generally comprises a mixture of different substrates. Different micro-organisms may be required for an efficient removal of all those components. Secondly, a heterogeneous population can also develop as biogenic material is available in the system, due to the lyses of micro-organisms, e.g. in deeper parts of the biofilms. Also, the presence of predators like protozoa and metazoa (e.g. nematodes, ciliates, worms), is known (Cooney, 1981). Thus the biofilm may no longer be homogeneous, which means that porosity, density, activity, etc. may vary throughout the film.

As micro-kinetics are often investigated for pure cultures of bacteria only, the application of those results may also be limited. When modelling the macro-kinetics of the environmental bioprocesses, much simplifications are necessary, due to the complex nature of the characteristics of immobilised non-septic biofilms, as described above.

Figure 5. A pH gradient in a biofilm of *Hyphomicrobium* GJ21 degrading dichloromethane

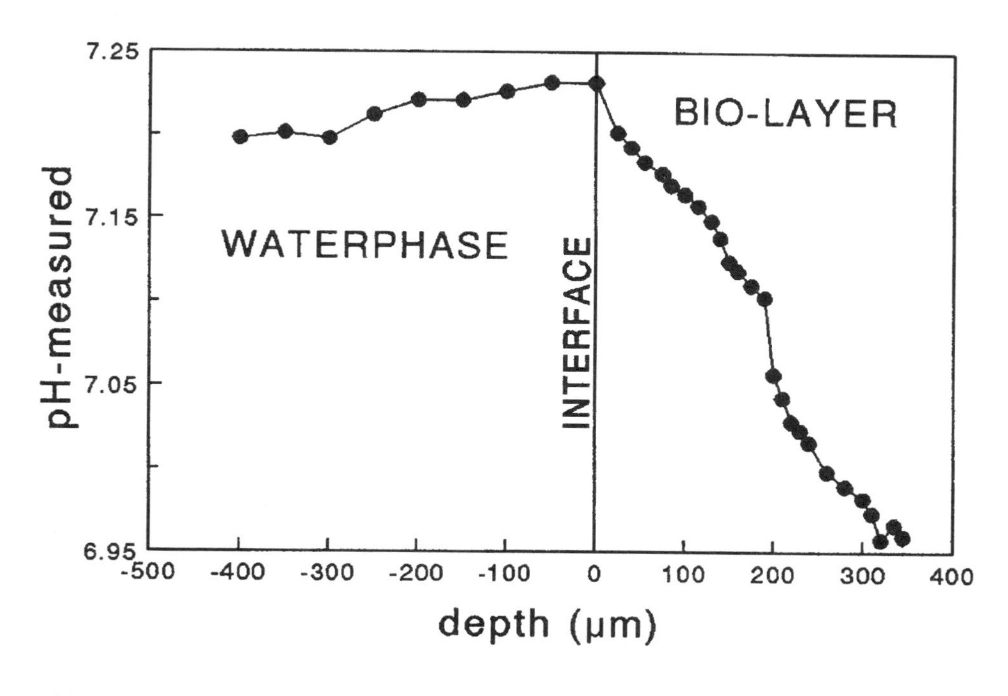

Source: Measurements by de Beer, 1990.

Biofilters

Three groups of biological waste gas purification systems are known, which can be distinguished (Table 3) by the behaviour of the liquid phase (which is either continuously moving or stationary, present in the contact apparatus) and of the micro-organisms (which are either freely dispersed in the aqueous phase or immobilised on a carrier material). Biofilters reflect the easier mode of operation as compared to other biological purification systems. In a biofilter (Figure 6) the waste gas is forced to rise through a simple structured packed bed of materials, in which a suitable microbial population develops over the course of time, or by inoculation of the material with suitable microbial strains. It is generally assumed that the constituent particles of the packing material are surrounded by a wet biolayer. The volatile compounds and oxygen present in the waste gas are transferred from the gas phase into this biofilm, where the microbial degradation takes place. The packing material normally consists of small particles (d < 10 mm), hence a high specific area (300-1 000 m^{-1}) and an excellent mass transfer are established. As inorganic nutrients (phosphate, nitrogen, sulphur, etc.) must also be supplied by the carrier material, mostly natural materials (like compost, peat, etc.) are applied. Extra addition of nutrients have sometimes shown to increase the conversion rate of a biofilter. Don (1986) reported that the removal efficiency of a toluene eliminating biofilter could be increased from 50 per cent to 95 per cent at an inlet concentration of 100 mg/m^3 and a gas velocity of 100 m/h, by the addition of inorganic nutrients to the filter material.

Table 3. Distinctions between different biological waste gas purification systems

Microbial flora	Aqueous phase	
	Mobile	Stationary
Dispersed	Bioscrubber	*
Immobilised	Trickling filter	Biofilter

Note: * = It is not possible to have a system with a stationary liquid phase ad dispersed microbial flora.

Source: Author.

From the conventional soil and compost filters, which were used in the early 1950s, a high pressure drop (Figure 7) and a non-homogeneous structure were known to exist (Don, 1986; Eitner, 1989). In order to reduce energy consumption, the height of such a filter bed amounted to 0.5-1.0 m, while the initial gas loads applied amounted to 5-10 m^3/m^2h (Don, 1986). Fairly long residence times (up to several minutes) were needed in order to achieve a high removal efficiency. This initiated intensive research projects for the development of better filter materials. Nowadays, mainly compost and peat are applied, or mixtures of these materials with wooden chips, heather, bark, polystyrene or lava particles (VDI-Berichte, 1989 [Verein Deutschen Ingenieure]; Ottengraf and Van den Oever, 1986). The latter materials are added to in order to create a stable filter bed structure over the course of time, and to decrease the pressure drop (< 500 Pa/m at superficial gas loads up to 100-500 m^3/m^2h) (Figure 7). Due to the optimised conditions in present biofilter systems provided with sufficiently active packing materials, the superficial gas contact time is in the range of 10-30s. Micro-organisms generally applied in biofilter systems are mesophylic. The degradation should hence take place within the temperature range of 15-40 °C. Furthermore, it has been reported that the optimal water content of the packing material should be in the range of 40-60 per cent w/w. In order to preserve the microbial activity, it should at least exceed 40 per cent (Eitner, 1989; Bardtke, 1987).

Figure 6. Experimental set-up of a biofilter system

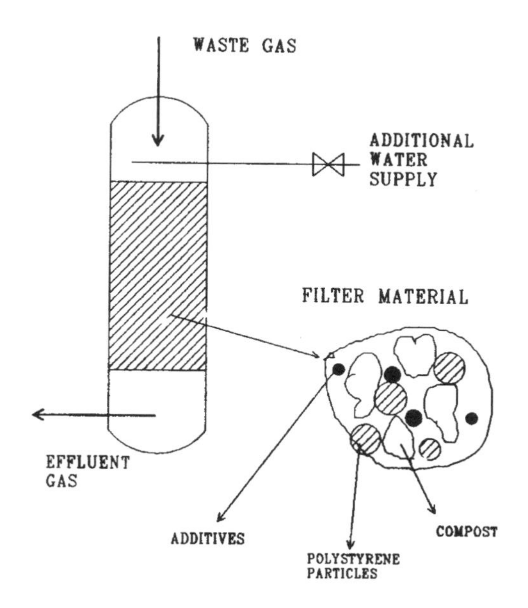

Notes: The filter material consists of e.g. compost, mixed with additives.

Source: Author.

Figure 7. The pressure drop of a filter bed versus the superficial gas load

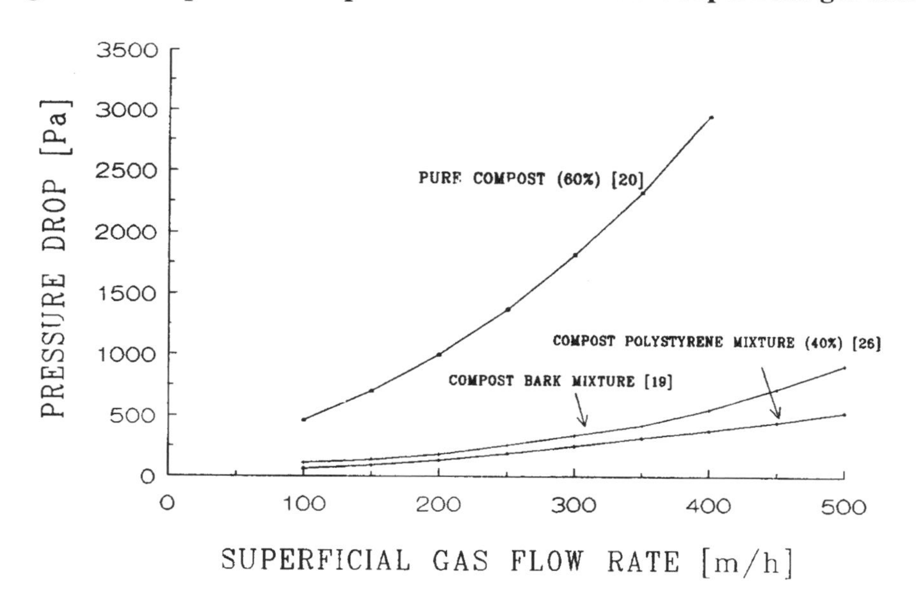

Notes: The percentages at each curve indicate the water content (bed height = 1 M).

Source: Don, 1986; Eitner, 1989; Ottengraf, 1981.

Water will evaporate from the filter bed, unless the inlet gas flow is completely saturated. This process will be enhanced by the heat generated by the microbial substrate degradation. In a pre-treatment section, therefore, the inlet gas may advantageously be humidified. As 100 per cent saturation can hardly be achieved, an additional (periodical) water supply to the filter at the top of the filter bed may sometimes be necessary. As the dry-out process generally starts at the gas inlet side, in those cases the filter may be advantageously operated down-flow. The water balance control of a biofilter is also connected to the lay-out of the filter.

Many different designs and forms have been presented in literature (VDI-Berichte, 1989; DECHEMA, 1987) and can be divided into 'open' and 'closed' biofilters. Open filters are generally subjected to changing weather conditions, thus to strongly changing temperature and humidity in time and place. This may result in setting and shrinking of the packing material, and an increasing formation of a non-homogeneous structure (e.g. channelling), etc.

Since a number of years, the interest in closed biofilters has increased, although these completely housed and insulated systems generally require somewhat higher capital costs. The biofilter performance can be better controlled and ageing phenomena of the packing material prevented, hence overdesign can be minimised. In this way, the microbial activity can be exploited for long periods of time (three to five years) with low operational cost and maintenance (Lith, 1990; Oude-Luttighuis, 1989).

Intensive experimental investigations on laboratory, semi-technical as well as full-scale have been carried out to determine the macro-kinetics of the filtration process and the values of the rate parameters for many volatile compounds discharged by industry. For biofilter systems, experimental results may be summarised as follows (Ottengraf, 1986; Ottengraf and Van den Oever, 1986; Ottengraf *et al.*, 1986):

- The macro-kinetics of the elimination processes in a biological filter bed can be described by an absorption process in a wet biolayer, surrounding the constituent particles, accompanied by a simultaneous biological degradation reaction.

- The elimination of nearly all of the compounds investigated, like alcohols, ketones, esters, aromatics, etc., in a biological filter bed follows zeroth order reaction kinetics down to very low concentration levels. This has been confirmed by batch investigations of the degradation process in aqueous solutions of the compounds concerned.

- At low gas phase concentration levels or low water solubility of the compounds concerned, the elimination rate of the filter bed may shift towards a diffusion controlled regime.

- Due to the predominantly zeroth order character of the elimination process, the degree of removal of any biodegradable compound may be close to 100 per cent at finite residence times of the gas phase in the filter bed.

The general path of the elimination capacity (EC) of a filter bed with height (H) as a function of the inlet gas phase concentration C_{go} at constant gas flow rate ω, is shown in Figure 8. EC is defined as:

$$EC = \frac{\omega}{H} \bullet (C_{go} - C_{ge})$$

Three operational regimes may be generally distinguished:

i) Below point A in the graph, the organic load (w/H) × C_{go} to the filter is so low that practically 100 per cent conversion of the compound is achieved.

ii) If the gas phase is higher than C_{gb} (Figure 8), the load is so high that the maximum elimination capacity EC_{max} of the filter is reached. The system is in the so-called reaction limited regime (Ottengraf *et al.*, 1986), the activity of the biolayer surrounding the packing particles is fully utilised and no reaction free zone exists (see also curve 1, Figure 9).

iii) If anywhere in the filter bed the concentration C_g equals C_{gb},two zones can be distinguished in the filter bed. If the filter is operated down-flow, the upper part of the bed is still in the reaction limited regime, the lower part however in the diffusion-controlled regime. The transition between both regimes occurs at the critical gas phase concentration C_{gb} shown in Figure 9. Below this critical concentration (as curve 3 in Figure 9 shows) a reaction-free zone exists in the biolayer due to the zeroth order character of the micro-kinetics of the elimination process. It can be shown that in a multi-stage filter, each stage having a height H, the final exit concentration may be found according to the construction shown in Figure 8. Starting with a concentration C_{go} in the waste gas, the exit concentration from the first stage amounts to C_{g1} being the inlet concentration for the second stage. The exit concentration of this stage amounts to C_{g2} etc. In this way the number of stages (or in other words, the total height of the filter bed) may be calculated to achieve a desired degree of conversion.

Figure 8. Schematic path of the elimination capacity of a biofilter as a function of the gas inlet concentration, at constant gas flow rate ω

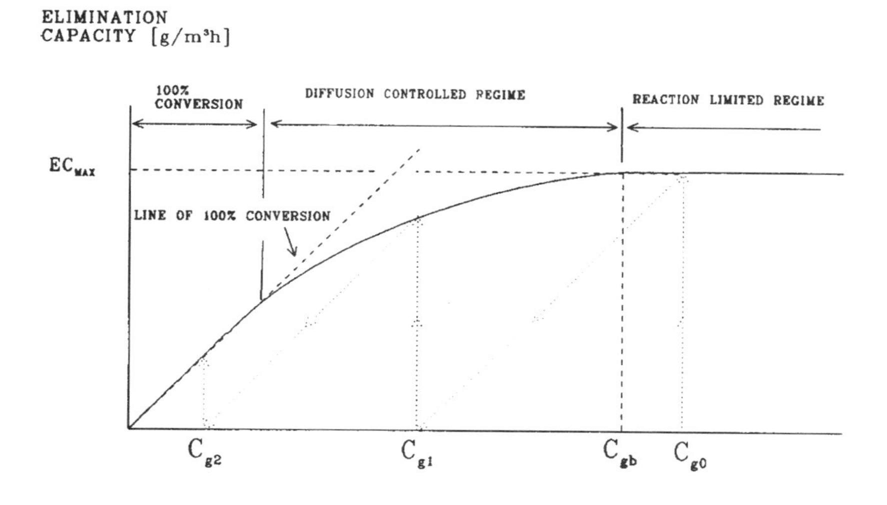

Source: Ottengraf and Oever, 1986.

Biofilters are applied nowadays in many branches of industry. Table 4 lists a number of full-scale applications for the removal of many different mixtures of volatile organic and inorganic substrates. From this table it will be clear that degrees of conversion of over 90 per cent can normally be reached. The specific costs of biofiltration are generally of the order of magnitude of Hfl 0.5-3 per 1 000 m³ of waste gas treated (see Table 1).

Figure 9. Biophysical substrate penetration model according to zeroth order kinetics

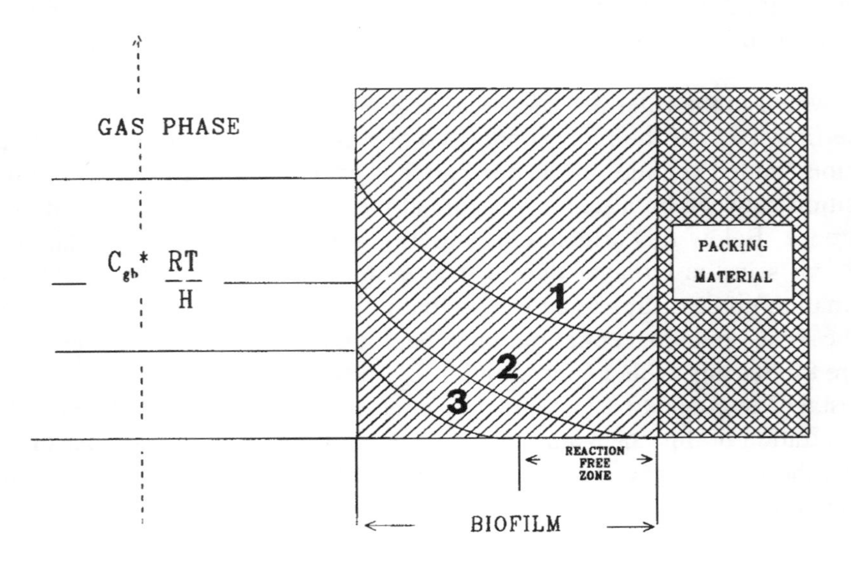

Source: Diks, 1992.

Table 4. Examples of full-scale biofilter applications

Application	Gas Flow [m³/h]	Elimination of	Number of filter stages	Total superficial residence time(s)	Efficiency (%)
Gelatine production	35 000	odour / n.s.	0.6-1	12-21	70-93
Cacoa & choc. processing	10 000	odour / n.s.	2	22	99
Fishmeal factory	40 000	odour / 230 mgC/m³	1	20	50-90
Tobacco industry	30 000	odour / NH_3 (1.5 mg/m³) nicotine (3.5 mg/m³)	2	14	95
Waste water treatment	10 000	odours / H_2S (10 mg/m³) acetone (8 mg/m³)	2	29	90-95
Flavour & fragrance ind.	25 400	odours (10^5 o.u./m³)	2	22	98
Paint production	11 700	org. solvents (1 800 mg/m³)	2	38	90
Pharmaceutical plant	75 000	org. solvents (aromatics, aliphatics, chlorinated comp.)	3	108	80
Photo film production	140 000	org. solv.(400 mg/m³)	2	30	75
Food processing industry	9 000	odour from oil (10^5 o.u./m³)	2	20	93
Ceramics production	30 000	ethanol	1	8	98
Metal foundry	40 000	benzene (9 mg/m³)	1	30	80

Notes: n.s. = no specification
o.u. = odour units
Height of each filter stage = 1m.

Sources: Hereth, 1987; Lith, 1990; Liebe, 1989; Koch, 1989; Maier, 1989.

Biological trickling filters

In the biological purification of waste gases, problems may arise if acid metabolites are produced during the biological degradation. If this process takes place to such an extent that the pH buffering capacity of the filter material is effective for only a relatively short period of operation, the presence of a flowing liquid phase in the system is required for the continuous neutralisation of the acids produced, as well as for the drainage of neutralisation products from the system. This situation is encountered in the degradation of halogenated hydrocarbons, ammonia, hydrogen sulphide, etc. For example, the degradation of dichloromethane yields hydrochloric acid:

$$CH_2Cl_2 + O_2 \xrightarrow{\text{Hyphomicrobium sp.}} CO_2 + 2\,HCl$$

Figure 10 shows the inhibiting influence of the pH upon the microbial activity of the micro-organism *Hyphomicrobium sp.* (Diks and Ottengraf, 1991).

Figure 10. The relative activity of *Hyphomicrobium* GJ21 from a trickling filter versus the pH in the liquid phase

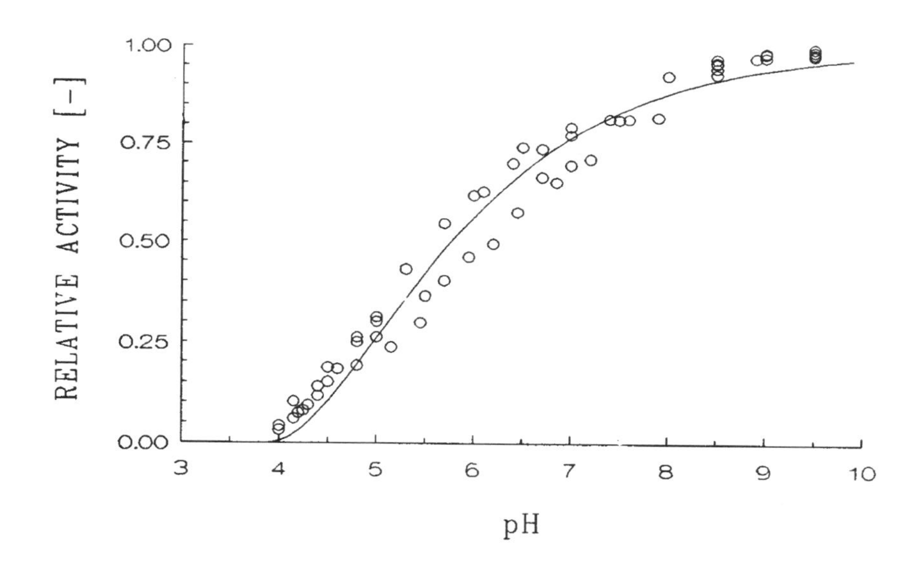

Source: Diks, 1992.

The problems of acidification and neutralisation can easily be solved by the application of a biological trickling filter (Figure 11). In this system, a water phase is continuously re-circulated over a packed bed of a carrier material, on which biofilm is immobilised. The contaminants in the waste gas are absorbed in the liquid phase and transferred to the biolayer. Simultaneously, the acids produced are removed from the filter bed, while the pH value of the liquid is controlled by adding an alkaline solution (e.g. NaOH). As the neutralisation product NaCl also inhibits the biological activity, as shown in Figure 12, a continuous refreshment of the liquid phase takes place, thus keeping the

NaCl concentration below inhibiting levels (< 200 mM). The gas flow is forced to rise through the bed co- or counter- currently to the liquid phase. The packed bed consists of packing elements (e.g. Pall-rings, Novalox-saddles or a structured packing of corrugated sheets) made of inert materials (e.g. glass, plastics, ceramics). On the surface of these carriers, the biofilm is present, which generally develops naturally over the course of time, after a trickling filter system is inoculated.

Figure 11. Experimental set-up of a biological trickling filter system

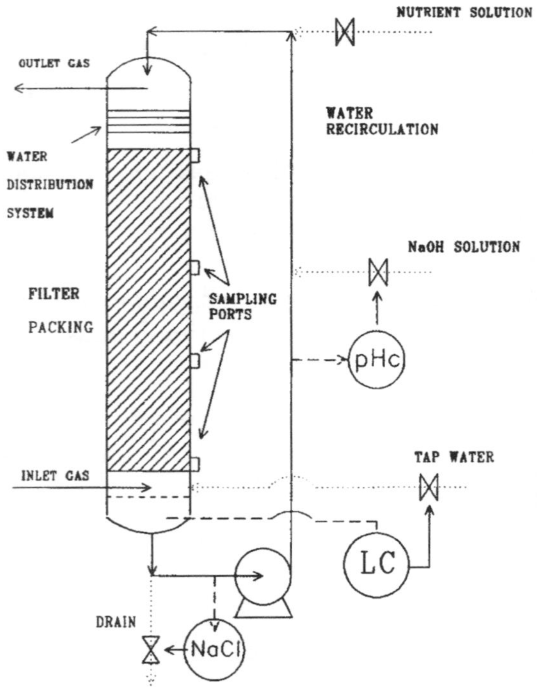

Notes: The packing material consisted of conventional packing elements.

Source: Diks, 1992.

Other phenomena depending on the liquid flow rate should also be taken into account, as they may strongly determine the efficiency of the trickling filter (Diks and Ottengraf, 1991). In the first place, the liquid flow rate strongly determines the degree of the wetting of the packing material. Within the range of the liquid flow rate mentioned, the wetted area in a packed bed of dumped elements ($a_s \approx$ 100-350; element size > ½") is normally less than 40 per cent of the specific area available (Perry and Green, 1987). Somewhat higher degrees of wetting may be found for structured packings. The total rate of mass transfer as well as the conversion rate of the system is proportional to the total biofilm area, and hence, proportional to the wetted area.

Figure 12. The influence of the NaCl concentration on the degradation rate of dichloromethane by *Hyphomicrobium* GJ21

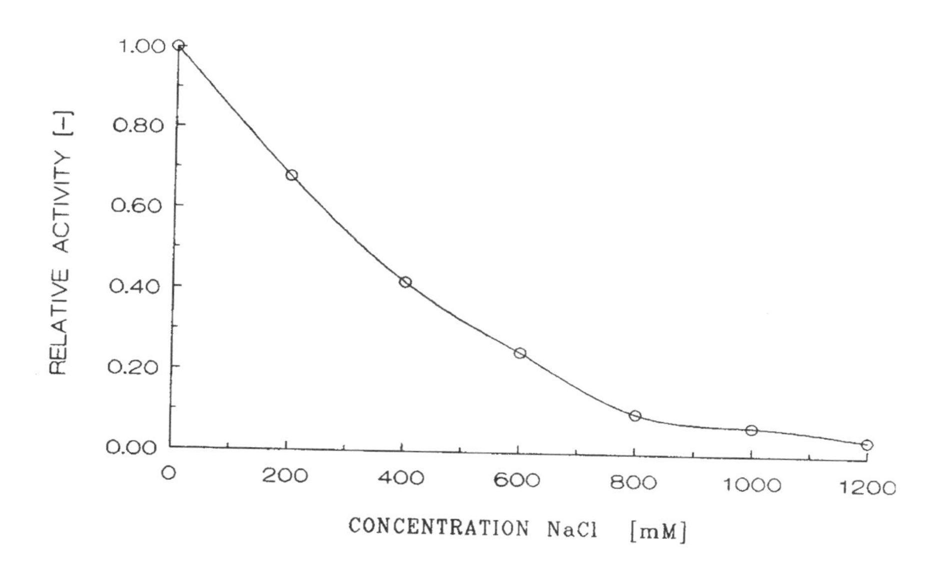

Source: Diks, 1992.

Secondly, the production of acid results in the existence of an axial pH gradient, as the liquid is generally neutralised only before it enters the column. A pH drop from pH=8 to pH=4 has, for example, been observed during the degradation of dichloromethane in a 2.7 m biological trickling filter at a low liquid flow rate (2 m/h) and a high inlet gas concentration (8 g/m³). This pH-gradient, which also depends on the concentration of dissolved buffering components (phosphates, carbonates, ammonium, etc.), will result in a decreased elimination performance due to the pH-dependency of the biological reaction rate (Figure 10). In Figure 13, the trickling filter performance for the removal of dichloromethane is shown at different liquid flow rates. The influence of the above-mentioned phenomena is clearly shown.

From the phenomena mentioned above, a high superficial liquid flow seems preferable. However, high liquid flow rates enhance the energy costs of the process and increase undesired sloughing of attached biomass. The optimal value of the liquid flow rate should therefore be found by experiments.

Finally, some remarks should be made about the risk of clogging the filter bed. This phenomenon is well-known from waste water treatment plants, where low rate trickling filters are applied, using dumped packings with characteristic sizes of 40-75 mm (Bishop and Kinner, 1986). If the average organic load exceeds certain limits (generally 0.08-1 kg/m³d at hydraulic loads of 0.04-1.6 m/h), serious clogging may result from the extensive growth of biomass and the retention of suspended solids from the waste water. At our own experiments on the elimination of 1,2-dichloroethane and toluene from waste gases in a trickling filter, clogging was also observed. Within a few weeks, a packed bed of ½" ceramic novalox saddles was fully grown with biomass. However, for the elimination of dichloromethane, such a packed bed proved to be stable for considerably longer periods of time (years). This different behaviour was suspected to be caused by the morphology of biofilms formed by the different microbial strain applied. In the case of toluene and 1,2-dichloroethane, the formation of a network of long films and filaments between the packing elements was observed. However, the biofilms of *Hyphomicrobium* in the degradation of dichloromethane were characterised by an aggregation of small flocs. It may thus be concluded that also the morphology of a biofilm may strongly determine the successful application of the trickling filter.

Figure 13. The influence of the liquid flow rate on the elimination capacity in a biological trickling filter

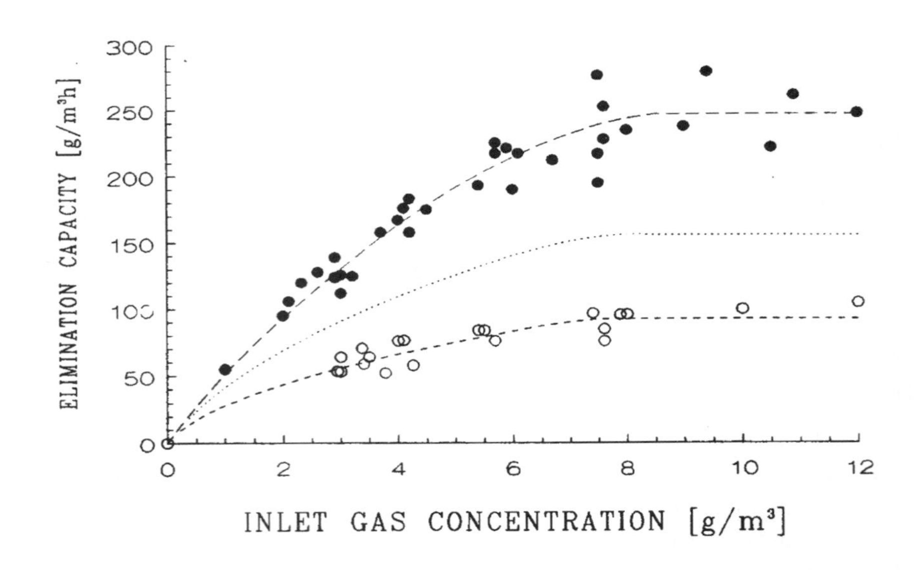

Notes: u_l = 1.8 m/h (O, dotted line)
u_l = 3.6 m/h (O, solid line)
u_l = 7.2 m/h (O, dashed line)

Source: Diks, 1992.

The number of full-scale applications of a biological trickling filter for the purification of waste gases is very limited as far as the authors know. However, from laboratory scale investigations (Diks and Ottengraf, 1991), it may be concluded that this technique is very promising and offers real prospects for a practical application.

Bioscrubbers

As indicated in Table 3, the common aspect of trickling filters and bioscrubbers is the mobility of the liquid phase. However, in a biotrickling filter the biomass is immobilised on a carrier material, while in a bioscrubber (Figure 14), the biomass is freely suspended in the liquid phase. A bioscrubber normally consists of a scrubber section, in which the mass transfer between gas and liquid phase takes place. This section may be a packed bed, similar to a trickling filter, but different designs have also been presented (VDI-Berichte, 1989; Kohler, 1986; Schmidt, 1986). Regeneration of the liquid phase takes place in a regeneration tank by the suspended micro-organisms. An additional oxygen supply in this compartment may be necessary, if the concentration of the substrates is so high, that the liquid itself does not contain enough oxygen for a complete degradation.

After the regeneration, the water is returned to the scrubber section. In order to increase the efficiency, biomass is also often re-circulated to the scrubber compartment, where the mass transfer rate is increased by the simultaneous biological reaction. For example, the application of biomass suspensions containing 0,2-1 kgTSS/m³ (TSS = Total Suspended Solids) are reported for a venturiscrubber (Schmidt, 1986), while concentrations up to 60 kgTSS/m³ have been used in a packed bed absorber (Beck-Gasche, 1989). The latter application was tested on laboratory scale only. Clogging of the packed bed scrubber may be achieved by the application of a high liquid velocity (> 20 m/h) or by using specific packing elements (Schippert, 1989*a*).

Figure 14. The experimental set-up of a bioscrubber

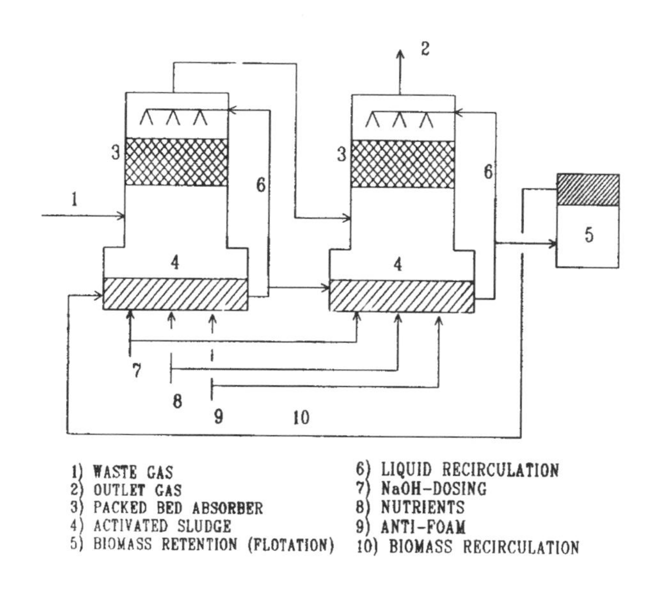

Source: Diks, 1992.

In a bioscrubber, a superficial gas flow rate of 0.5-1 m/s is normally applied. Degrees of conversion of over 90 per cent can be reached in a bioscrubber for compounds with a relatively low value of the Henry-coefficient (< ca. 50 [Pa m³/mol]). At much higher values of the Henry-coefficient the required liquid flow rate, hence, the energy consumption for liquid recirculation, becomes too high (Schippert, 1989*b*). This problem can be partly solved by allowing the biological degradation to take place in the scrubber section, either by suspended biomass, or by immobilised biomass as done in a trickling filter.

Another solution for this problem has also recently been presented. This concerns the addition of a high boiling organic solvent to the liquid phase (Schippert, 1989*b*; Lebeault, 1989). For a successful operation, the organic solvent must have a very low water solvability, a very low Henry-coefficient of the compound to be absorbed, and a high boiling point, thus a low partial pressure at operating conditions. Besides, it must neither be toxic for the micro-organisms, nor biodegradable. Organic fluids applied for this purpose mostly are silicone fluids or phtalates. A high mass transfer rate can be realised in the scrubber compartment by the high absorption capacity of the solvent. In this way, both the reactor and the energy consumption can be reduced. The substrate concentration in the organic phase may be 100 to 1 000 times higher than those in the aqueous phase. Thus, in the regeneration compartment the compounds, which have mainly been absorbed in the organic phase, are transferred to the aqueous phase where the microbial degradation takes place. Due to the buffering capacity of the organic phase, high substrate concentrations in the aqueous phase, hence, toxic effects, can be avoided (Lebeault, 1989). For bioscrubbers, a small number of full-scale applications are known, thus, a comparison with, for example, trickling filters is still quite difficult.

Environmental aspects of biofiltration: the emission of micro-organisms

As mentioned earlier, a clear trend exists nowadays to apply biofiltration on a much broader scale in different branches of industry, and the interest in this technique is rapidly increasing. This necessitates the evaluation of possible risks of working with biologically active materials, and the consequent establishment of guidelines. Particularly in the food processing, the pharmaceutical and the fermentation industry, some fear exists for an increased contamination of raw materials and products with undesirable micro-organisms during handling processes, due to significantly high microbial emissions from a nearby installed biofilter.

Until today, little has been known on the subject of contamination of biologically-treated air by micro-organisms. Data dealing with the concentration of micro-organisms in an effluent gas from a biological system are very limited. However, in recent years a few investigations have been carried out on the subject (Ottengraf and Konings, 1991; Klages *et al.*, 1987) determining the number of micro-organisms in effluent gases from full-scale as well as laboratory-scaled biological waste gas purification systems. From the evaluation of the results, a number of important conclusions have been drawn.

Figure 15. Capture of micro-organisms in a biofilter is mainly caused by inertial deposition

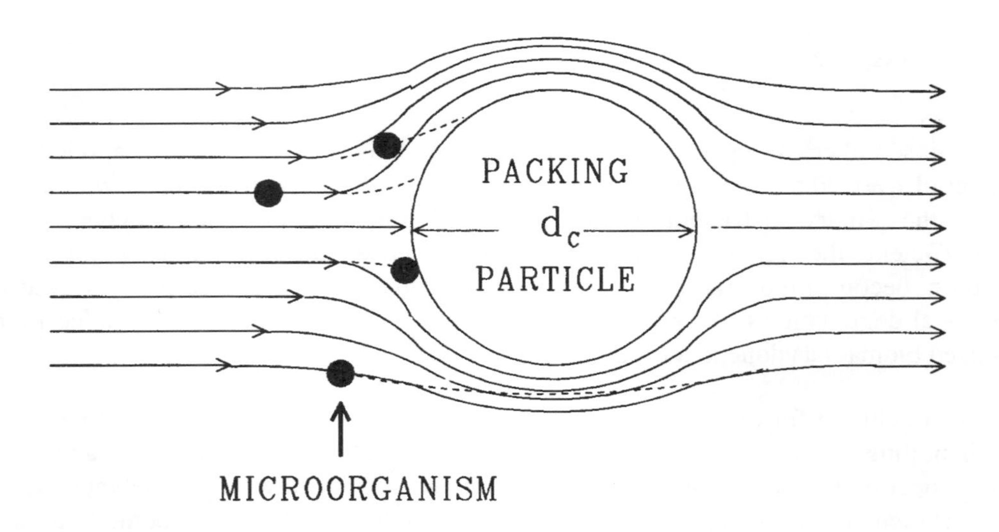

Source: Ottengraf and Konings, 1991.

In general, it has been found that the number of microbial germs (mainly bacteria, and fungi to a smaller extent) in the effluent gas of different full-scale biofilters varies from 10^3 to 10^4 m^{-3}, which is of the same order of magnitude as the numbers encountered in indoor air, and slightly higher than encountered in open air.

On the basis of experimental investigations, it was also found that a biofilter considerably reduces the concentration of micro-organisms of waste gases contaminated with a high number of germs. Moreover, from laboratory experiments it followed that the effect of the gas velocity on the discharge process of micro-organisms from biofilters is due to a simultaneous process of capture (Figure 15) and emission (Figure 16). A model, including these phenomena, has been developed, which is able to describe the experimental data rather well.

It can be concluded that the risks of any microbial contamination are generally not increased by the installation of a biological waste gas purification system on the location concerned.

Figure 16. The biolayer surrounding a compost particle is thought to be an exhaustible source of micro-organisms

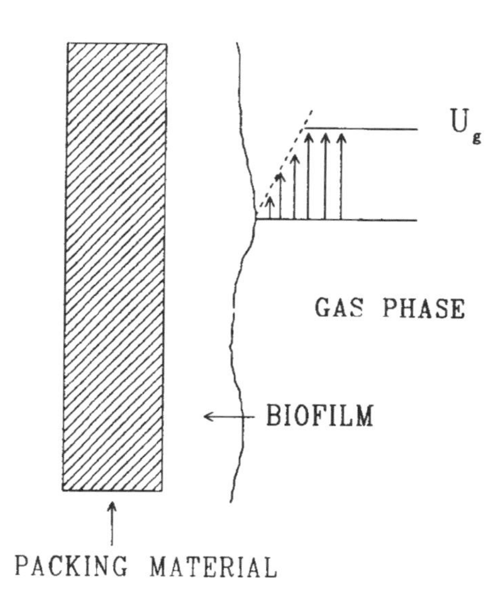

Source: Author.

Summary and conclusions

From the above-presented overview on biological waste gas purification systems, it will be clear that as far as biofilters are concerned, this system can be considered as a state-of-art technology. It nowadays is an important alternative to many other physical and chemical methods of waste gas purification, being safe to operate at relatively low costs. Concerning the bioscrubbers and the biological trickling filters, these systems do provide good prospects for the near future in the area where the biofilter application is limited. In specific cases, e.g. the removal of dichloromethane, the biological trickling filter has proved to operate successfully, while good results have also been reached for different compounds using a bioscrubber.

However, much research is still needed in order to understand and optimise the processes from a microbiological, as well as from an engineering, point of view.

List of symbols

α : Ratio of the concentration at the biofilm interface and the Monod constant
a_w : Specific wetted area [m^2/m^3]
C_{go} : Inlet gas phase concentration [g/m^3]
C_{gi} : Concentration at interface biofilm - gas phase [g/m^3]
δ : Biofilm thickness [m]
D : Effective diffusion coefficient in biofilm [m^2/s]
E_{act} : Activation energy for the biological reaction [kJ/mol]
EC : Elimination capacity, or the amount of substrate degraded per unit of reactor volume and time [g/m^3h]
H : Height of a filter bed [m]
Ko : Volumetric reaction rate in biofilm [$g/m^3_{reactor}h$]
K_s : Monod constant [g/m^3]
μ_{max} : Maximum growth rate [day^{-1}]
T : Absolute temperature [K]
u_g : Superficial gas velocity [m/h]
u_l : Superficial liquid velocity [m/h]
ϕ_0 : $\delta \bullet \sqrt{\dfrac{K_0}{D \bullet C_{gi}}}$ = Zeroth order Thiele number
ω : Gas load = u_g

Questions, comments, and answers

Q: How long can the packing material stay in the filter? Does it need to be replaced because of clogging, for example?

A: This depends on the material. Nutrients are not lost to an important degree. The average life of a compost filter is two to three years, but high quality materials can last five to seven years. Clogging is controlled by the limit conditions - sites and limiting nutrients.

Q: Were there any pathogens detected in the emissions of micro-organisms?

A: Yes, pathogens were found, both at the inlet and outlet.

REFERENCES

BACH, H. (1923), *Gesundheits-Ingenieur* 46, 38, pp. 370-376.

BARDTKE, D. (1987), "Fundamental microbiological principles of biological waste gas treatment", Proceedings of the International Meeting on Biological Treatment of Industrial Waste Gases, DECHEMA, Heidelberg, 24-26 March.

BECK-GASCHE, B. (1989), *Untersuchungen zum Einsatz und Modellierung von Biowäschern*, Fortschritt-Berichte der VDI-Zeitschriften, Reihe 15, nr. 68, VDI-Verlag, Düsseldorf.

de BEER, D. (1990), "Microelectrode studies in biofilms and sediments", Ph.D. thesis, University of Amsterdam, Amsterdam, the Netherlands.

BISHOP, P. and N. KINNER (1986), "Aerobic fixed film processes", in *Biotechnology*, H.-J. Rehm and G. Reed (eds.), Vol. 8, Ch. 3, pp. 113-176, VCH-Verlaggesellschaft, Weinheim.

de BONT, J., M.J.A.W. VORAGE, S. HARTMANS and W.J.J. VAN DEN TWEEL (1986), "Microbial Degradation of 1,3-Dichlorobenzene", *Appl. Environ. Microbiol.* 52, pp. 677-680.

BRUNNER, W., D. STAUB and T. LEISINGER (1980), "Bacterial Degradation of Dichloromethane", *Appl. Environ. Microb.* 40, pp. 950-958.

COONEY, C.L. (1981), "Growth of Microorganisms", in *Biotechnology*, H.-J. Rehm and G. Reed (eds.), Vol. 1, Ch. 2, VCH-Verlaggesellschaft, Weinheim.

DECHEMA, Proceedings of the International Meeting on Biological Treatment of Industrial Waste Gases, Heidelberg, 24-26 March.

DIKS, R.M.M. (1992), *The removal of Dichloromethane from waste gases in a biological trickling filter*, Ph.D. Thesis, Technical University of Eindhoven, The Netherlands.

DIKS, R.M.M. and S.P.P. OTTENGRAF (1991), "Verification Studies of a Simplified Model for the Removal of Dichloromethane from Waste Gases Using a Biological Trickling Filter", *Bioproc. Eng.* 6, Part I, pp. 93-99; Part II, pp. 131-140.

DON, J. (1986), *The Rapid Development of Biofiltration for the Gas Purification of Various Waste Gas Streams*, VDI-Berichte 561, pp. 63-73, VDI-Verlag, Düsseldorf.

EITNER, E. (1989), *Vergleich von Biofiltermedien anhand Mikrobiologischer und Bodenphysikalischer Kenndaten*, VDI-Berichte 735, pp. 191-213, VDI-Verlag, Düsseldorf.

HARRIS, N.P. and G.S. HANSFORD (1976), "A Study of Substrate Removal in a Microbial Film Reactor", Water Research 10, pp. 935-943.

HARTMANS, S., J.A.M. DE BONT, J. TRAMPER and K.C.A.M. LUYBEN (1985), "Bacterial Degradation of Vinylchloride", *Biotechnology Letter* 7, pp. 383-388.

HARTMANS, S., A. SCHMUCKLE, A. COOK and T. LEISINGER (1986), "Methylchloride: Natually Occurring Toxicant and C-1 Growth Substrate", *J. Gen. Microbiol.* 132, pp. 1139-1142.

HERETH, H. (1987), Proceedings of the International Meeting on Biological Treatment of Industrial Waste Gases, DECHEMA, Heidelberg, 24-26 March.

JÄGER, B. and J. JAGER (1978), "Geruchsbekämpfung in Kompostwerken am Beispiel", *Müll und Abfall* 5, pp. 48-54.

JANSSEN, D.B., A. SCHEPER, L. DIJKHUIZEN and B. WITHOLT (1985), "Degradation of Halogenated Aliphatic Compounds by *Xanthobacter autotrophicus* GJ10", *Appl. Environ. Microb.* 49, pp. 673-677.

KLAGES, S. and M. HATAMI (1987), "Vergleichender Mikrobiologische Untersuchungen an Biofiltern zur Abgasreinigung", Ph.D. thesis, University of Stuttgart, Stuttgart, Germany.

KOCH, W. (1989), VDI-Berichte 735, VDI-Verlag, Düsseldorf.

KOHLER, H. (1986), "Biowäscher zur minimierung von Organischen Gasförmigen Emissionen", VDI-Berichte 561, pp. 169-190, VDI-Verlag, Düsseldorf.

LEBEAULT, J.M. (1989), "Biological Purification of Waste Gases and Waters in a Multiphasic Bio-Reactor", Abstract Forum for Appl. Biotechnol., Gent, Belgium, p. 20.

LIEBE, H. (1989), VDI-Berichte 735, VDI-Verlag, Düsseldorf.

van LITH, C. (1990), Clairtech B.V., Veenendaal, the Netherlands, personal communication.

MAIER, G. (1989), VDI-Berichte 735, VDI-Verlag, Düsseldorf.

MAURER, P.G. (1979), "Systemstudie zur Erfassung und Verminderung von Belästigenden Geruchsenmissionen", *BMFT Forschungsbericht* T, pp. 79-114.

OLDENHUIS, R., L. KUJIK, A. LAMMERS, D.B. JANSSEN and B. WITHOLT (1989), "Degradation of Chlorinated and Non-Chlorinated Aromatic Solvents in Soil Suspensions by Pure Bacterial Cultures", *Appl. Microbiol. Biotechnol.* 30, pp. 211-217.

OTTENGRAF, S.P.P. (1986), "Exhaust Gas Purification", in *Biotechnology*, H.-J. Rehm and G. Reed (eds.), Vol. 8, Ch. 12, pp. 425-452, VCH-Verlaggesellschaft, Weinheim.

OTTENGRAF, S.P.P. and J.H.G. KONINGS (1991), "Emission of Microorganisms from Biofilters", *Bioproc. Eng.* 7, pp. 89-96.

OTTENGRAF, S.P.P. and A.H.C. VAN DEN OEVER (1986), "Theoretical Model for a Submerged Biological Filter", *Biotechnol. Bioeng.* 19, pp. 1411-1417.

OTTENGRAF, S.P.P., J.P.P. MEESTERS, A.H.C. VAN DEN OEVER and H.R. ROZEMA (1986), "Biological Elimination of Volatile Xenobiotic Compounds in Biofilters", *Bioproc. Eng.* 1, pp. 61-69.

OUDE-LUTTIGHUIS, H. (1989), VDI-Berichte 735, VDI-Verlag, Düsseldorf.

PERRY, R.H. and D.W. GREEN (1987), *Perry's Chemical Engineering Handbook,* 6th ed., Mcgraw-Hill, New York.

PRÜSS, M. and H. BLUNK, (1941), German patent number 710954, applied 1934.

REINEKE, W. and H.J. KNACKMUSS (1984), "Microbial Metabolism of Haloaromatics: Isolation and Properties of a Chlorobenwene-Degrading Bacterium", *Appl. Environ. Microbiol.* 47, pp. 395-402.

SCHIPPERT, E. (1989*a*), *Biowäscher nach Einer Dosenlackieranlage*, VDI-Berichte 735, pp. 78-88, VDI-Verlag, Düsseldorf.

SCHIPPERT, E. (1989*b*), *Das Biosolv-Verfahren von Keramchemie zur Absorption von schwer wasserlöslichen Lösemitteln*, VDI-Berichte 735, pp. 161-177, VDI-Verlag, Düsseldorf.

SCHMIDT, F. (1986), European patent number 0-133-222.

SCHOLTZ, R., L.P. WACKETT, C. EGLI, A.M. COOK, and T. LEISINGER (1988), "Dichloromethane Dehalogenase with Improved Catalytic Activity Isolated from Fast-Growing Dichloromethane Degrading Bacterium", *J. Bacteriol.* 170, pp. 5698-5704.

SCHRAA, G., M.L. BOONE, M.S.M. JETTEN, A.R.W. VAN NERVEN, P.J. COLBERG and A.J.B. ZEHNDER (1986), "Degradation of 1,4-Dichlorobenzene by *Alcaligenes sp.* strain A175", *Appl. Environ. Microbiol.* 52, pp. 1374-1381.

VDI-BERICHTE 735 (1989), *Tagung Biologische Abgasreinigung*, VDI-Verlag, Köln, 23-24 May.

van den WIJNGAARD, A., D.B. JANSSEN and B. WITHOLT (1989), "Degradation of Epichlorohydrin and Halohydrins by Three Bacterial Cultures Isolated from Fresh Water Sediment", *J. Gen. Microb.* 135, pp. 2199-2208.

BIOPREVENTION OF AIR POLLUTION

by

Ryuichiro Kurane
National Institute of Bioscience and Human Technology (NIBH)
Tsukuba, Japan

Introduction

The last 20 years have seen a growing realisation that acidifying air pollutants, resulting largely from the combustion of fossil fuels such as petroleum, can affect the functioning of forest and soil ecosystems. As the costs of disposal, waste treatment, and regulatory compliance go up, environmental aspects of products and processes will necessarily become more important to society.

Some OECD Member countries are already moving aggressively to integrate "clean" products and processes into their industrial strategies, and international trade will increasingly be influenced by environmental concerns. As a competitive strategy, "green design concepts" can help manufacturing generate less waste and reduce production costs at the same time. As an environmental strategy, green products and process design concepts offer new ways to address many environmental problems such as acid rain.

Crude oil has 250 kinds of sulfur (S) compounds, and their concentration in oil ranges from 0.5 per cent to 5 per cent. In all countries, present regulations limit sulfur concentrations in petroleum to 0.5 per cent. The removal of sulfur and/or nitrogen from petroleum and petroleum distillates leads to cleaner-burning fuels and lower sulfur and nitrogen emissions. Currently, hydrodesulfurisation, a catalytic, energy-intensive thermochemical process, is the only method for desulfurising petroleum distillates. It is very effective against mercaptans and related organic sulfur compounds. However, it is not effective for thiophen structure compounds such as dibenzothiophen (DBT) and related organosulfur compounds contained in oil.

Serious pollution problems owing to acid rain are due to the difficulty of removing organosulfur compounds such as DBT. There are, for example, serious acid rain problems in areas such as the Great Lakes (on the US-Canadian border), the Scandinavian peninsula, and the Biel forest in Germany.

Regulations in the United States, Japan and other OECD countries mandate or will mandate reducing sulfur concentrations in diesel and other fuels from 0.5 per cent to 0.05 per cent.

Figure 1 shows the relationship between the removal rate of S compounds in oil and energy consumption when the current thermochemical method is used. To reduce the sulfur level from 0.5 per cent to 0.05 per cent will require a huge amount of energy and will result in large CO_2

emissions; these will contribute to the warming of the Earth, another serious global environmental problem.

Figure 1. The relationship between the removal rate of S and energy consumption in current method

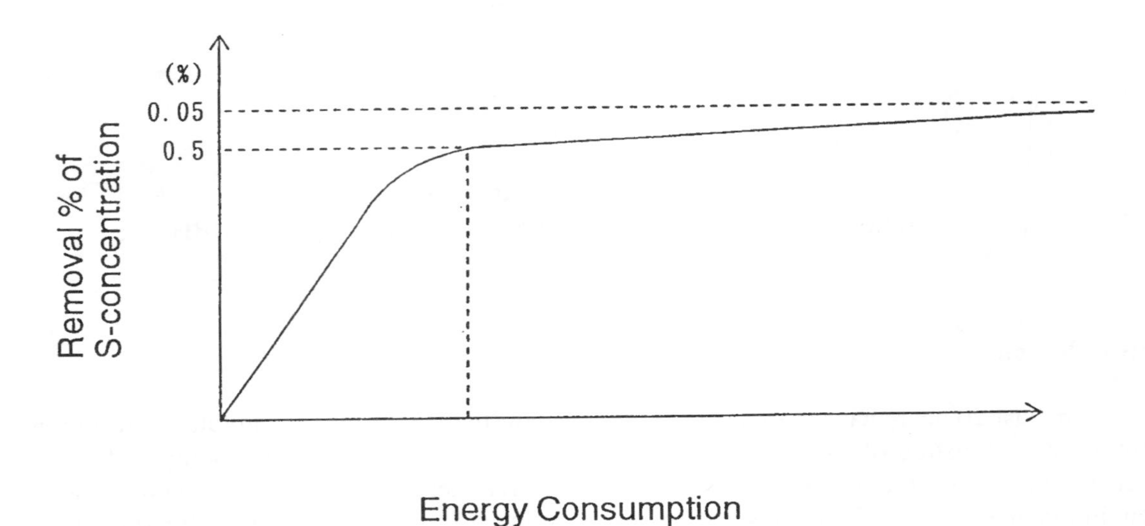

Source: Author.

It is possible that the use of biodesulfurisation in conjunction with thermochemical desulfurisation can solve the problem (Figure 2). If micro-organisms, such as bacteria, with the ability to degrade or modify DBT and related compounds are available, they may be able to remove the organically bound sulfur in DBT from petroleum distillates. Because such micro-organisms can cleave recalcitrant S compounds such as DBT, their use would make it possible to remove almost 100 per cent of the organosulfur compounds using the current thermochemical method. Thus, environmentally friendly petroleum with a low sulfur content could be obtained.

However, biocatalysts such as microbes are normally hydrophilic and cannot live in an organic solvent such as petroleum. When hydrophilic micro-organisms that are able to degrade or modify DBT are applied in a normal water phase system, very little DBT is degraded or modified. In addition, the oil must be collected and dehydrated, and this is expensive. Thus, it is very important to carry out R&D on micro-organisms for biodesulfurisation that can survive in organic solvents. A bacterial mutant that survives in an organic solvent has been successfully created.

At present, gas from automobiles and factories is emitted into the air, and the result is serious air pollution and acid rain, which damages forest and soil ecosystems. However, expenditures to restore forest and soil damaged by acid rain are negligible at present. In the future, in order to protect the environment, the costs of remediation will be huge. If we are able to prevent gas emissions into the air by using environmentally friendly petroleum with lower sulfur content, total costs, including remediation costs, will be much reduced.

Figure 2. Solution method: Application of biodesulfurization to currently used desulfurization method

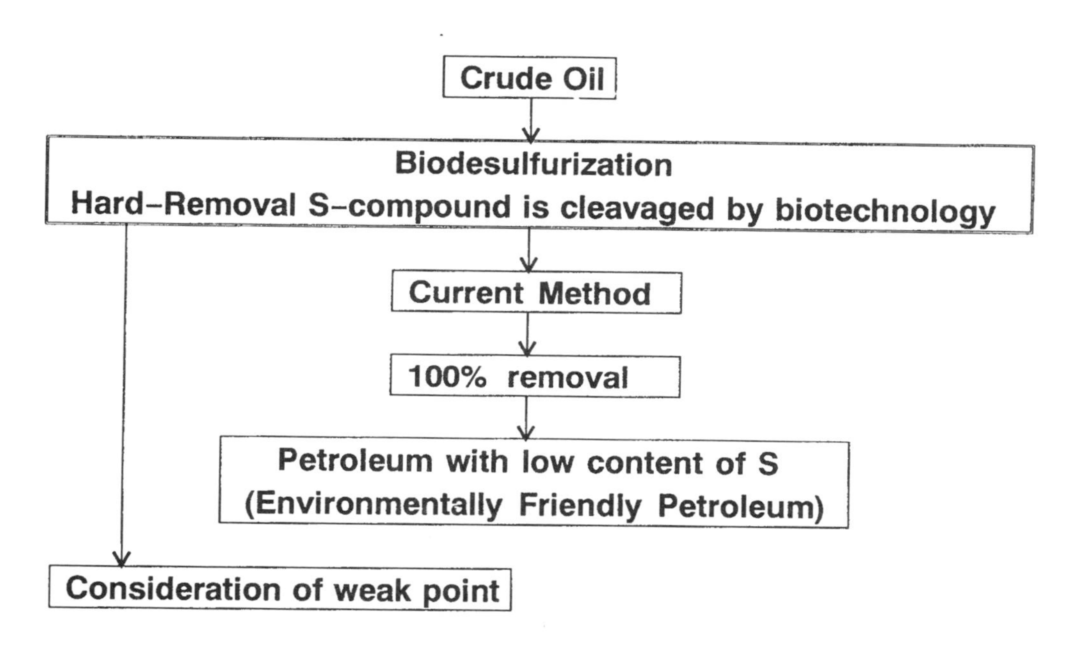

Source: Author.

There have been many reports of aerobic microbial degradation of DBT, which has been used as a model compound for recalcitrant organic sulfur compounds in fossil fuels such as petroleum and coal. The microbial degradation pathways of DBT in aerobic condition is shown are Figure 3.

DBT has been widely used as a model compound for the organic sulfur component of fossil fuels (Monticello and Finnerty, 1985). Many studies have reported on transformation, degradation, and desulfurisation of DBT. Kodama *et al.* (1973) reported on a ring destructive pathway of DBT metabolism in two *Pseudomonas* species. This carbon destructive pathway is very similar to the naphthalene degradative pathway (Foght and Westlake, 1990; Gallagher, 1992). Although the carbon destructive pathway may be useful in biodegradation of DBT in the environment, it would not be valuable for the desulfurisation of fossil fuels for three reasons. First, ring attack occurs at positions 2 and 3 of DBT positions that frequently have alkyl or aryl substitutions. DBT substituted at these positions is not a substrate for the Kodama pathway (Gallagher *et al.*, 1991). Second, the carbon destructive pathway reduces the energy content of the fuel. Third, the major product is 3-hydroxy-2-formyl-benzothiophene. Very little DBT is degraded sufficiently to result in desulfurisation.

Isbister and Koblynski first described the *Pseudomonas* sp. CB-1 that can accomplish the sulfur-specific metabolism of DBT (Isbister and Koblylnski, 1985). The products identified as intermediates in the desulfurisation of DBT were DBT sulfoxide, DBT sulfone, and 2,2'-dihydroxybiphenyl. *Pseudomonas* sp. CB-1 was lost before the metabolic pathway could be fully characterised. Van Afferden *et al.* (1990) have reported on a *Brevibacterium* sp. that desulfurises DBT by a destructive pathway resulting in benzoate and sulfite formation, both of which are oxidised further. *Arthrobacter* sp. K3b reacts similarly to the *Brevibacterium*; when DBT sulfone is used as the substrate, sulfite and benzoate are produced (Dahlberg, 1992).

Figure 3. Microbial degradation pathways of DBT

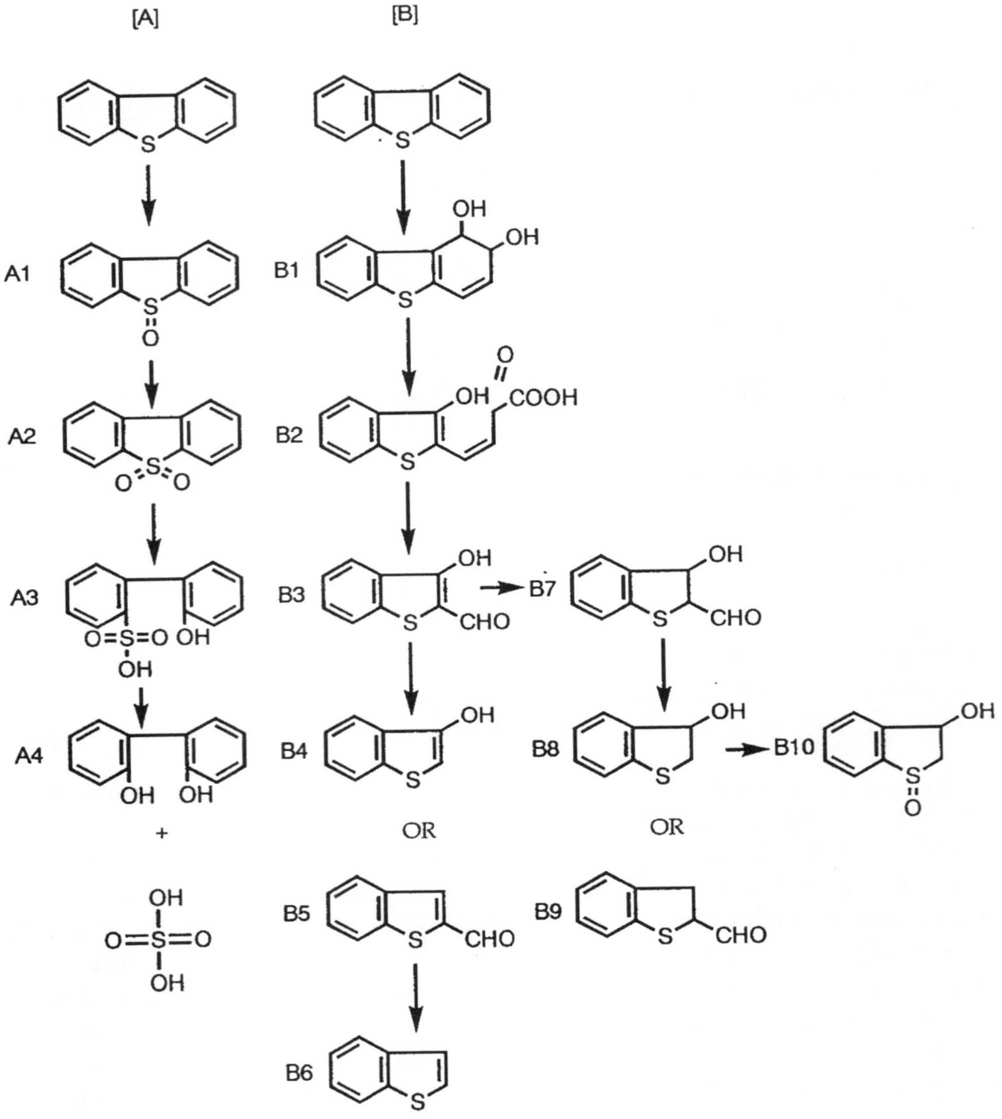

Source: Author.

Kilbane and Bielaga isolated *R. rhodochrous* IGTS8 using organic sulfur compounds as the sole source of sulfur (Kilbane and Bielaga, 1990). *R. rhodochrous* has been shown to produce 2-phenylphenol as the product of DBT desulfurisation. Also, these authors identified DBT sulfoxide and DBT sulfone as products of desulfurisation. Omori *et al.* (1992) reported a *Corynebacterium* sp. that desulfurised DBT, resulting in 2-phenylphenol and sulfite; DBT sulfoxide and DBT sulfone were found, but no other intermediates. Sulfinate and sulfonate compounds have been postulated as intermediates in DBT desulfurisation but have only been isolated recently. Kilbane and Bielaga (1991) reported finding a sultone formed during desulfurisation of thianthrene by *R. rhodochrous* IGTS58.

Two types of pathways for DBT metabolism are now recognised. These pathways are the ring-destructive pathway, represented by the Kodama (Kodama *et al.*, 1973) and Van Afferden (Van Afferden *et al.*, 1990) pathways, and the sulfur-specific pathway (Isbister, 1986; Kilbane and Bielaga, 1990; Omori *et al.*, 1992). The second pathway, sometimes called the 4S pathway, was said to involve sequential metabolism of DBT to sulfoxide, sulfone, sulfonate, and ultimately, to sulfate. Other investigators reported finding the sulfoxide, sulfone, sulfate, and a desulfurised product (biphenyl) (Isbister and Koblynski, 1985; Kilbane and Bielaga, 1990). Recently, two additional intermediates in the sulfur-specific path were identified as their cyclic forms, dibenz [c,e][1,2]-oxathiin-6-oxide (sultine) and dibenz [c,e][1,2]-oxathiin-6, 6-dioxide (sultone).

Aerobic microbes such as *Rhodococcus* are able to degrade DBT rapidly, but unfortunately require oxygen. In the oil refinery industry, aeration with oxygen and air is feared owing to the possibility of explosions. Moreover, anaerobic microbes cannot survive in the presence of oxygen, and building an anaerobic facility, such as anaerobic reactor tank that shuts out oxygen, is extremely costly.

Therefore, microaerobic micro-organisms able to degrade or modify DBT and related compounds under conditions of oxygen that will not provoke dangerous explosions (less than 2 per cent oxygen) may be the most likely candidates for biodesulfurisation of petroleum distillates in the near future.

Results

The chemical structure of recalcitrant organic sulfur compounds is shown in Figure 4. Alkyl dibenzothiophen is more difficult to remove than DBT. It is not a commercial chemical, and we therefore synthesised 4,6-dimethyl dibenzothiophen as shown in Figure 5. The micro-organisms were screened, and DBT was used as sole carbon and sulfur source (Kurane and Tubata, submitted).

Figure 4. Hard-removal organic sulfur compound contained in oil

Source: Author.

Several gram-positive and gram-negative bacteria able to degrade DBT and alkyl dibenzothiophen (4,6-DM-DBT) under microaerobic conditions, especially under conditions of N_2 gas exchange, were obtained from soils. The degradation rate of these sulfur compounds after cultivation for 28 days are presented in Table 1.

Figure 5. Synthesis of 4.6-Dimethyl dibenzothiophene

Dibenzothiophene + $(CH_3)_2SO_4$ (Dimethyl sulfate) —BuLi / Hexane→ 4,6-Dimethyl dibenzothiophene M.W. 212.32

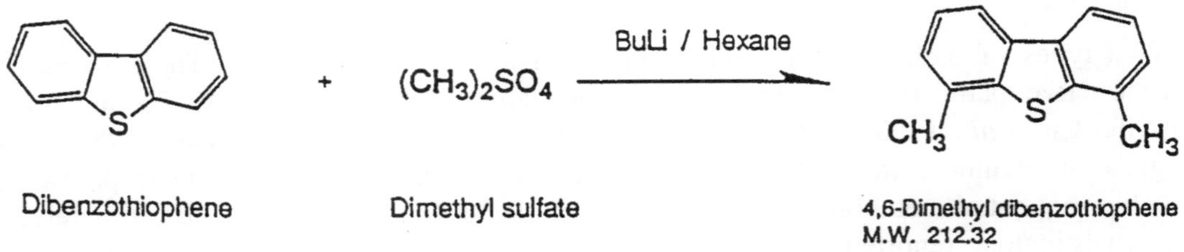

Source: Author.

Table 1. Degradation rate of Dibenzothiophene (DBT) and Alkyl Dibenzothiophene (4,6-DM-DBT) by DBT-utilising bacteria under N_2 gas exchange condition

Strain No.	DBT Growth (OD660) Degradation (%)	4,6-DM-DBT Growth (OD660) Degradation (%)
No. 39	0.145 15.5%	0.114 9.6%
No. 53	0.131 15.4%	0.122 9.6%
No. 59	0.145 11.6%	0.151 18.2%
No. 66	0.176 21.1%	0.219 19.3%
No. 71	0.202 18.9%	0.081 9.1%
No. 76	0.130 12.3%	0.051 8.1%
No. 89	0.117 15.1%	0.068 8.8%

Note:

1. Shaking at 30°C for 28 days, substrate concentration 1 000 ppm.

Source: Author.

Biophysical, biochemical and chemotaxonomy characteristics of strain 54, 69, and 89 were investigated, as shown in Table 2. These three strains belong to the genus of *Pseudomonas fluorescens, Pseudomonas putida,* and *Pseudomonas* sp. Among the strains obtained in our screening, *Pseudomonas putida* No. 69 has the greatest ability to degrade DBT under conditions of N_2 gas exchange.

Table 2. Characteristics of Strain Nos. 54, 69, 89

	No. 54	No. 69	No. 89
Gram reaction	-	-	-
Cell size (µm)	0.7 x 1.8	0.7 x 1.5	0.7 x 1.6
Morphology	Rods	Rods	Rods
Motility	+	+	+
Flageller arrangement	Polar	Polar	Polar
OF-test	O	O	O
Oxidase test	+	+	+
Catalase	+	+	+
Lipase	+	+	+
Nitrate reduction	+	+	+
Protocatechuate ortho cleavage	+	+	+
Growth at 4°C	+	-	-
Growth at 41°C	-	-	+
Fluorescent pigment	+	-	+
Pyocianin production	-	-	+
Urease	-	-	-
Lysine decarboxylase	+	+	+
Arginine dehydrolase	+	+	+
DNase	-	-	-
Gelatine liquefaction at 20°C	+	-	+
Starch hydrolysis	-	-	-
PHB accumulating	-	-	-
Utilisation of:			
Nicotinate	-	+	-
Succinate	+	-	+
Propionate	+	+	+
D-Malate	+	-	+
D-Glucose	+	+	+
Lactose	-	-	-
Sucrose	+	+	-
D-Galactose	+	-	+
D-Sorbitol	+	-	-
Arabinose	+	+	-
Mannose	-	+	-
Ramnose	+	-	-
Treharose	+	-	-
Maltose	-	-	-
D-Xylose	+	-	-
DNA G+C content	60.1%	57.4%	59.6%
Quinone type	Q-8	Q-8	Q-8
Cellular fatty acids:			
12:0	+	+	+
14:0	+	+	+
16:0	+	+	+
16:1	+	+	+
18:1	+	+	+
12:0 2-OH	+	+	+
16:0 2-OH	-	-	-

Notes:

1. Strain No. 54: *Pseudomonas fluorescens.*
2. Strain No. 69: *Pseudomonas putida.*
3. Strain No. 89: *Pseudomonas* sp.

Source: Author.

Figure 6 shows the time course of DBT degradation by *Pseudomonas putida* No. 69 under conditions of N_2 gas exchange. After three weeks of cultivation, about 200 mg of DBT were degraded per litre. *Pseudomonas putida* No. 69 can also degrade about 100 mg of 4,6-DM-DBT per litre after three weeks of cultivation.

Figure 6. Time course of DBT degradation by Gram Negative Bacterium Strain No.69 under N_2 gas exchange condition

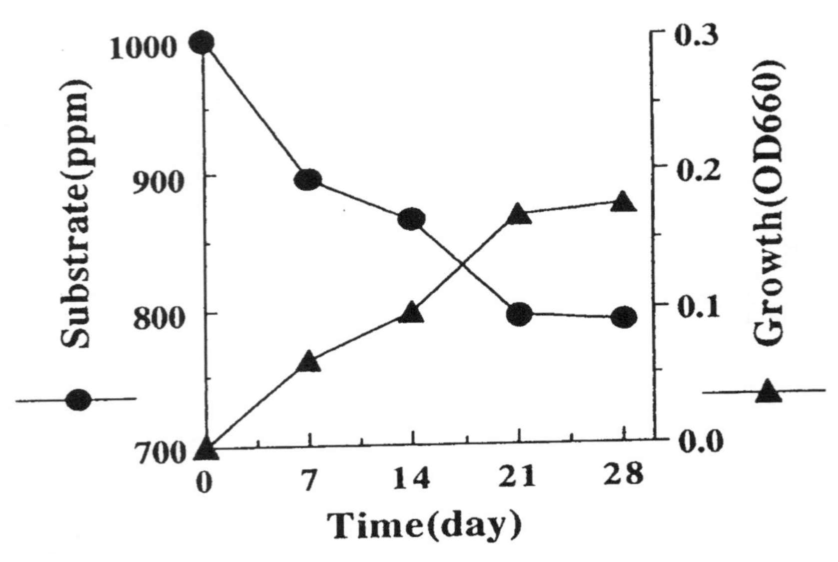

Source: Author.

Three kinds of DBT degradation products (A, B, C) were obtained by *Pseudomonas putida* No. 69 under conditions of N_2 gas exchange using the TLC separation method. Product B was analysed further.

Figure 7 shows the chemical structure of product B. The degradation pathway of DBT by *Pseudomonas putida* No. 69 under conditions of N_2 gas exchange is now under study; *Pseudomonas putida* appears to have a different degradation pathway from the aerobic microbial degradation pathways under microaerobic conditions.

Many organic solvents such as petroleum are highly biotoxic and kill most micro-organisms at low concentrations, sometimes 0.1 per cent ($^v/_v$). According to Horikoshi *et al.* (1989), growth of micro-organisms depends on polarity of the solvent (parameter log P). The parameter log P, where P is the partition coefficient of a given solvent in an equimolar mixture of octanol and water, is used as a quantitative index of solvent polarity. The smaller the number of log P, the higher the biotoxicity. The log P of n-octane is 4.9, that of toluene 2.8, and that of n-heptanol 2.4; thus, an n-heptanol-resistant mutant can survive easily in petroleum.

Figure 7. Chemical structure of Product B

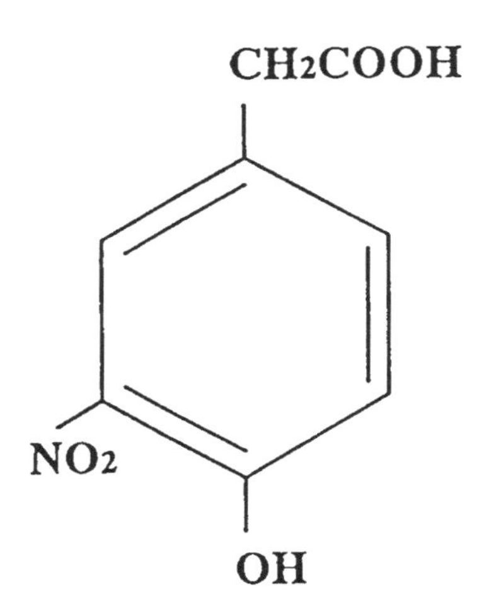

Source: Author.

Figure 8. Time course of cell growth (O.D. at 660nm of Organic Solvent Sensitive Parent *Pseudomonas putida* No.69 (×) and Organic Solvent Resistant Mutant *Pseudomonas putida* No.69-3 (O) in the presence of 50 per cent 1-Hepthanol

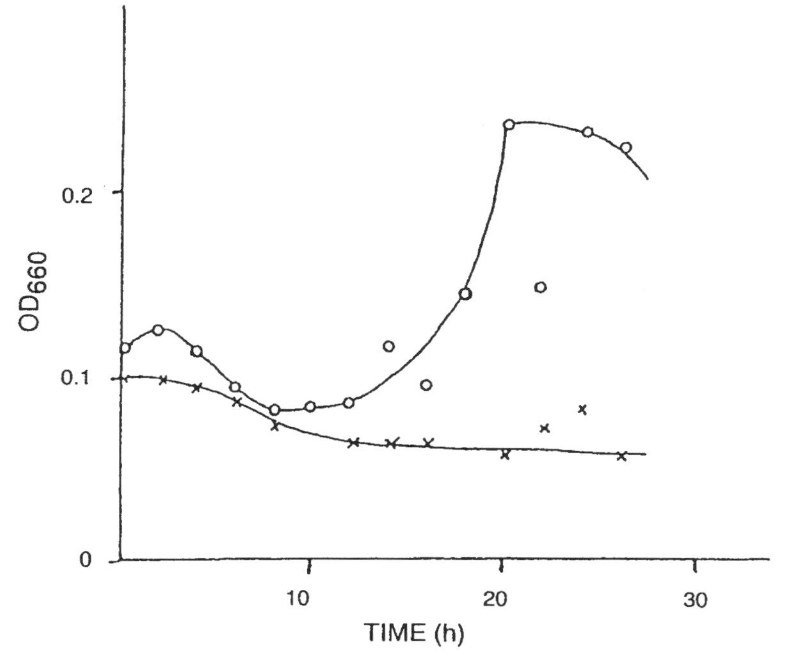

Source: Author.

Using evolutionary techniques, we have been able to create a mutant of *Pseudomonas putida,* which normally lives only in water; it can degrade DBT under conditions of N_2 gas exchange and will survive in a 50 per cent heptanol organic solvent system. The *Pseudomonas putida* mutant No. 69-3 was able to survive in a concentration of over 99 per cent heptanol. Figure 8 shows the time course of the cell growth of the organic solvent sensitive parent, *Pseudomonas putida* No. 69, and the organic solvent resistant mutant, *Pseudomonas putida* No. 69-3, in the presence of 50 per cent n-heptanol.

Figure 9 shows DBT degradation by *Pseudomonas putida* No. 69-3, in the presence of various organic solvents under conditions of N_2 gas exchange.

Figure 9. DBT degradation by *Pseudomonas putida* No.69-3 in the presence of organic solvents under microaerobic conditions

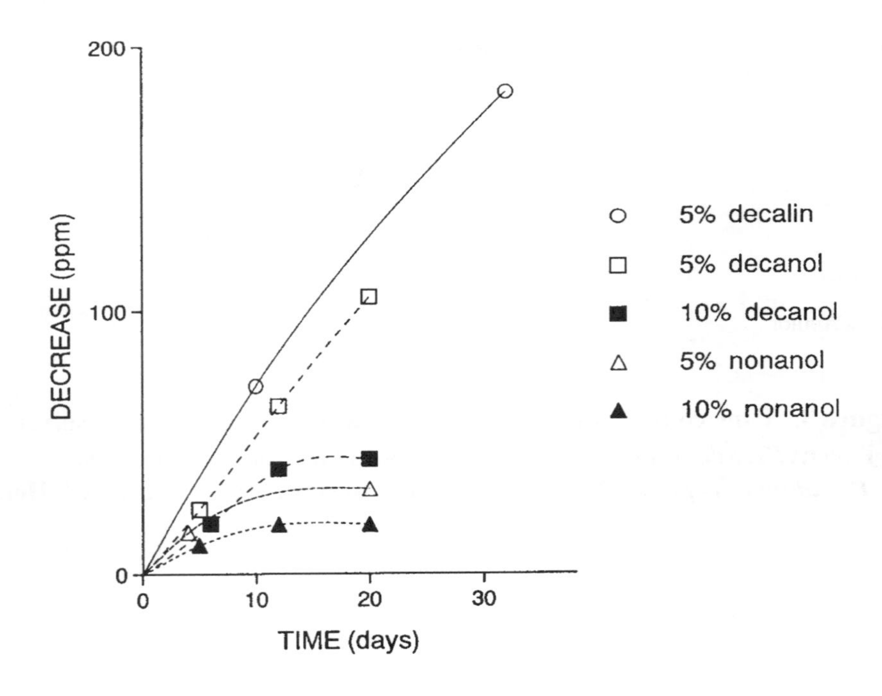

Source: Author.

The cellular fatty acid composition of *Pseudomonas putida* No. 69 has saturated fatty acids with even carbon number. The fatty acid structure of the cell membrane of *Pseudomonas putida* No. 69-3 has changed greatly, as the saturated fatty acids with even carbon number are changed to unsaturated fatty acids, branched fatty acids, and saturated fatty acids with an odd carbon number. The fluidity of the cell membrane is changed, and this may be the one of factors that renders the strain resistant to organic solvents.

Conclusion and prospects

This bioprevention technology, when coupled with currently used thermochemical desulfurisation methods, would provide an effective method for removing recalcitrant organic sulfur compounds from petroleum, thereby resulting in reduced sulfur emissions and cleaner air.

Questions, comments, and answers

Q: An important step is to get the right organism and you have succeeded in doing this. Is it now going to full application or are there other bottlenecks?

A: The bottlenecks are the microaerobic capability and the organic solvent system. Most effort has been in the evolution of molecular engineering which is very effective to create new micro-organisms. First we found an organism which degrades DBT and then we applied this technique.

Q: Is the process shape known - is it batch or continuous?

A: So far we have done two years work on the development of the biocatalyst. There will be another two years before we reach the process engineering stage.

Q: Is DBT one of the most prominent S compounds?

A: The thiophen structure is the hardest to remove. DBT was used as a model compound because it is commercially available. Other related compounds are even harder to remove using conventional methods.

Q: The DBT molecule is mineralised - what is the end product of decomposition?

A: The organism can degrade it completely to CO_2- the S is in water as sulphate and S-amino acids.

Q: The fact that you can adapt organisms to grow in solvents will be useful for production purposes. Is it generally so easy to do?

A: Yes, it is easy using our technique of molecular engineering.

REFERENCES

DAHLBERG, M.D. (1992), "Desulferization of coal", paper presented at the Third International Symposium on the Biological Processing of Coal, 4:-7 May, Clearwater Beach, FL., Electric Power Research Institute, Palo Alto, CA.

FOGHT, J.M. and D.W.S. WESTLAKE (1990), "Expression of dibenzothiophene – degradative gene in two *Pseudomonas* species", *Can. J. Microbiol.* 36, pp. 718:-72.

GALLAGHER, J.R. (1992), "Sulfur specific pathway", paper presented at the Third International Symposium on the Biological Processing of Coal, 4:-7 May, Clearwater Beach, FL, Electric Power Research Institute, Palo Alto, CA.

GALLAGHER, J.R., S.D. DENOME, P.M. MONROE and K.D. YOUNG (1991), "Microbial desulferization of dibenzothiophene", Abstr. Q:-80, p. 289, *Abstracts of the 91st General Meeting*, Am. Soc. Microbiol., Washington, DC.

INOUE, A., and K. HORIKOSHI, (1989), "A *Pseudomonas* thrives in high concentration of toluene", *Nature* 388, pp. 264:-265.

ISBISTER, J.D. and E.A. KOBLYNSKI (1985), "Microbial desulferization of coal", in Y.A. Attia (ed.), *Processing and Utilization of High Sulfur Coals,* pp. 627:-641, Elsevier, Amsterdam.

KILBANE, J.J. and B.A. BIELAGA (1990), "Desulferization of coal: the microbial solution", *Chemtech.* 20, pp. 747:-751.

KILBANE, J.J. and B.A. BIELAGA (1991), "Molecular Biological Enhancement of Coal Biodesulfurization", Final Report, US DOE contract DE:-AC22:-88PC8891.

KODAMA, K., K. UMEHARA, K. SHIMIZU, S. NAKATANI, Y. MINODA and K. YAMADA (1973), "Identification of microbial products from dibenzothiophene and its proposed oxidation pathway", *Agr. Biol. Chem.* 37, pp. 45:-50.

KURANE, R. and T. TSUBATA (1996), "Selection of an organic solvent:-resistant bacterium from an organic solvent:-sensitive bacterium and characterization of its fatty acids", submitted to *Can. J. Micro.*

MONTICELLO, D.J. and W.R. FINNERTY (1985), "Microbial desulferization of fossil fuels", *Ann. Rev. Microbiol.* 39, pp. 371:-389.

OMORI, T., L. MONNA, Y. SAIKI and T. KODAMA (1992), "Desulferization of dibenzothiophene by *Corynebacterium* sp. strain 5Y1", *Appl. Environ. Microbio.* 58, pp. 911:-915.

VAN AFFERDEN, M., S. SCHACHT, J. KLEIN and H.G. TRUPER (1990), "Degradation of dibenzothiophene by *Brevibacterium* sp. DO", *Arch. Microbiol.* 153, pp. 324:-328.

APPLICATION OF CLOSED BIOFILTRATION SYSTEMS FOR TREATMENT OF INDOOR ATMOSPHERES

by

S. Keuning
Bioclear b.v., Groningen, the Netherlands

P.G. Paul
Stork-Comprimo b.v., Amsterdam, the Netherlands

R.A. Binot
ESA-ESTEC, Noordwijk, the Netherlands

Introduction

The development of a closed biological air filter (BAF) was initiated in 1988 by the European Space Agency (ESA) and the Netherlands Agency for Aerospace Programs (NIVR) in order to determine whether this technique could be adapted for removal of trace gas contaminants in space applications. The system was also expected to reduce the risk of microbial contamination onboard manned spacecraft. This development is a joint activity of Stork-Comprimo and Bioclear, and is supported by the universities of Wageningen and Groningen since the BAF program, from the beginning, required the close integration of complementary disciplines, namely: microbiology bio-engineering, mathematical modelling and data acquisition, and the development of analytical procedures.

The principle of a biological air filter is based on the capability of certain micro-organisms to completely convert gaseous contaminants to water, carbon dioxide and salts by oxidation. In a manned space vehicle, contamination is caused by the release of natural biological products by the crew or by material off-gassing. Unless these contaminants are removed from the confined atmosphere, they will accumulate and rapidly reach toxic concentrations.

Out of the several hundreds of gaseous contaminants expected due to material off-gassing or biological activity, approximately 100 contaminants, belonging to 14 major chemical classes, have been studied for their biodegradability. Based on present knowledge, and providing a correct selection of micro-organisms and operating conditions can be achieved, nearly all contaminants expected aboard will be susceptible to biodegradation (Binot and Paul, 1989; Paul, 1992).

The classical physico-chemical approach involving a combination of oxidation catalysts with pre- and post- adsorption and chemisorption filters suffers from a number of important drawbacks. These include the high susceptibility to poisoning by certain natural and xenobiotic compounds, and the risk of creating chemical decomposition products which are possibly more difficult to remove or more toxic than the initial contaminants.

In comparison with physico-chemical systems, the biological air filter is expected to have several advantages, inherent of the living nature of its "biological catalyst", such as adaptability to unexpected contaminants, and ability to recover normal efficiency after accidental poisoning. At the moment, earth applications are also considered like systems for building air conditioning and removal of specific industrial emissions.

For waste air streams containing a broad range of contaminants with various water solubilities and present in low concentrations, engineered versions of a traditional process known as the compost filter are usually very successfully applied. In their simplest form, however, these filters are not acceptable for use in space due to their generally low volume efficiency, the poor confinement of the active micro-organisms inside the filter, and the often unknown composition of the microbial populations used. On the other hand, their high resistance to poisoning, their important capability to adapt to new contaminants and fluctuating concentrations, their very low maintenance requirements and very long life-time (years), along with, generally, an improvement of performances with time and a capacity to recover after accidental events, were sufficient arguments to start investigating the potentials for space applications.

BAF development

In the first phase of the project, the theoretical feasibility and the characteristics and requirements of a BAF system were investigated. It was found that many micro-organisms have biodegradation capacity for a great number of space cabin contaminants. Although it is believed that the BAF could not replace the whole trace contaminant and particulate control systems, the BAF offers the potential for a simple, low energy means for the removal of easily biodegradable compounds (especially the metabolic products from the crew), and ultimately, of many xenobiotic contaminants.

In the second and third phase of the BAF project, applied research to biodegradation and the reactor concept was carried out. The efficient operation of a BAF is largely dependent on the kinetics properties of the microbial community or "biocatalyst" in the biofilter device. To investigate the complex interaction effects of a mixed population of micro-organisms with specific biodegradation capabilities, experiments were performed with pure and mixed bacterial cultures in which the number of contaminants and bacterial strains were gradually increased. The results of the biodegradation studies obtained so far show good prospective for the applicability in the BAF. Table 1 shows the results of an experiment where six different strains efficiently degraded a mixture of six different components over a period of more than 50 days.

Table 1. Elimination of space cabin contaminants in a continuous flow bioreactor by a mixed bacterial population

Organism	Substrate	Elimination
Pseudomonas GJ40	toluene	> 99.9 %
Pseudomonas GJ31	chlorobenzene	> 99.9 %
Pseudomonas GJ8	p/m-xylene	> 99.9 %
Hyphomicrobium DM20	dichloromethane	> 99.8 %
strain ARD1	dichloromethane	
strain BCG22	acetone	> 99.4 %

Source: Keuning *et al.*, 1991.

To support the testing of the experimental BAF-specific analytical tools are under development like:

- identification techniques for the microbial strains in the BAF in order to control the BAF population and to prevent any pathogenic organism to enter the BAF (Van der Waarde *et al.*, 1994);

- specific tools for analysing the composition of biofilms in laboratory membrane bioreactors.

Experimental BAF membrane reactor

An experimental BAF membrane reactor has been designed. It is characterised by a large interfacial area and a good separation between the liquid and the gas phases in the bioreactor. The BAF uses a hydrophobic porous membrane which has the dual function of supporting formation of an active microbial biofilm on one side of the membrane, and of assuring phase separation between the aqueous biofilm phase and the flowing air. This latter function is especially important in microgravity applications. Contaminants are progressively removed by the creation along the membrane of a continuous dynamic equilibrium between diffusion and degradation. High degradation rates along the entire length of the reactor are achieved by selecting organisms with high affinity for the contaminants, namely organisms with low K_s values. K_s is the contaminant concentration which results in half the maximum specific microbial degradation rate. K_s must be below the toxicological limit for contaminants aboard spacecraft (SMAC). See Figure 1.

Figure 1. Experimental BAF membrane reactor

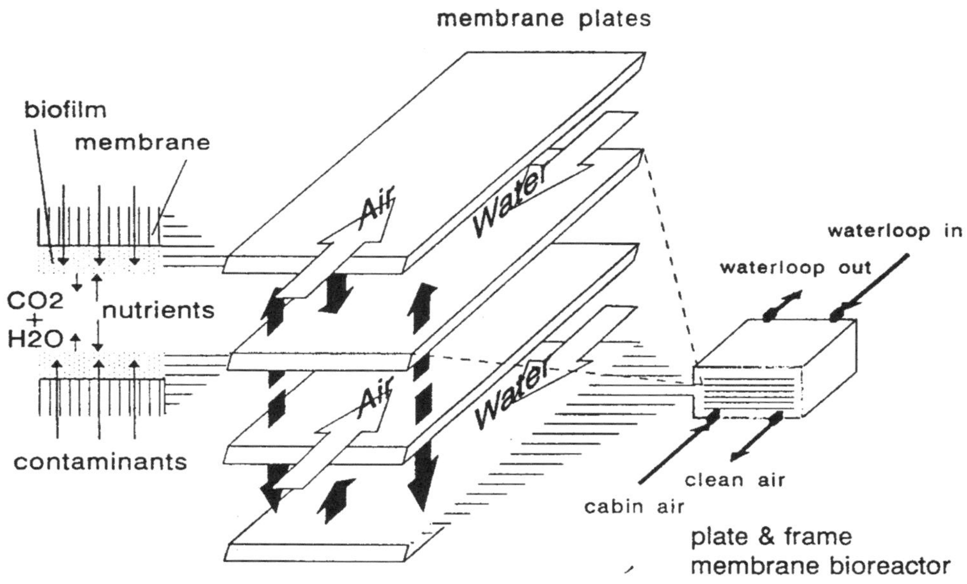

Sources: Breukers *et al.*, 1994.

A mathematical model for simulations of the biophysical process of removal of contaminants by a defined mixed population of micro-organisms has been developed and is progressively optimised and extended for membrane bioreactor systems (Breukers *et al.*, 1994).

Based on experience gained by a first experimental prototype, a considerably improved experimental BAF (Module-02) was designed and tested with very satisfactory results. This advanced experimental module and a supporting analytical testbed now support an extensive program of quantitative and qualitative tests and also life-tests. Results from these tests are used to further optimise the system, for both space and earth applications (Binot *et al.*, 1994).

Conclusions

Closed biological air filters are not only feasible for use in space. They are also potentially applicable on the ground -- in applications ranging from treatment of indoor atmospheres, building air conditioning and odour removal, to treatment of specific industrial gas streams. The use of membrane separation techniques and engineered populations of known non-pathogenic organisms makes them intrinsically safe. Mathematical models have been validated experimentally and are used to guide future designs.

Questions, comments, and answers

Q: What is the area of the membrane surface and the mean retention time of molecules in contact with immobilised organisms? How high is the inlet gas concentration?

A: The module shown has a total surface of 2 000 cm^2. The total depends on the number of modules but with the capillary membrane this will increase many times. The flow is 3 m^3 air/hour. The concentration is 0-several hundred ppm, which is significantly lower than industrial applications and the biomass is near resting (substrate limiting). There is little growth and clogging.

Q: It is surprising that you could leave the system for six months without substrate and then start it up so quickly. You would expect that most of the organisms would have disappeared.

A: After the full shut down, there was not an immediate recovery, but a regeneration of the filter with a few weeks of re-adaptation. There are enough organisms left alive to recolonise.

Q: Is there a selection pressure for toluene degraders at such a low level of toluene?

A: There is no pressure. We have a closed system inoculated with specific strains and there are a few toluene degraders always present.

Q: Can you compete with biotrickling filters here on earth? The efficiency is not as high and isn't the cost higher than, say, compost filters?

A: We don't intend to compete with conventional systems. We expect very specific uses such as air conditioning systems - imagine this in a car, for example.

Q: Have you compared the economics of use in a car with physical absorbers such as activated carbon?

A: This is a similar situation to industrial uses. In biological systems, you do not have a rest material, and the process is therefore cheaper in the long term. Also, many volatile compounds including chlorinated compounds are adsorbed very poorly by activated carbon.

REFERENCES

BINOT, R.A. and P.G. PAUL (1989), "BAF - An Advanced Ecological Concept for Air Quality Control", Proceedings 19th Intersociety Conference on Environmental Systems, San Diego, SAE Technical Paper Series no. 891535.

BINOT, R.A., R.J. BREUKERS, P.G. PAUL and D. JAGER (1994), "BAF-EXEMS '92: Testing of the biological air filter for air quality control during a manned space mission simulation", Proceedings of the Fifth European Symposium on Space Environmental Control Systems, Friedrichshafen, Germany, SAE Technical Papers Series no. 941343.

BREUKERS, R.J.L.M., L.M.M.M. MEYLINK, P.G. PAUL and R.A. BINOT (1994), "Mathematical modelling of the membrane biological air filter - BAF", Proceedings of the Fifth European Symposium on Space Environmental Control Systems, Friedrichshafen, Germany, SAE Technical Papers, Series No. 941315.

KEUNING, S., D. JAGER, P.G. PAUL and R.A. BINOT (1991), "Biodegradation study with space cabin contaminants to determine the feasibility of Biological Air Filtration (BAF) in space cabins", Proceedings of the Fourth European Symposium on Space Environmental Control Systems, Florence, 21-24 Oct. 1991, (ESA SP-324, Dec. 1991).

PAUL, P.G. (1992), "Biofiltration - an odour suppression", *Hydroc. Techn. Int.* 1992, pp. 229-321.

VAN DER WAARDE, J.J., D. JAGER, S. KEUNING, P.G. PAUL and R.A. BINOT (1994), "BAF - Detection of microbial contamination and identification of biofilm bacterial strains in a biological air filter", Proceedings of the Fifth European Symposium on Space Environmental Control Systems, Friedrichshafen, Germany, SAE Technical Papers Series no. 941341.

Parallel Sessions/B1 and B2

B1.1. FOCUS AND TRENDS IN R&D: SOIL & AIR/OFF-GAS

PROSPECTS FOR GMOs: DESIGNING RECOMBINANT BACTERIA FOR ENVIRONMENTAL RELEASE

by

J. M. S. Romero, A. Haro, S. Panke, A. Cebolla and V. de Lorenzo
Centro de Investigaciones Biológicas, CSIC, Madrid, Spain

Introduction

In recent years, the explosive development of modern molecular genetics and its application to non-enteric bacteria, such as soil and root-colonising *Pseudomonas*, has made it possible to construct, in the laboratory, various strains with novel phenotypes for use in biodegradation of environmental pollutants or for biological control in agriculture. So far, however, these strains have rarely left the laboratory, so that the tremendous power of modern molecular biology for these areas has not been yet fully realised. This is due in part to regulations relating to large-scale validation. The major bottleneck, however, lies with the difficulty of producing GMOs with sufficient predictability, in terms of performance and ecological behaviour, for their application in the field. It is one thing to have a bacterium with an interesting phenotype in a Petri dish under well-defined laboratory conditions, and quite another to achieve the expression of the same phenotype under the conditions prevailing in the field. There is here a significant difference with other biotechnological processes for which the working conditions can be manipulated by the operator at will.

Fortunately, there have been significant advances in the design of genetic tools specifically tailored for construction of GMOs destined for environmental release. This communication addresses some of the problems facing the molecular geneticist when designing GMOs and proposes some potential solutions.

Insertion and deletion of DNA segments in the chromosome of bacteria

The modularity of the genetic complement of bacteria implies the existence of natural systems for assembling the various sequence components of the different genes and gene clusters. Catabolic operons are frequently included within transposons which, in many cases, can exchange DNA segments through transposon-determined site-specific recombination systems. What lessons should we learn from these processes? In an effort to imitate in the laboratory some of the events underlying the appearance of new phenotypes of environmental interest, use is made of a series of transposon vectors derived from Tn5; these have become available since 1990 and permit the insertion of heterologous DNA segments into the chromosome of target bacteria and their stable inheritance through a process which resembles natural mechanisms of acquisition or loss of adaptation-related phenotypes. Transposon vectors make it possible to engineer, with very few manipulations, stable recombinant phenotypes in the laboratory. Their simple design has made them increasingly popular for constructing GMOs destined for uncontained applications. These vectors have found broad use for manipulations of a variety of Gram-negative bacteria of environmental interest such as *Pseudomonas*. Besides allowing insertion of heterologous genes into the chromosome of various Gram-negative strains, transposon vectors also allow the use of native promoters for the expression of recombinant genes, even in the absence of any information on promoter structure or regulation in the native host. In a further step to develop tools for metabolic engineering inspired by natural processes of DNA insertion/excision, various site-specific recombination systems have been exploited to generate predetermined deletions within DNA segments inserted in the chromosome of Gram-negative bacteria that permit the eventual addition of heterologous DNA segments devoid of any phenotypic marker to the genome of the target strain. The resultant micro-organism, endowed with a novel phenotype determined by the DNA segment of interest, does not carry any of the genetic determinants used for its construction, i.e. it is indistinguishable from a natural strain.

Heterologous expression in response to environmental signals

Gene expression is the result of transcriptional and post-transcriptional events. In spite of the many broad host-range systems available, expression of a recombinant phenotype may act unpredictably for expressing heterologous activity in micro-organisms poorly characterised genetically, as is the case in many Gram-negatives of environmental interest. Transposon vectors allow, however, for probing and parasiting indigenous promoters for the expression of recombinant genes even in the absence of any information on promoter structure or regulation in the native host. The most straightforward procedure for having a gene or operon transcribed in response to a pre-determined signal is to construct a specialised mini-Tn5 transposon in which the promoterless gene or gene cluster is placed next to the I or O termini of the mobile element. The strain of interest is then mutagenised with such a transposon, and insertions that happen to have occurred next to a promoter regulated by that signal are screened, for instance, with immunological procedures. In many cases, however, construction of specialised transposons of this sort is impractical, and some information on promoter strength and its regulation may be needed prior to engineering expression of a heterologous gene placed artificially downstream. The range of promoter strengths available in a given Gram-negative strain can be assessed with mini-Tn5 derivatives carrying a variety of convenient reporter genes such as *lacZ, xylE* or *luxAB*. The resulting insertions not only provide a repertoire of promoter strengths, as revealed by the activities of the corresponding reporter products, but they also offer a genetic "hook" to place the expression of heterologous genes under the same regulator of the reporter gene, through homologous recombination of specialised plasmid vectors with transposon sequences inserted into the chromosome of target bacteria.

In some cases, starvation might be considered a universal signal potentially useful for driving gene expression when no other known signal can be used for the purpose. Promoters responsive to carbon, nitrogen, iron and phosphate starvation are, in principle, adequate building blocks for expression of heterologous genes in the field. Furthermore, some specialised genetic probes are available for identifying promoters which are preferentially active when cells have ceased to grow. Such types of promoters may have interesting applications in GMOs destined for environmental release which need to perform under nutrient starvation and/or very low growth rates.

Engineering biological activities on the surface of bacteria

A number of simple procedures are being developed to use permissive sites of outer membrane proteins for exposure to the external medium of peptides with some biological activity, i.e. chelation of metal ions. A favourite carrier for such peptides is the LamB protein of *E. coli* (maltose and lambda phage receptor), which can be expressed in a variety of heterologous Gram-negative hosts such as *Pseudomonas.* The amino acid sequence around position 153 tolerates genetic insertion of up to 200 extra residues which, due to the topology of the protein, become exposed to the external medium. When derivatives of the *lamB* gene are transcribed, the epitope is presented on the surface of the bacterial cells in a configuration available to specific antibodies. This simple scheme has been used for three purposes: i) antigenic labelling of GMOs to follow their fate in microbial ecosystems; ii) non-disruptive monitoring of gene expression by single cells in complex ecosystems; and iii) engineering of bacteria with superior metal-binding abilities. The use of surface determinants as attachment sites of bacteria on specific solid matrixes, as well as on other bacteria, will soon become the basis of engineering rationally stable consortia with distinct structures and constructing micro-organisms with a built-in tropism for certain ecological niches.

Outlook

We have learnt so much in recent years about the molecular basis of bacterial adaptation to novel environmental conditions that we can imitate to a considerable degree the natural processes by which micro-organisms evolve new properties of environmental interest. Therefore, use of the most recent genetic strategies available leaves little room for the somewhat arbitrary distinction between recombinant and natural bacteria, since the former basically mimic faithfully the natural processes of DNA shuffling between distant locations of the bacterial genome. All these recent developments have virtually solved the problem of genetic stability of recombinant genes, as well as that of the use of antibiotic gene markers as selection determinants, the latter being one of the features of recombinant organisms which, to date, cause most concern among the public. In this context, the potential problems for the use of GMOs are no greater and no less than those posed by the use of "natural" micro-organisms in bioremediation, i.e. colonisation ability, survival, and proper expression of the phenotypes of interest.

Questions, comments, and answers

Q: Can these organisms be distinguished from natural ones? Does the strain have any of the properties concerning risk assessment, e.g. is there any transferability in the final strain?

A: The way organisms evolve the ability to degrade new compounds is by the acquisition and loss of DNA sequences. By doing the same in the laboratory we are imitating what is going on in nature. All possibilities of transferring have been eliminated by the design of the transposon. It is possible to engineer even less transferability if desired.

REFERENCES

BRAZIL, D., L. KENEFICK, M. CALLANAN, A. HARO, V. DE LORENZO, D. N. DOWLING and F. O'GARA (1995), "Construction of a rhizosphere *Pseudomonas* with potential to degrade polychorinated biphenyls and detection of *bph* gene expression in the rhizosphere", *Applied and Environmental Microbiology* 61, pp. 1946-1952.

CEBOLLA, A., C. GUZMÁN and V. DE LORENZO (1996), "Non-disruptive detection of activity of catabolic promoters of *Pseudomonas* with an antigenic surface reporter system", *Applied and Environmental Microbiology* 62, pp. 214-220.

DE LORENZO, V. (1992), "Genetic engineering strategies for environmental applications", *Current Opinion in Biotechnology* 3, pp. 227-231.

DE LORENZO, V. (1994), "Designing microbial systems for gene expression in the field", *Trends in Biotechnology* 12, pp. 365-371.

DE LORENZO, V. (1994), "Genetic strategies to engineer expression systems responsive to relevant environmental signals", F. O'Gara, D. Dowling, B. Boesten (eds.), pp. 91-101, *Molecular Ecology of Rhizosphere Microorganisms.* VCH, Weinheim.

DE LORENZO, V., I. CASES, M. HERRERO and K.N. TIMMIS (1993), "Early and late responses of TOL promoters to pathway inducers: Identification of growth-phase dependent promoters in *Pseudomonas putida* with *lacZ-tet* bicistronic reporters", *Journal of Bacteriology* 175, pp. 6902-6907.

DE LORENZO, V. and J. PÉREZ-MARTÍN (1996), "Regulatory noise in prokaryotic promoters: How bacteria learn to respond to novel environmental signals", *Mol. Microbiol.* (in press).

DE LORENZO, V. and K.N. TIMMIS (1992), "Specialized host-vector systems for the engineering of *Pseudomonas* strains destined for environmental release", in *Pseudomonas: Molecular Biology and Biotechnology*, pp. 415-428, American Association of Microbiology, Washington, D.C.

DE LORENZO, V. and K.N. TIMMIS (1994), "Analysis and construction of stable phenotypes in Gram-negative bacteria with Tn*5* and Tn*10*-derived mini-transposons", *Methods in Enzymology* 235, pp. 386-405.

DÍAZ, E., M. MUNTHALI, V. DE LORENZO and K. N. TIMMIS (1994), "Universal barrier to lateral spread of specific genes among micro-organisms", *Molecular Microbiology* 13, pp. 855-861.

KRISTENSEN, C., L. EBERL, J. M. SÁNCHEZ-ROMERO, M. GISKOV, S. MOLIN and V. DE LORENZO (1995), "Site-specific deletions of chromosomally located DNA segments with the multimer resolution system of broad-host-range plasmid RP4", *J. Bacteriol.* 177, pp. 52-58.

RAMOS, J. L., E. DÍAZ, D. DOWLING, V. DE LORENZO, S. MOLIN, F. O'GARA, C. RAMOS and K. N. TIMMIS (1994), "Behavior of bacteria designed for biodegradation", *Bio/Technology* 12, pp. 1349-1356.

POSSIBILITIES AND LIMITATIONS FOR THE MICROBIAL DEGRADATION OF POLY-CYCLIC AROMATIC HYDROCARBONS (PAH) IN SOIL[1]

by

B. Mahro
Institut für Technischen Umweltschutz, Bremen, Germany; and
Arbeitsbereich Biotechnologie II, Technische Universität Hamburg-Harburg, Hamburg, Germany

A. Eschenbach and G. Schaefer
Arbeitsbereich Biotechnologie II, Technische Universität Hamburg-Harburg, Hamburg, Germany

M. Kästner
Universität Jena, Institut für Microbiologie, Jena, Germany

Introduction

Polycyclic aromatic hydrocarbons (PAH) are compounds with a basic structure of two or more condensed aromatic rings. Due to their mutagenic potential, a number of PAHs are considered as important environmental pollutants. They may occur at high concentrations, especially on the ground of former gasworks or cooking plants, or as components of tar oil spills. According to the so-called "Holland List", which is probably the most widely applied list of acceptable and non-acceptable concentration standards in Europe, PAHs are considered as non-tolerable if they exceed an overall PAH concentration in soil of 200 mg PAH/kg of soil or, for particular toxic compounds, [i.e. benzo(a)pyrene], if a threshold value of 10 mg BaP/kg of soil is exceeded.

The hope that it might be possible to clean up such PAH-contaminated soils by biological means is based on observations that a number of micro-organisms are able to degrade and mineralise PAH, at least under laboratory conditions. At the present time, one can distinguish three different biochemical pathways of microbial PAH degradation (Mahro *et al.*, 1994):

- ***Type 1 - Complete mineralisation:*** intracellular; complete degradation of the ring systems. Theoretically there should be no formation or accumulation of metabolites. CO_2 as main product.

[1] This research was funded by the Deutsche Forschungsgemeinschaft (SFB 188, B1) and by the Deutsche Bundesstiftung Umwelt (DBU) in cooperation with the Freie und Hansestadt Hamburg. Parts of the work were carried out in co-operation with the research groups of Professor R. Stegmann (TUHH) and Dr. R. Wienberg (private research laboratory). We also gratefully acknowledge the continuous support and advice of Professor V. Kasche, the head of the Biotechnology Section in Hamburg-Harburg.

- ***Type 2 - Cometabolic transformation:*** initial oxygenation intracellular; partial oxidation of the rings. Partly oxidised metabolites are usually accumulated. CO_2 as possible product.

- ***Type 3 - Non-specific radical oxidation:*** extra-cellular; initial oxidation by formation of radicals, non-specific further reaction of the oxidation products. CO_2 as possible product.

The biochemistry of these reactions has been extensively reviewed by Cerniglia (1992), and by Barr and Aust (1994). This report will focus on the following questions:

1) Does the PAH biodegradation that was observed in the liquid medium also occur in soil?

2) Which pathways contribute to PAH depletion in soil?

3) How can one influence, steer, and optimise these pathways in soil, and especially in soil from contaminated sites ?

Results

To answer the above-mentioned questions, we investigated different soil batches that had been supplemented with one or more model PAH compounds, and one or more PAH-degrading microbial isolates. The selected components were mixed, and the change of the extractable PAH concentration was subsequently followed over time. A typical result obtainable with this approach is pictured in Figure 1.

Figure 1. Influence of the addition of *Mycobacterium* sp. VF1 on the degradation of phenanthrene/fluoranthene in a batch of loamy sand

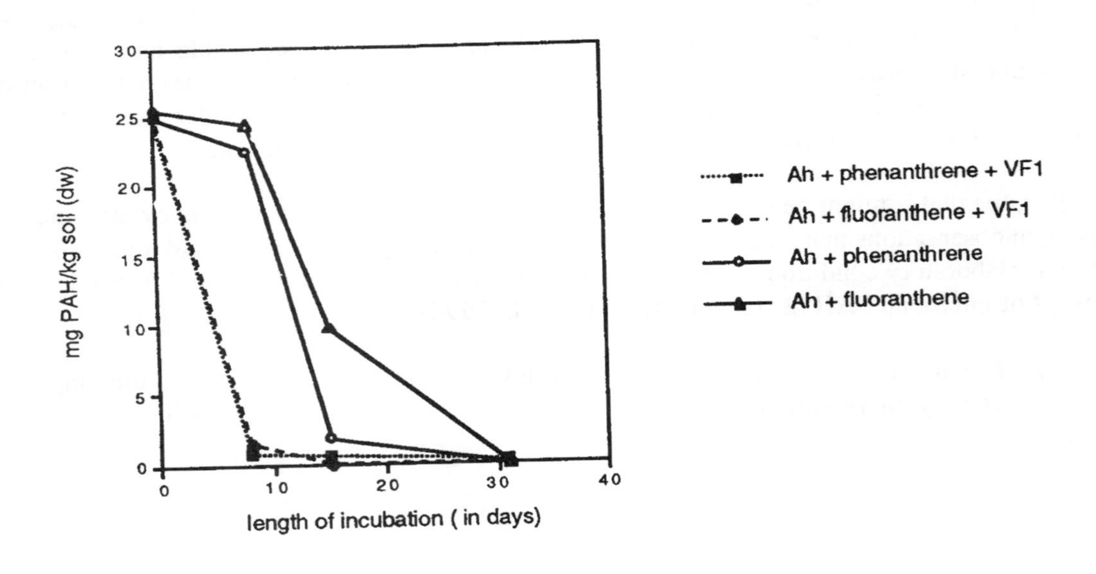

Source: Mahro, 1995.

The initial concentration of PAH dropped rapidly in the supplemented soil batch, indicating a significant microbial PAH degradation. One has to be aware, however, that the apparent analytical disappearance of PAH in soil cannot automatically be equated with real PAH degradation. The disappearance of the extractable amount of PAH in soil may sometimes also be caused by an insufficient experimental analytical procedure (Eschenbach *et al.*, 1994). Thus, it was observed in our laboratory that PAH that could not be detected anymore after an organic extraction with acetone became re-extractable if the soil was treated a second time with a methanolic sodium hydroxide solution. This indicated that not all non-detectable PAH had actually been degraded. A considerable amount of PAH molecules might be strongly adsorbed in humic cavities and may thereby become inaccessible to organic solvents. The humic structure needed to be destroyed by alkaline hydrolysis in order to release this fraction of un-degraded PAH (Eschenbach *et al.*, 1994).

But on top of that, even if the extraction and analytical procedure has been optimised, another part of the soil PAH may also become irreversibly bound to the soil matrix. One can easily demonstrate this by the use of ^{14}C-labelled PAH, as can be seen in Figure 2.

Figure 2: Fate of ^{14}C-labelled PAH in unsupplemented soil from a contaminated site

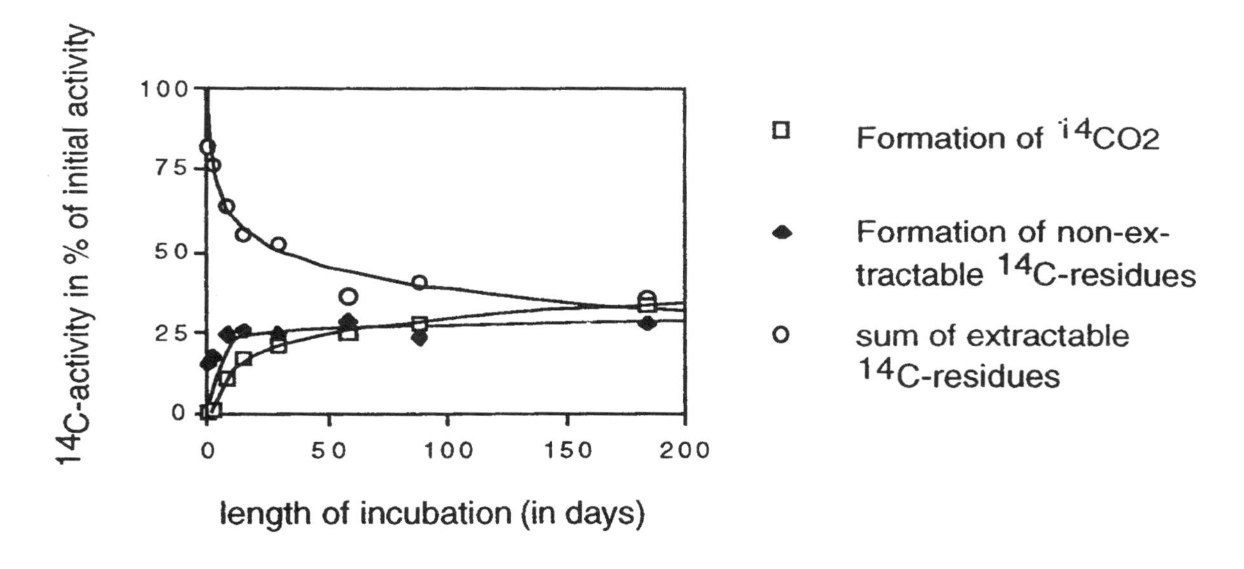

Source: Mahro, 1995.

The apparent disappearance of the PAH that was found after the more rigid two-step extraction procedure was due to real degradation (i.e. formation of CO_2), obviously only to some extent. Another part of the disappearance of PAH in the soil can also be caused by the formation of non-extractable residues in the soil. The fact that xenobiotics may stick irreversibly to the soil matrix is well-known for pesticides, and agro-scientists have called this phenomenon "the formation of bound residues" (Bollag *et al.*, 1992; Calderbank, 1989; Führ, 1987).

Though the phenomenon seems somewhat worrying at first sight, it isn't necessarily problematic -- as long as the PAH stay irreversibly bound in the soil matrix. On the contrary, since one cannot avoid the binding of PAH anyway, it seems reasonable to try intentionally making use of the irreversible binding process of these PAHs in soil. A possible rationale behind this idea is provided in the following model (Figure 3).

Figure 3. Presumed PAH degradation pathways in soil

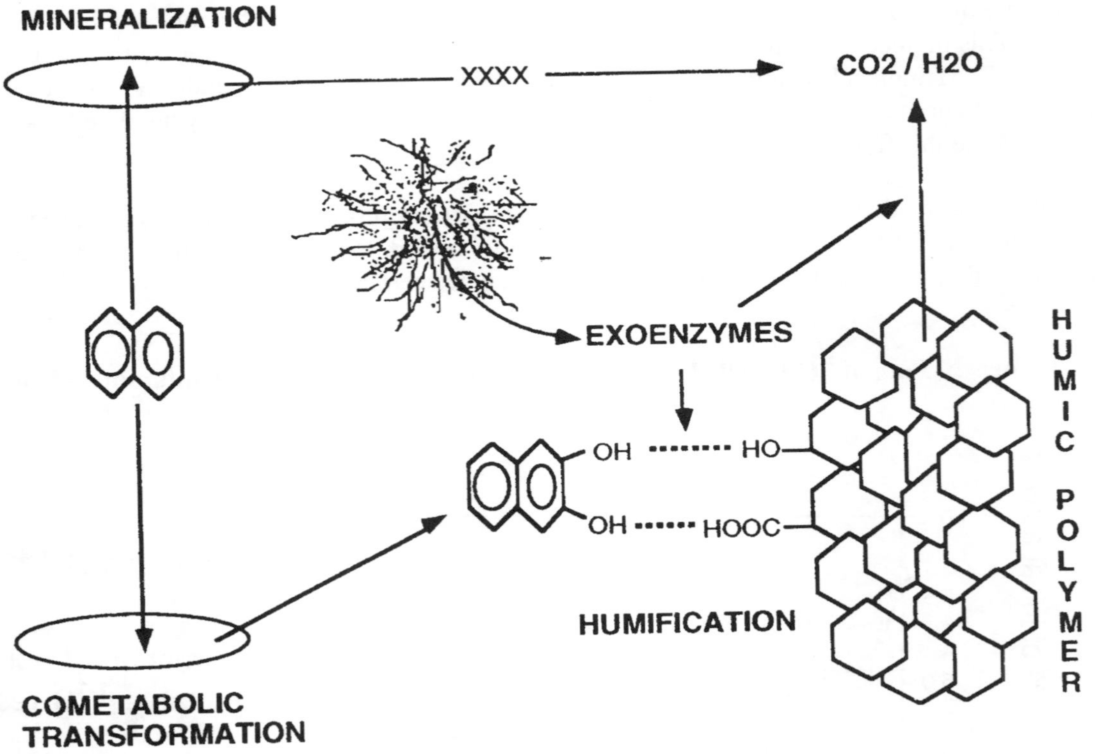

Source: Mahro, 1995.

The model is based on the assumption that the PAH degradation in soil may proceed along two different pathways. The first pathway is the well-known PAH mineralisation pathway, as it was described briefly above. This pathway may be stimulated especially by the presence of PAH-mineralising bacteria. The second pathway is based on the assumption that not all PAHs are fully degraded, and that the partially degraded products, i.e. metabolites, can be excreted in the cells environment. These carboxylic, phenolic or chinoid metabolites may subsequently get linked with the appropriate functional groups in the humic polymer, supposing that there is a sufficient amount of enzymes or radicals which trigger and stimulate these processes.

The relevance of the presumed binding of PAH in soil has meanwhile also been proven by Richnow *et al.* (1994). The authors were able to show by ^{18}O-labelling experiments that at least the formation of ester bonds between PAH metabolites and humic substances does really take place. The formation of bound residues is therefore not only a characteristic feature of many pesticides, but also of PAH.

The preceding remarks on the microbial degradation of PAH in soil have shown that the disappearance and biodegradation of PAH in soil can be due to more pathways than in liquid systems. Figure 4 summarises the major depletion pathways for PAH in soil.

Figure 4. Possible depletion pathways for PAH in soil

possible fate of PAH in soil

1 CO_2

2 transformation in partially oxidized metabolites (detectable in soil extract)

3 adsorption or fixation by chemical or physical interactions with organic soil substance

4 adsorption or fixation by inclusion in soil- or tar oil particles

5 elution or transport by dissolved organic matter (DOM)

6 evaporation (also as metabolite)

7 sorption or uptake in biomass including plants

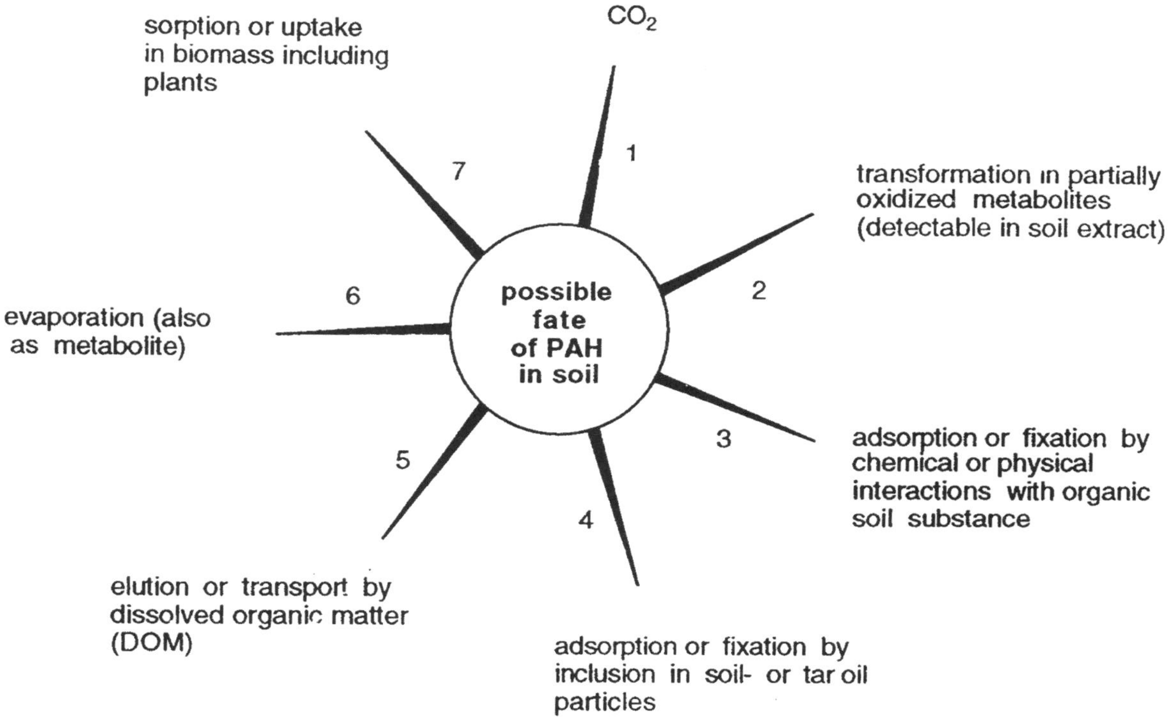

Source: Mahro, 1995.

In order to reduce the toxic potential of PAH in soil, one has therefore to evaluate more closely:

a) how one can use the whole diversity of these depletion pathways to reach this goal; and

b) which specific features must be taken into account if one is working with soil from contaminated sites.

To address the first question more closely we evaluated whether it is possible to preferentially stimulate the binding of PAH in the humic matrix (i.e. humification of PAH in soil). Two different biological additives (i.e. ripe compost and the white rot fungus *Pleurotus ostreatus*) were tested in this context. Ripe compost was selected since it is a well-known carrier of mainly cellulose and lignin degrading streptomycetes and fungi, while the white rot fungus *Pleurotus ostreatus* is known for its ligninolytic capacity and the fact that it may also be cultivated on a large scale. It was found that the use of *Pleurotus ostreatus* did stimulate the PAH-depletion in soil, but a specific stimulation of the humification process was not observed. In most experiments the binding that took place without the fungus was almost as high as the one in the supplemented soil. Control experiments with sterile soil however showed that the irreversible binding of the PAH in soil was mainly a biogenic process. In that context another important observation was the fact that the extent of humification was also dependent on the PAH molecule itself. While anthracene could disappear to a great extent along the humification pathway, naphthalene was mainly degraded in soil along the mineralisation pathway. The most problematical observation in this context, however, was that benzo(a) pyrene could not be really degraded either way.

The addition of ripe compost to the PAH-contaminated soil had a similar impact on the depletion of PAH in soil as with the addition of the white rot fungus *P. ostreatus*. Compost could also stimulate both pathways, i.e. the mineralisation and the binding of PAH in the soil matrix (Kästner *et al.*, 1995) and accelerate the initial PAH turnover rate (Kästner and Mahro, 1996).

The last point which shall be considered now in more detail is the question concerning which specific features must be taken into account if one is working with soil from contaminated sites. A comparison of the kinetics of the microbial PAH degradation in artificially contaminated soil on one hand, and of the microbial degradation in soil from sites that have been contaminated for a long time (Figure 5) on the other hand, reveals two characteristic differences. Thus it is usually relatively easy to observe a complete and fast biodegradation if the PAH degradation is measured with PAH-degrading bacteria in artificially contaminated soil systems (see Figure 1). However, in most of our investigations with soil from contaminated sites it was not possible to reduce the extractable PAH concentration in soil completely to zero, even if the soil was incubated for 18 months (data not shown).

Figure 5: Kinetics of microbial degradation in soil from a tar-oil contaminated site

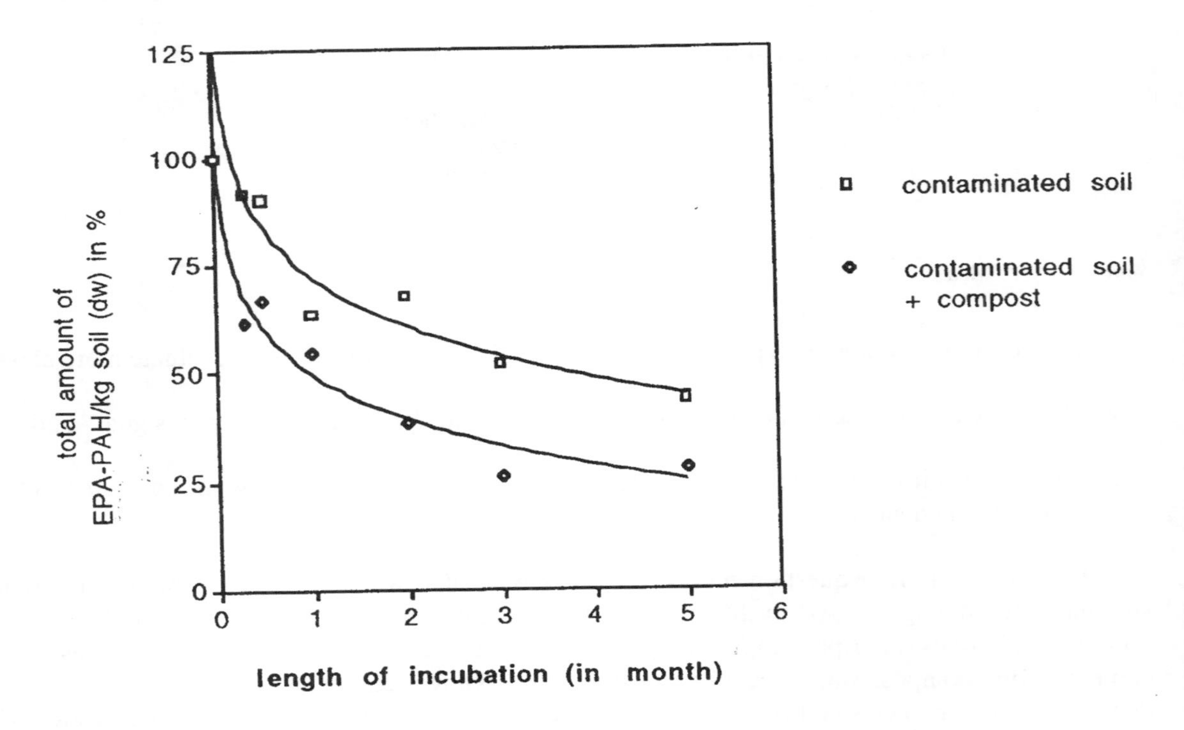

Source: Mahro, 1995.

The reason for the striking difference between the biodegradation in soil from contaminated sites and soil that has been artificially contaminated can be discovered if one considers the experimental set-up of both experiments. The PAH degradation in scientific experiments is usually recorded right after the mixing of the component soil, bacteria, and PAH. At this point in time, most of the PAH molecules are still available for bacteria in the aqueous phase or only loosely adsorbed in soil particles (see left graph of Figure 6).

Figure 6. Differences in bioavailability in artificially contaminated soil and soil from contaminated sites

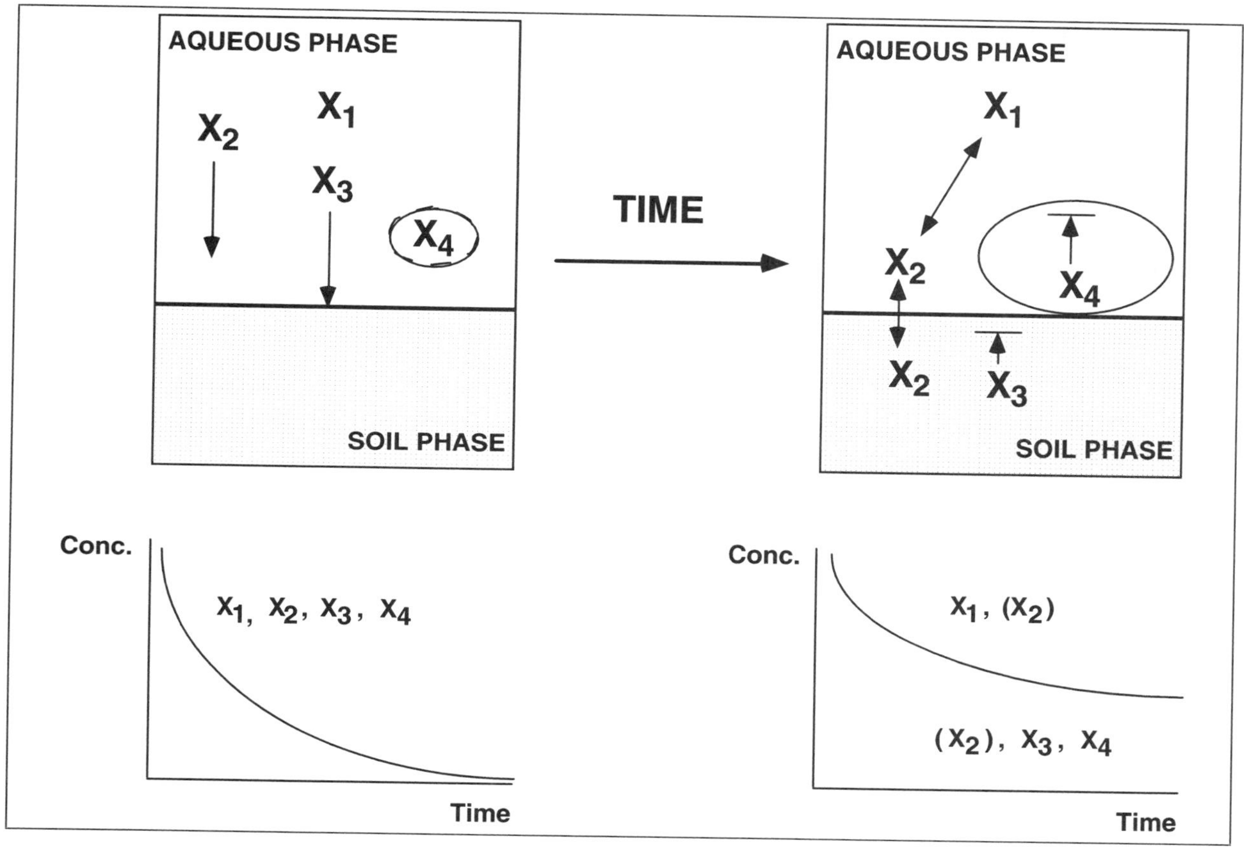

Note: The left part of the figure represents the situation in the artificially contaminated soil. All PAH X_1-X_4 are still freely accessible for micro-organisms, and can be easily degraded.

The right part of the figure reflects the situation in the soil that had been contaminated for a long time. X_1-X_4 represent different situations of PAH-bioavailability:

- X_1 = PAH freely accessible in the aqueous phase.
- X_2 = PAH adsorbed to soil surfaces; availability dependent on solubility of the molecule and the equilibrium.
- X_3 = PAH irreversibly bound, humified or trapped within the humic matrix or soil particles.
- X_4 = PAH polymerised in tar particles or in solidified surface films on top of the soil particles.

Source: Author.

The right side of Figure 6 shows the situation in the aged soil material. The PAH had enough time to adsorb at, or to diffuse into the soil particles themselves; it may have become trapped in soil or tar clumps, or became polymerised at the surface, or within the tar particle itself (see Luthy *et al.*, 1993). This means that the mass transfer most likely is the major limiting factor in biodegradation of PAH in soil from contaminated sites. The apparent residual level of PAH is therefore not indicating any limit of biodegradation, but is an indication of the big difference in the time-scale for mass transfer in or from the soil particle into the aqueous phase on one hand, and for biodegradation on the other.

In addition, one has also to realise that the "scientific way" to mix the PAH into the soil (i.e. mostly by adding tiny drops with a syringe) is also very unlike the way PAHs get into the ground on, for example, tar oil contaminated sites. It is significantly different for a micro-organism whether it has to penetrate dense and cm- or mm-sized tar or soil particles, or whether the substrate is "served at the table" in well-dissolved liquid portions in the aqueous phase. It will therefore be an important point in future bioremediation research to design experiments that mimic more closely the situation in the real world -- the bioremediation of a contaminated site!

Conclusions

1) The mere analytical disappearance of PAH in soil does not indicate a great deal about the actual fate of the PAH in the soil matrix. Mass balance techniques should be included in the experimental strategy.

2) PAH degradation in soil is not just the formation of CO_2 and water. Humification is a significant depletion pathway as well.

3) The humification pathway might become part of an intentive strategy to get rid off PAHs in soil. This requires additional evaluation of:

   - the type of PAH binding in soil;
   - the toxicity of bound PAH residues;
   - the long-term fate of the bound PAH residues;
   - the mobilising effects of compost on other soil compounds (especially metals).

4) Mass transfer into the accessible pore space and bioavailability in the aqueous phase are the major limiting factors of biodegradation. There is no need for the addition of more or less sophisticated micro-organisms to soil as long as mass transfer rates are slower than the biodegradation rates.

5) Future research on soil bioremediation should therefore focus more on the implementation of process engineering techniques, such as:

   - separation of uncontaminated soil particles;
   - reduction of soil or tar oil particle sizes;

– application of means to separate or dissolve tar particles.

6) There is a need to design more reality-like experiments especially addressing problems like ageing of contaminations in soil, impact of oily matrices on PAH degradation, degradation of complex mixtures of xenobiotics, degradation of PAH at the interfacial film between tar particles and the aqueous phase.

Questions, comments, and answers

Q: There are a number of treatments proposed for improving the availability of PAHs, such as the use of solvents, surfactants or additives. Were these used and did they improve bioremediation? The IGT process uses Fenton's Reagent to chemically oxidise PAH before biotreatment.

A: So far, we used no other dissolving reagents in our bioremediation experiments. We have used Fenton's Reagent in another context to stress the soil material in order to mobilise the non-extractable PAH residues.

Q: The mass transfer limitation is acknowledged, but are there other limitations to the degradation of more complex PAHs? Is it all due to poor mass transfer?

A: In fact, no one has yet discovered organisms which can completely mineralise PAH with more than four rings; therefore there may indeed be some biochemical barriers. But due to the poor solubility of these PAH, it is hard to distinguish biological mass transfer limitations.

REFERENCES

BARR, D.P. and S.D. AUST (1994), "Mechanisms white rot fungi use to degrade pollutants", *Environ. Sci.Technol.* 28, pp. 78 A-87 A.

BOLLAG, J.M., C.J. MYERS, R.D. MINARD (1992), "Biological and chemical interactions of pesticides with soil organic matter", *The Science of the Total Environment* 123/124, pp. 205-217.

CALDERBANK, A. (1989), "The occurrence and significance of bound pesticide residues in soil", *Rev. Environ. Contam. Toxicol.* 108, pp. 71-103.

CERNIGLIA, C.E. (1992), "Biodegradation of polycyclic aromatic hydrocarbons", *Biodegradation* 3, pp. 351-368.

ESCHENBACH, A., M. KÄSTNER, R. BIERL, G. SCHAEFER and B. MAHRO (1994), "Evaluation of a new and more effective method to extract polycyclic aromatic hydrocarbons from soil samples", *Chemosphere* 28, pp. 683-692.

FÜHR, F. (1987), "Non-extractable pesticide residues in soil", in *Pesticide science and biotechnology*, R. Greenhalgh and T.R. Roberts (eds.), pp. 381-389, Blackwell Sci. Publ., Oxford.

KÄSTNER, M. and B. MAHRO (1996), "The effect of compost on the biodegradation of polycyclic aromatic hydrocarbons (PAH) in soil", *Appl. Microbiol. Biotechnol.* 44, pp. 668-675.

KÄSTNER, M., S. LOTTER, J. HEERENKLAGE, M. BREUER-JAMMALI, R. STEGMANN and B. MAHRO (1995), "Fate of ^{14}C-labeled Anthracene and Hexadecane in compost manured soil", *Appl. Microbiol. Biotechnol.* 43, pp. 1128-1135.

LUTHY, R.G., A. RAMASWAMI, S. GHOSHAL and W. MERKEL (1993), "Interfacial films in coal tar non-aqueous phase liquid water systems", *Environ. Sci.Technol.* 27, pp. 2914-2918.

MAHRO, B. (1995), "Vergleichende Untersuchung zur Wirkung der Zugabe biologisch aktiver Supplemente auf den mikrobiellen Abbau polyzyklischer aromatischer Kohlenwasserstoffe (PAK) im Boden", Habilitation Thesis, Technische Universität Hamburg-Harburg.

MAHRO, B, G. SCHAEFER and M. KÄSTNER (1994), "Pathways of microbial degradation of polycyclic aromatic hydrocarbons in soil", in *Bioremediation of chlorinated and polycyclic aromatic hydrocarbon compounds,* R.E. Hinchee *et al.* (eds.), pp. 203-217, Lewis Publishers, Boca Raton, Florida.

RICHNOW, H.H, R. SEIFERT, J. HEFTER, M. KÄSTNER, B. MAHRO and W. MICHAELIS (1994), "Metabolites of xenobiotica and mineral oil constituents linked to macromolecular organic matter in polluted environments", *Org Geochem* 22, pp. 671-681.

MODELLING AND OPTIMISATION FOR ENVIRONMENTAL BIOTECHNOLOGY

by

G. Lyberatos, M. Kornaros, C. Zafiri and U. Zissi
Department of Chemical Engineering, University of Patras, Patras, Greece

Brief historical review

Many environmental technologies (activated sludge, nitrification/denitrification systems, anaerobic digestion units, sanitary landfills, composting facilities, etc.) are largely based on biological processes. In most instances spontaneously grown cultures of micro-organisms are responsible for the consumption and/or bioconversion of various polluting substances. These environments, contrary to most biotechnological processes that aim at the production of pharmaceuticals, solvents, enzymes, etc. are usually characterised by mixed and undefined microbial species, and by complex, time-varying and undefined substrate media (liquid and/or solid wastes). The prevailing microbial populations are a function of (a) the starting inoculating cultures, (b) the waste composition, and (c) the process operating parameters (temperature, pressure, pH, stirring rates, aeration rates and hydraulic and solid retention times). Often the cultures are enriched by special enrichment cultures that are capable of specific bioconversions and biodegradations, or are capable of higher performance when compared to spontaneously grown microbes. These cultures may then in many cases be maintained, in others, however, they are unable to withstand the competition within the 'open environment'. In any case one has to deal with complex growth media and mixed microbial cultures that are characterised by a wide wealth of interactions (competition, commensalism, ammensalism, mutualism, predation etc.).

The significance of understanding the factors that lead to a specific prevalence of a microbial culture with a particular bioconversion-biodegradation performance may not be underestimated. Most available mathematical models for processes like the activated sludge are based on crude lumping assumptions. Thus, the overall organics concentration is described by overall empirical parameters such as chemical oxygen demand (COD), biochemical oxygen demand (BOD), total organic carbon (TOC), etc. These parameters hardly describe the nature of the waste, and hardly reflect the consequences of its discharge or the effectiveness of alternative treatment possibilities. For instance, a waste containing a hardly biodegradable aromatic compound may yield, when tested, a very low value of BOD, leading erroneously to the possible conclusion that the waste is rather weak. On the other hand, such heuristic aggregate measures of pollution levels are not a linear function of the concentration of the various organics the waste contains. In addition, the usual assumption is that micro-organisms may be lumped into groups such as total biomass, total heterotrophic biomass, acidogenic biomass, etc. It is clear that the kinetics of growth for mixed cultures are hardly described by such a lumped approach. The consequence of this approach is that one obtains rather simple-to-use models that are, however, very process-specific, and have limited validity and use for process design. One major risk, and the most adverse consequence of such model development is that one may be led to attributing certain macroscopic observations erroneously. The only reason why design based on such models does not fail is that usually one overdesigns in order to secure the required effluent requirements.

The question that naturally arises from the previous discussion is: What type of research is needed in order to elucidate the key biological processes that take place? What type of research is needed to develop models that would be useful for designing and optimising environmental technology processes based on microbial metabolism? We will attempt to address these questions, expressing our opinion, and invite the scientific community to join in this discussion.

First of all, it is necessary to identify the key microbial species that are responsible for a particular bioconversion observed in a waste treatment environment. A substantial body of work has been done in this area, and for many processes we now have a clearer picture of the key species that are responsible. Thus we know, for instance, that *Acinetobacter* sp. is responsible for biological phosphorus removal (Fuhs and Chen, 1975; Deinema *et al.*, 1980); *Nitrosomonas* and *Nitrobacter* operate sequentially in oxidising ammonia to nitrite and nitrate respectively (Focht and Verstraete, 1977); *Pseudomonas* species, among others are responsible for denitrification (Tiedje, 1988); many microbial populations have been isolated from activated sludge and anaerobic digestion processes (Metcalf & Eddy, 1991).

Once the key microbial species have been identified, their growth and bioconversion kinetics as a function of the growth medium characteristics, and the operating parameters (temperature, pH, etc.), need to be determined. A rather limited body of literature exists in this area, although a fair amount of research has been done on bioconversion rates observed in wastewaters. We feel that pure culture studies of pertinent microbial growth and bioconversion on synthetic well-defined media are absolute necessities. Examples of such studies are outlined in the subsequent sections.

Having the kinetics of individual species, one should next be concerned with the behaviour of mixed cultures grown on mixed substrates. In this area, there have been a number of theoretical studies, and rather few experimental studies which have been conducted with species that are not pertinent for waste treatment processes, revealing a great wealth of possible behaviours.

Having a mathematical model that adequately predicts the behaviour of defined mixed cultures in defined synthetic wastes, one may then use optimisation theory techniques in order to assess the optimal design and operating strategies as functions of feed characteristics.

The final step should then be the testing of such modelling and optimisation results against real waste situations. An attempt to explain deviations from the predicted behaviour should be made, leading to possible reconsiderations of the 'model system' defined.

Kinetic modelling of *Pseudomonas denitrificans* growth and denitrification under aerobic, anoxic and transient operating conditions

Biological denitrification combined with nitrification is one of the most feasible and cost-effective processes for nitrogen removal from potable water and wastewaters. Denitrification is defined as the dissimilatory reduction, carried out by heterotrophic facultative aerobic bacteria, of one or both of the ionic nitrogen oxides (nitrate and nitrite) to gaseous nitrogen products. The process proceeds through a series of four steps, from nitrate to nitrogen gas (Payne, 1973):

$$NO_3^- \rightarrow NO_2^- \rightarrow NO\uparrow \rightarrow N_2O\uparrow \rightarrow N_2\uparrow$$

Denitrification is considered to be an anoxic process, occurring when nitrate or nitrite is used instead of oxygen as the terminal electron acceptor, and requires an organic or inorganic substrate for energy (electron donor) and cell synthesis. A wide variety of solid, liquid, and gaseous substrates have been reported to be used by denitrifying organisms. Complete nitrogen removal from wastewaters can be accomplished using several different process configurations (Rittmann and Langeland, 1985). Almost all of the most recently installed systems are basically single-sludge pre-denitrification, which means that their principle of operation is the recirculation of the nitrified mixed liquor to the anoxic basins in order to be denitrified. In any one of the systems achieving biological nitrogen removal the heterotrophic bacteria, which are responsible for the conversion of nitrogenous oxides to nitrogen gas, are alternately exposed to aerobic and anoxic conditions.

Kinetic data for the establishment of a mathematical model were taken from batch experiments that were carried out under strictly anoxic, strictly aerobic and transient conditions of growth, with a pure culture of a representative denitrifying bacterium, *Pseudomonas denitrificans* (ATCC 13867), in a defined synthetic medium (Kornaros and Lyberatos, 1994; Kornaros *et al.*, 1995). The basic medium used for the growth contained L-glutamic acid as the sole carbon, energy and nitrogen source. Nitrate and nitrite were added to the basic synthetic medium according to the desirable condition of growth. Batch experiments were performed in a 2.5l, stirred at 500 rpm. The temperature and the pH in the fermentor were automatically controlled at 25±0.1 °C and 7.1 respectively.

Kinetic model

Considering the biological reduction of nitrates to molar nitrogen as a two-step process (production of nitrite-nitrogen as an intermediate), an overall kinetic denitrification model was developed. A batch system with *Ps. denitrificans* cells could be defined by the following differential equations:

- cell growth:

$$\frac{dx}{dt} = \left[\mu_{aer} + \mu_{an1}\left(n_1, s_g, c_o\right) + \mu_{an2}\left(n_2, s_g, n_1, c_o\right)\right]\cdot x$$

- reduction of nitrate-nitrogen to nitrite-nitrogen:

$$\frac{dn_1}{dt} = -\frac{1}{Y_{n1}}\mu_{an1}\left(n_1, s_g, c_o\right)\cdot x$$

- generation of nitrite-nitrogen and reduction of nitrite-nitrogen to nitrogen gas:

$$\frac{dn_2}{dt} = \frac{1}{Y_{n1}}\mu_{an1}\left(n_1, s_g, c_o\right)\cdot x - \left[\frac{1}{Y_{n2}}\mu_{an2}\left(n_2, s_g, n_1, c_o\right) + v_{an}\left(n_2, s_g, c_o\right)\right]\cdot x$$

– organic carbon utilisation:

$$\frac{ds_g}{dt} = -\frac{1}{Y_{x/s}}\mu_{aer}\cdot x - Y_{sn1}\left[\frac{1}{Y_{n1}}\mu_{an1}\left(n_1, s_g, c_o\right)\right]\cdot x$$

$$-Y_{sn2}\left[\frac{1}{Y_{n2}}\mu_{an2}\left(n_2, s_g, n_1, c_o\right) + v_{an}\left(n_2, s_g, c_o\right)\right]\cdot x$$

– mass balance for oxygen:

$$\frac{dc_o}{dt} = k_L\alpha\left(c_o^* - c_o\right) - \frac{1}{Y_{x/o}}\mu_{aer}\cdot x$$

– aerobic specific growth rate:

$$\frac{d\mu_{aer}}{dt} = \alpha_{max\,2}\frac{K_{I3}}{K_{I3} + n_2}\left[\mu_{t,aer}\left(s_g, c_o, n_1, n_2\right) - \mu_{aer}\right],$$

$$\text{for } \mu_{t,aer}\left(s_g, c_o, n_1, n_2\right) > \mu_{aer}$$

or

$$\mu_{aer} = \mu_{t,aer}\left(s_g, c_o, n_1, n_2\right),$$

$$\text{for } \mu_{t,aer}\left(s_g, c_o, n_1, n_2\right) \leq \mu_{aer}$$

where,

$$\mu_{t,aer}\left(s_g, c_o, n_1, n_2\right) = \mu_{max}\frac{s_g}{K_s + s_g}\frac{c_o}{K_o + c_o}\frac{K_{I1}}{K_{I1} + n_1}\frac{K_{I2}}{K_{I2} + n_2}$$

$$\mu_{an1}\left(n_1, s_g, c_o\right) = \mu_{m1}\frac{n_1}{K_{n1} + n_1}\frac{s_g}{K_{s1} + s_g}\frac{K_{Io1}}{K_{Io1} + c_o}$$

$$\mu_{an2}\left(n_2, s_g, n_1, c_o\right) = \mu_{m2}\frac{n_2}{K_{n2} + n_2}\frac{s_g}{K_{s2} + s_g}\frac{K_{In1}}{K_{In1} + n_1}\frac{K_{Io2}}{K_{Io2} + c_o}$$

$$v_{an}\left(n_2, s_g, c_o\right) = v_{n2}\frac{n_2}{K_{n2} + n_2}\frac{s_g}{K_{s2} + s_g}\frac{K_{Io3}}{K_{Io3} + c_o}$$

and

$$Y_{x/s} = Y_{x/s}^{max} - b_1\cdot n_1 - \frac{b_2\cdot n_2}{b_3 + n_2}$$

$$Y_{x/o} = \frac{Y_{x/s}}{Y_{o/s}^{max}}$$

where x is the cell mass concentration, n_1 and n_2 are the nitrate and nitrite concentrations respectively, s_g is the organic carbon (glutamate) concentration, and c_o is the dissolved oxygen concentration.

The denitrification model developed in this work was proved capable of describing with sufficient accuracy the cell growth and organic carbon utilisation rates, the nitrate and nitrite reduction rates under anoxic conditions, the cell growth rate, the organic carbon and dissolved oxygen utilisation rates under aerobic conditions, as well as the behaviour of the denitrifier under transient conditions of growth. The excellent predictive ability of the model is clearly shown in Figures 1 and 2 where batch cultures of *Ps. denitrificans* are changed from aerobic to anoxic conditions and vice versa.

Figure 1. Experimental and theoretical profiles of biomass, dissolved oxygen, nitrate-, nitrite- and total nitrogen concentrations for the aerobic growth of a batch culture with *Ps. denitrificans* cells previously grown under denitrifying conditions

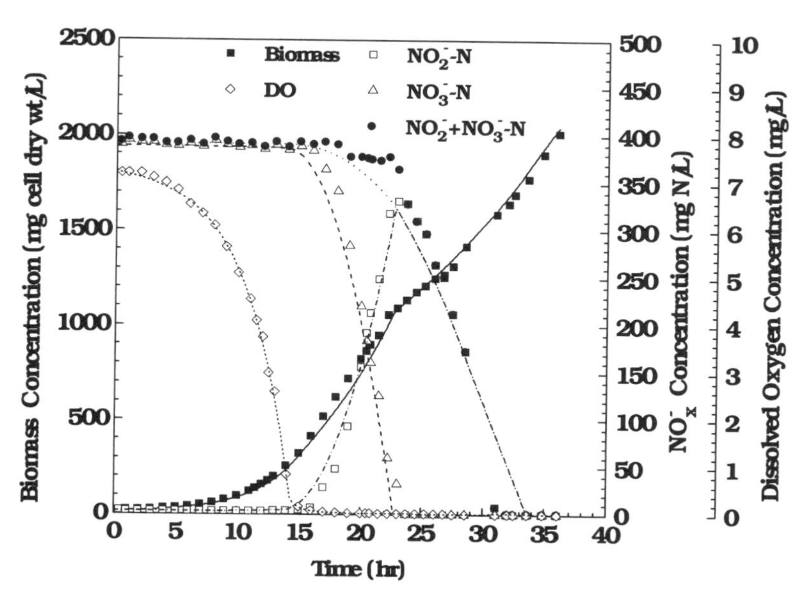

Source: Kornaros and Lyberatos, 1994.

Figure 2. Response of a batch culture in aerobic conditions to a change to anoxic conditions for 6.5 hr, back to aerobic conditions for 2.5 hr and then shift back to anoxic conditions

Experimental and theoretical profiles of biomass, nitrate-, nitrite- and total nitrogen concentrations for each growth phase

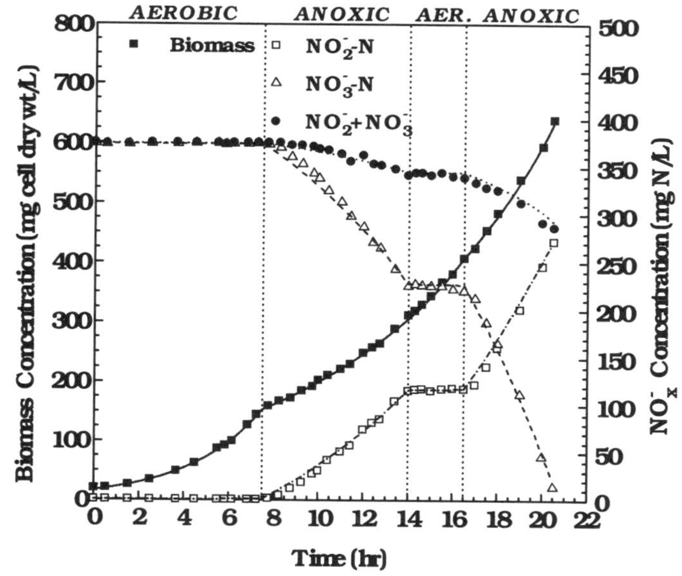

Source: Kornaros and Lyberatos, 1994.

Biological phosphorus removal using a pure culture of *Acinetobacter* sp.

Biological phosphorus removal from wastewaters is an important process for the protection of ground and surface waters from eutrophication (increase of algae concentration and concomitant oxygen depletion). It is a process in which phosphorus is taken up by bacteria able to store phosphate intracellularly as polyphosphate. Many investigators (Fuhs and Chen, 1975; Deinema *et al.*, 1980; Lötter, 1985) have isolated micro-organisms from activated sludge systems, achieving enhanced biological phosphorus removal. The most commonly isolated organisms have been members of the genus *Acinetobacter*. The enrichment of an activated sludge plant with a phosphorus removing bacterial population is achieved by sludge recirculation through anaerobic and aerobic zones. Under aerobic conditions, phosphate is not only utilised for cell synthesis, maintenance, and energy transport, but also stored as polyphosphate inside the cells. Under anaerobic conditions, the stored polyphosphate is hydrolyzed to orthophosphate which is released from the cells to the liquid. Phosphorus is removed from wastewater by wasting phosphorus-rich sludge from the system. Since the efficiency of a phosphorus removing wastewater treatment plant depends directly on the phosphorus content of the cells wasted from the system, the study and mathematical modelling of polyphosphate accumulation and utilisation rates under phosphate, and limiting and non-limiting growth conditions are of great importance.

Batch experiments were carried out using a pure culture of *Acinetobacter* sp (ATCC 11171). The bacterium was grown in a synthetic liquid medium with CH_3COONa (as carbon source) and KH_2PO_4 (as phosphorus source). Indicated amounts of phosphates and acetate were added to the basic mineral medium according to the desirable limiting conditions of growth and the pH was adjusted to 7.0. Initially, the bacterial cells were cultivated under aerobic conditions in a shaker bath for approximately 40 hours at 25 °C until all phosphate was exhausted from the culture medium. A portion of the preculture medium was transferred to the fermentor at 10 per cent of the working volume. Growth was monitored with a spectrophotometer. Filtered samples were analysed for orthophosphates and acetate by ion chromatography.

Kinetic model

For the aerobic growth of *Acinetobacter* sp. under phosphate- (soluble or intracellularly stored) and carbon-limited conditions:

- cell growth:

$$\frac{dx}{dt} = \mu_{max\,1} \frac{p}{K_p + p} \frac{c}{c + K_{c1} + \frac{c^2}{K_{c2}}} x$$

$$+ \mu_{max\,2} \frac{p_n/x}{K_{pn} + p_n/x} \frac{K_I}{K_I + p} \frac{c}{K_{c3} + c} x$$

- soluble phosphate utilisation:

$$\frac{dp}{dt} = -\left(\frac{1}{Y_{x/p}} + a_1\right) \mu_{max\,1} \frac{p}{K_p + p} \frac{c}{c + K_{c1} + \frac{c^2}{K_{c2}}} x$$

– acetate utilisation:

$$\frac{dc}{dt} = -\left(\frac{1}{Y_{x/c}} + v_1\right)\mu_{max\,1}\frac{p}{K_p + p}\frac{c}{c + K_{c1} + \frac{c^2}{K_{c2}}}x$$

$$-\left(\frac{1}{Y_{x/c}} + v_2\right)\mu_{max\,2}\frac{p_n/x}{K_{pn} + p_n/x}\frac{K_I}{K_I + p}\frac{c}{K_{c3} + c}x$$

– mass balance for polyphosphates :

$$\frac{dp_n}{dt} = a_1\,\mu_{max\,1}\frac{p}{K_p + p}\frac{c}{c + K_{c1} + \frac{c^2}{K_{c2}}}x$$

$$-\frac{1}{Y_{x/p}}\mu_{max\,2}\frac{p_n/x}{K_{pn} + p_n/x}\frac{K_I}{K_I + p}\frac{c}{K_{c3} + c}x$$

– mass balance for carbon reserves :

$$\frac{dc_n}{dt} = v_1\,\mu_{max\,1}\frac{p}{K_p + p}\frac{c}{c + K_{c1} + \frac{c^2}{K_{c2}}}x$$

$$+ v_2\,\mu_{max\,2}\frac{p_n/x}{K_{pn} + p_n/x}\frac{K_I}{K_I + p}\frac{c}{K_{c3} + c}x$$

where x is the cell mass concentration, p and c are the soluble phosphate and acetate concentrations respectively, p_n/x is the polyphosphate mass accumulated per unit cell mass and c_n/x is the carbon reserves mass accumulated per unit cell mass.

In order to calculate the amounts of polyphosphates and carbon reserve stored per dry cell mass, two more equations were used :

$$\frac{d\left(p_n/x\right)}{dt} = \frac{dp_n}{dt}\frac{1}{x} - \frac{p_n}{x^2}\frac{dx}{dt} \quad \text{and}$$

$$\frac{d\left(c_n/x\right)}{dt} = \frac{dc_n}{dt}\frac{1}{x} - \frac{c_n}{x^2}\frac{dx}{dt}$$

All the model's kinetic parameters were estimated exclusively from batch experimental data. The proposed model is capable of describing with sufficient accuracy the cell growth rate, the phosphate and acetate utilisation rates and the kinetics of polyphosphate and carbon storage per unit cell mass.

Figure 3. Experimental and simulated profiles of biomass, acetate and soluble phosphate concentrations

Aerobic growth of Acinetobacter sp. in a batch culture under phosphate limiting conditions

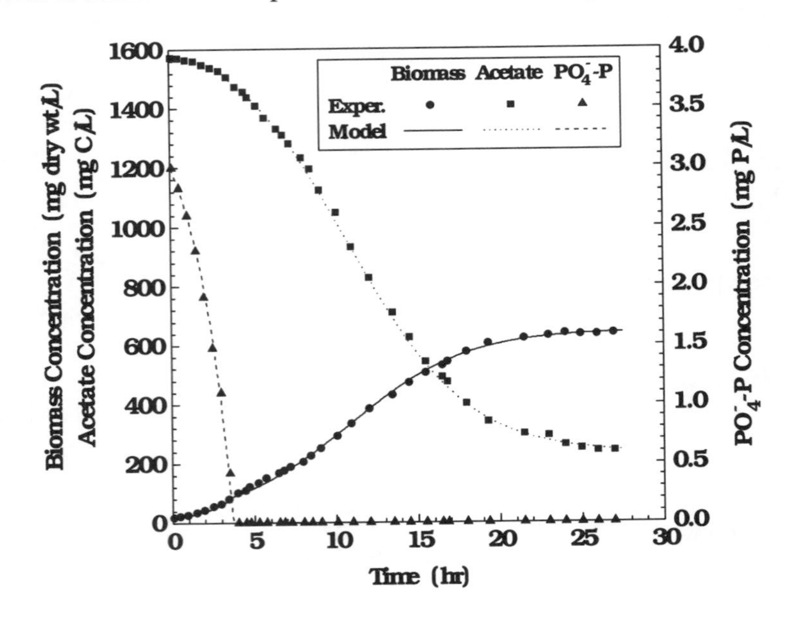

Source: Zafiri *et al.*, 1995.

Figure 4. Experimental and simulated profiles of polyphosphates and carbon reserves accumulated per unit cell mass

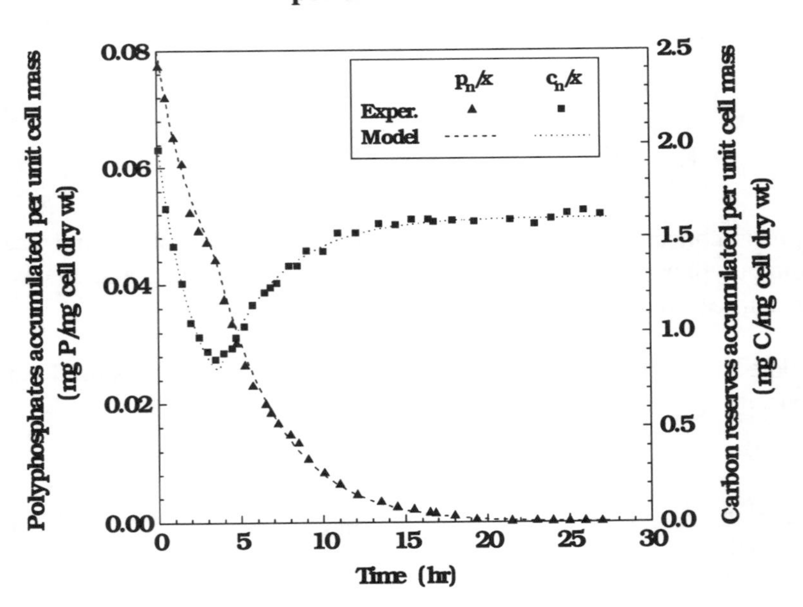

Source: Zafiri *et al.*, 1995.

The main conclusions of this work were as follows :

- The ability of a pure culture of *Acinetobacter* sp. to take up more phosphorus than is normally needed for cell growth and store it as polyphosphate when it is grown under aerobic conditions using acetate as sole carbon source was verified (Figures 3 and 4).

- The bacterium was also capable of accumulating intracellularly large amounts of organic carbon, especially when it was grown under polyphosphate-limiting conditions.

- A substantial decrease (about 60 per cent) of the maximum specific growth rate was observed when the cells were grown using phosphorus exclusively supplied from the hydrolysis of polyphosphates instead of the soluble phosphates.

- Substrate inhibition by acetate was observed when the bacterium was grown with soluble phosphates as the primer phosphorus source and could be very well described mathematically using Andrews kinetics. However, the acetate inhibitory effect on cell growth was not detected under polyphosphate-limiting conditions.

Degradation of p-aminoazobenzene by *Bacillus subtilis*

During the last few decades, a large number of synthetic compounds, called xenobiotics, has been introduced into the environment. The exogenous compound may serve as a substrate providing a source of nutrients or energy for the micro-organism, or may be cometabolized along with metabolism of another growth substrate (Alexander, 1981). Since aromatic azo and sulfo groups are not synthesised in nature, azo-dyes are considered to be xenobiotics. Dyes are regarded as pollutants because they are not readily degraded by conventional wastewater treatment systems (Pagga and Brown, 1986). In screening pure cultures for aerobic degradation of potential pollutants from dyestuff and the textile industry, Idaka *et al.* (1978) isolated a bacterium, *Aeromonas hydrophila* which is capable of aerobically degrading azo-dyes. Zimmermann *et al.* (1982) reported that azoreductase, the enzyme responsible for the initiation of degradation (azo-bond cleavage), from *Pseudomonas* KF46 can degrade azo-dyes aerobically.

The aerobic degradation of p-aminoazobenzene by *Bacillus subtilis* (ATCC 6051) was studied in our laboratory through batch experiments in order to investigate the effect of p-aminoazobenzene on the bacterial growth rate and elucidate the mechanism of dye degradation. The results proved that *Bacillus subtilis* cometabolizes p-aminoazobenzene in the presence of glucose as carbon source, producing aniline and p-phenylenediamine as the nitrogen-nitrogen double bond is broken. Based on the experimental observations, a kinetic model was developed, that describes well aerobic growth of *Bacillus subtilis*, carbon utilisation, p-aminoazobenzene degradation and product formation. The synthetic growth medium used contained 3 g/l glucose as carbon and energy source and ammonium chloride as nitrogen source. P-aminoazobenzene was added to the medium at a concentration of up to 20 mg/l, the solubility limit of the dye in water. The precultured cells were inoculated in a F-2000 Multigen Bench Fermentor with 1 500 ml working volume. All the batch experiments were carried out under aerobic conditions at 25 °C and pH=7.1. Glucose was determined using the rapid spectrophotometric procedure for determination of total carbohydrates. The concentration of p-aminoazobenzene was measured spectrophotometrically. Aniline and p-phenylenediamine were determined by high pressure liquid chromatography on a model 9010 Solvent Delivery System (Varian) and a model 9050 UV-VIS Detector (Varian).

Kinetic model

The following model was developed:

– biomass:

$$\frac{dX}{dt} = \mu(s_g, c_p) \cdot X$$

– glucose:

$$\frac{dS_g}{dt} = -\frac{1}{Y_{x/s}} \cdot \mu(s_g, c_p) \cdot X$$

– p-aminoazobenzene:

$$\frac{dC_p}{dt} = -Y^{appar}_{c_p/x} \cdot \mu(s_g, c_p) \cdot X - \beta_1(c_p) \cdot X$$

– aniline:

$$\frac{dP_1}{dt} = Y^{appar}_{p_1/x} \cdot \mu(s_g, c_p) \cdot X + \beta_2(c_p) \cdot X$$

– p-phenylenediamine:

$$\frac{dP_2}{dt} = Y^{appar}_{p_2/x} \cdot \mu(s_g, c_p) \cdot X + \beta_3(c_p) \cdot X$$

where

$$\mu(s_g, c_p) = \mu_{max} \cdot \frac{S_g}{K_s + S_g} \cdot \frac{K_I}{K_I + C_p}$$

$$Y^{appar}_{I/x} = \frac{C_p}{K_p + C_p} \cdot Y^{true}_{I/x} \qquad I = C_p, P_1, P_2$$

where Sg is the glucose concentration, X is the biomass concentration, C_p, P_1, and P_2, are p-aminoazobenzene, aniline and p-phenylenediamine concentration respectively. The system of equations (1)-(5) was integrated numerically with the determined values of K_s, K_I, μ_{max}, $Y_{x/s}$, $Y_{cp/x}$ $Y_{p1/x}$ and $Y_{p2/x}$, in order to predict cellular growth, glucose utilisation, p-aminoazobenzene degradation, aniline and p-phenylenediamine production. All model parameters were evaluated from the experimental data. Model predictions are compared with the experimental results in Figures 5-7.

Figure 5. Predicted and experimental biomass and glucose concentration vs. time
(C_{gluc0} = 3070 mg/l, C_{paab0} = 10 mg/l, T = 25 °C, pH = 7.1)

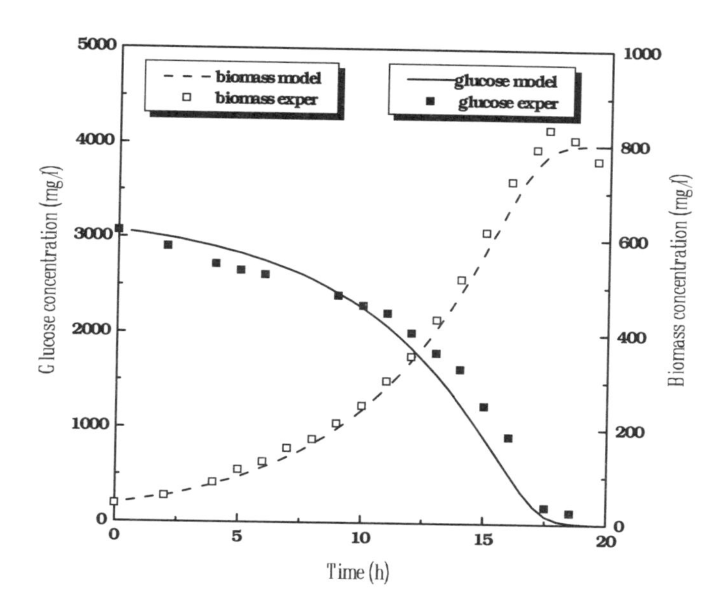

Source: Zissi *et al.*, 1995.

Figure 6. Predicted and experimental concentration profiles of p-aminoazobenzene and aniline with time

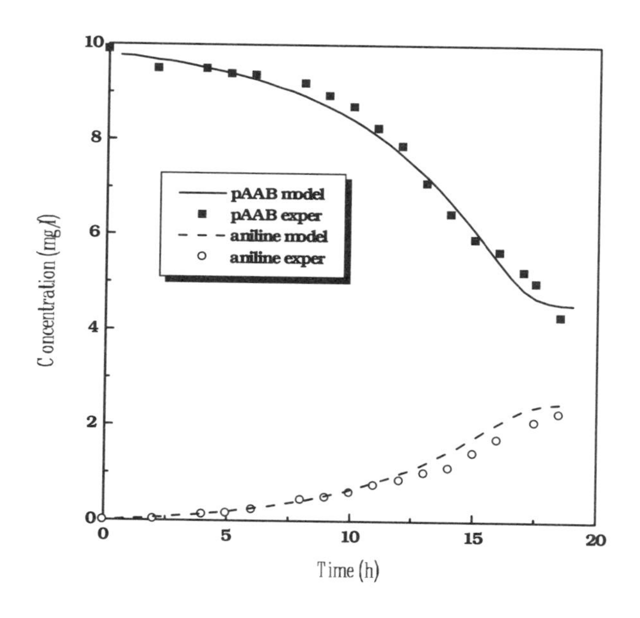

Source: Zissi *et al.*, 1995.

Figure 7. Predicted and experimental concentration profiles of p-aminoazobenzene and p-phenylenediamine with time

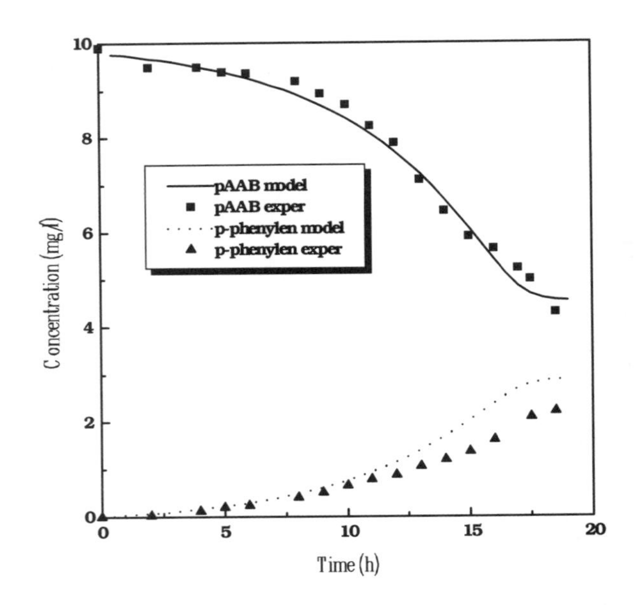

Source: Zissi *et al.*, 1995.

Questions, comments, and answers

Q: Why are the azo dyes not completely converted? Where is the bottleneck of this limited conversion? Can the model identify this?

A: The microbe has an azo-reductase and will go so far with co-metabolism, but needs another organism to complete the breakdown. If practical rates are the same as those predicted by the model, then organism growth is the bottleneck.

REFERENCES

ALEXANDER, M. (1981), "Biodegradation of Chemicals of Environmental Concern", *Science* 211, pp. 132-138.

DEINEMA, M.H., L.H.A. HABETS, J. SCHOLTEN, E. TURKSTRA and H.A.A.M. WEBERS (1980), "The accumulation of polyphosphate in *Acinetobacter* spp.", *FEMS Microbiology Letters* 9, pp. 275-279.

FOCHT, D.D. and W. VERSTRAETE (1977), "Biochemical ecology of nitrification and denitrification", *Advances in Microbial Ecology* 1, pp. 135-214.

FUHS, G.W. and M. CHEN (1975), "Microbiological basis of phosphate removal in the activated sludge process for the treatment of wastewater", *Microbial Ecology* 2, pp. 119-138.

IDAKA, E., T. OGAWA, H. HORITZU and M. TOMOYEDA (1978), "Degradation of azo-compounds by *Aeromonas hydrophila* var 24B", *Journal of the Society of Dyers and Colourists* 94, pp. 91-94.

KORNAROS, M. and G. LYBERATOS (1994), "Kinetic modelling of *Pseudomonas denitrificans* growth and denitrification under aerobic, anoxic and transient operating conditions", (Submitted for publication in *Water Research*).

KORNAROS, M., C. ZAFIRI and G. LYBERATOS (1995), "Kinetics of denitrification by *Pseudomonas denitrificans* under growth conditions limited by carbon and/or nitrate or nitrite", (Accepted for publication in *Water Environment Research*).

LÖTTER, L.H. (1985), "The role of bacterial phosphate metabolism in enhanced phosphorus removal from the activated sludge process", *Water Science and Technology* 17, pp. 127-138.

METCALF & EDDY, INC. (1991), *Wastewater Engineering: Treatment, Disposal and reuse* (3rd ed.), McGraw-Hill, New York.

PAGGA, U. and D. BROWN (1986), "The degradation of dyestuffs. II. Behavior of dyestuffs in aerobic biodegradation tests", *Chemosphere* 15, pp. 479-491.

PAYNE, W.J. (1973), "Reduction of nitrogenous oxides by microorganisms", *Bacteriological Reviews* 37, pp. 409-452.

RITTMANN, B.E. and W.E. LANGELAND (1985), "Simultaneous denitrification with nitrification in single-channel oxidation ditches", *Journal of Water Pollution Control Federation* 57, pp. 300-308.

TIEDJE, J.M. (1988), "Ecology of denitrification and dissimilatory nitrate reduction to ammonium", in A.J.B. Zehnder, ed., *Biology of Anaerobic Microorganisms*, Wiley, New York.

ZAFIRI, C., M. KORNAROS and G. LYBERATOS (1995), "Kinetics of polyphosphate accumulation by a pure culture of *Acinetobacter* sp. under phosphate and carbon limiting conditions of growth", (Submitted for publication in *Biotechnology and Bioengineering*).

ZISSI, U., G. LYBERATOS and S. PAVLOU (1995), "Biodegradation of p-aminoazobenzene by *Bacillus subtilis* under aerobic conditions", (Submitted for publication in *Water Research*).

ZIMMERMANN, T., H. KULLA and T. LEISINGER (1982), "Properties of purified Orange II azoreductase the enzyme initiating azo-dye degradation by *Pseudomonas KF46*", *European Journal of Biochemistry* 129, pp. 197-203.

APPLICATION OF MOLECULAR BIOLOGY TO REAL TIME MONITORING IN BIOREMEDIATION[1]

by

Gary S. Sayler, Udayakumar Matrubutham, Robert Palmer and Christine Kelly
The Center for Environmental Biotechnology, The University of Tennessee
Knoxville, Tennessee, United States

Background

Most environmental pollutants can be transformed or destroyed by a variety of aerobic and anaerobic biochemical actions of micro-organisms. These actions may be metabolically specific and can allow organisms to grow on the pollutants as a source of carbon and energy. In other cases degradation of the pollutants may be non-specific or co-metabolic in nature with ultimate destruction of the pollutant being a fortuitous event.

In many cases, rather complete mechanistic understanding of the biochemistry, genetics and molecular biology of degradation does exist. This understanding permits the development of advanced technology for monitoring the biological degradation process (bioremediation) in the environment. The reasons for developing monitoring technology for bioremediation process are numerous and can be summarised as follows:

- verifying that biodegradation is functionally active in a bioremediation process;
- insuring that the contaminated environment is capable of sustaining microbial activity leading to bioremediation;
- measuring the presence and removal of specific toxicants during the bioremediation process;
- evaluating the biological availability and movement of sorbed contaminants during bioremediation;
- quantifying the rates and dynamics of bioremediation processes under natural and accelerated conditions;
- assessing the persistence, activity and relative risk of microbial agents used in bioremediation;
- controlling the bioremediation process to achieve optimal efficacy and cost effectiveness.

[1] This work was supported by the U.S. Department of Energy Grant Number DE-FG05-94ER61870.

Many molecular methods are available for sensitive and highly selective monitoring of microbial populations in bioremediation. However, current technology for on-line real time monitoring of microbial processes in bioremediation is much more limited. The focus of this report is to examine the use of reporter gene technology for the purpose of on-line monitoring of biodegradation processes and the potential for bioremediation process control and optimisation.

Reporter gene technology seeks to make gene expression more directly assayable and quantifiable. This is accomplished through coupling genes for easily assayable enzymes or proteins to genes responsible for the process of interest, in this case biodegradation. When the genes for biodegradation are active, expressed, so too are the reporter genes which can be more readily measured than the specific genes of biodegradation. A variety of potential reporter genes exists such as *lacZ*, *gus*, GFP, CAT, CDO, and *lux*. In most cases, an exogenous substrate or light source must be supplied to visualise and quantify gene expression. This may be difficult or impossible to accomplish for on-line monitoring in bioremediation.

The genes for bacterial bioluminescence (*lux* in *Vibrio fischeri*) offer unusual potential for on-line measuring. This potential is due to the fact that an entire cassette of *lux* genes, when used as reporter genes, permits creation of a whole cell biosensor that is capable of signalling gene expression independent of any other manipulation of the cells or the external environments. Light produced is direct and can be sensitively measured by a variety of light sensing technology.

A well developed molecular system for *lux* reporter technology in bioremediation is available for naphthalene and polyaromatic hydrocarbon degradation. This system is based on the transposon Tn3341 insertion of a *lux* gene cassette into a salicylate inducible, plasmid encoded pathway for naphthalene biodegradation in *Pseudomonas fluorescens* strain HK44 (King *et al.*, 1990). This strain has received approval from the U.S. Environmental Protection Agency for large scale environmental test use in bioremediation process monitoring and control.

Figure 1. Biochemical reactions involved in the bioluminescence

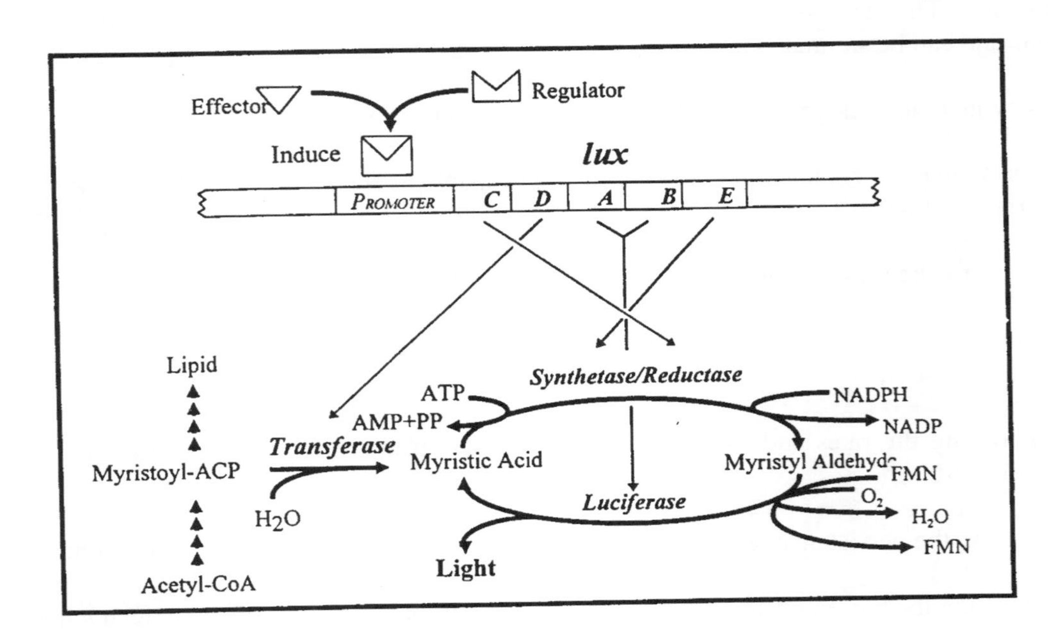

Source: Author.

Figure 1 represents a generalised model of *lux* catabolic gene fusions and the resulting physiological activity to produce bioluminescent light as a reporter of degradation. *Lux* fusions of this molecular design are inducible and positively controlled by regulatory proteins and effectors that bind to and permit expression from the catabolic gene promoters.

Induced bioluminescence from encapsulated HK44 in groundwater

Immobilisation of cells in alginate-$SrCl_2$ matrix has been advantageous in the design and application of HK44 as a biosensor (Heitzer *et al.*, 1994). Experiments were conducted to determine the sustainability of the matrix in long-term application.

The alginate-cell matrix was in the form of beads and were dispensed into 3 ml of sterile groundwater (GW) in vials. The GW consisted in mg l^{-1}: $CaCl_2$, 166; $MgCl_2.6H_2O$, 85; $BaCl_2.2H_2O$, 1.8; $SrCl_2.6H_2O$, 0.6; $FeSO_4.7H_2O$, 25; and KNO_3, 17 and buffered to provide different pH. The beads were incubated at 28°C, at static conditions. As a normal incubation control, a set of vials with beads in 0.1x YEPG medium was used. The cells were induced for bioluminescence on days 1, 7, 14, 21, 28 and 35 with two types of induction solutions. The induction solutions contained 100 mg l^{-1} of sodium salicylate in sterile distilled water or in YEP medium. The bioluminescence response was monitored every 15 minutes for 5 hours after induction. The light signals were recorded as nanoamperes of current with the help of a photomultiplier (PMT) and a liquid light guide. The salicylate concentration was determined at the beginning and end of induction (Heitzer *et al.*, 1992). At the end the beads were dissolved and spread plated on YEPG agar with 14 mg l^{-1} of tetracycline. The colony forming units were counted after 36 hours.

Figure 2. Induced bioluminescence during long-term incubation in groundwater

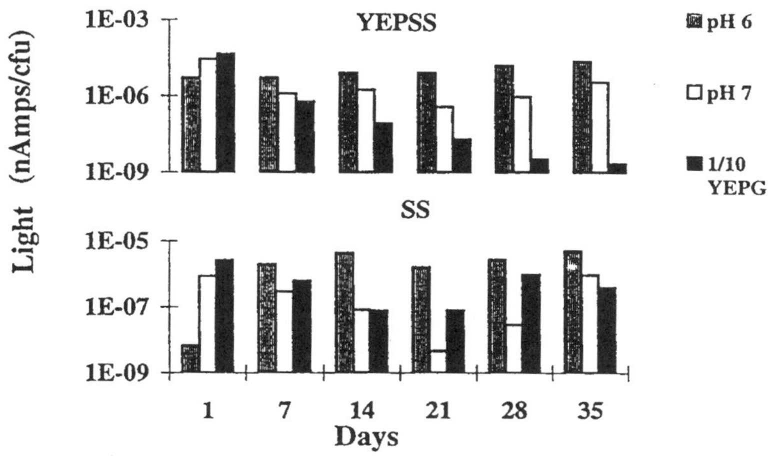

Source: Author.

The maximum bioluminescence response observed within the 5 hour period, during the 6 different induction periods are shown in Figure 2. There was no response from cells incubated at pH 3-5 GWs, irrespective of the induction solution. The signals in pH 6 GW were consistently stable for all induced periods. In pH 7 GW a declining trend in response was observed until day 21, however, the actual magnitude was not highly variable. The study demonstrated the robustness of the bioreporter bacteria and the bioluminescent signal during an extended time.

Fibre-optic mediated continuous light monitoring in groundwater

Fibre optic wave guides were examined for their utility in transmitting the light signals in soil environments to a PMT. Glass optical fibres with a radius of 1-mm and 0.4 numerical aperture, (Fiberguide Industries, NJ). were cut in two meter lengths, bundled and packaged inside a heat-shrink plastic tube. The ends were polished to form flat smooth surfaces, one end was placed in series to a PMT and the other end close to the alginate beads in a stainless steel module. The module was placed in a sand bed and GW with naphthalene was pumped at 1 ml minute^{-1}. The flow was controlled by the effluent tubing level such that a constant volume was always retained on the sand bed. As a control, GW without naphthalene was pumped in a identical format. Bioluminescent signals were averaged every 15 minutes and recorded in the computer.

The flow of GW through the module provided naphthalene to the cells. The response in nanoamperes and the concentration of naphthalene in the GW are shown in Figure 3. The specific signals illustrated that fibre optics are useful in continuous monitoring of the biological light signals. With the increase or decrease of naphthalene the signals varied. The light detection and monitoring

Figure 3. Fibre optic mediated light transmission

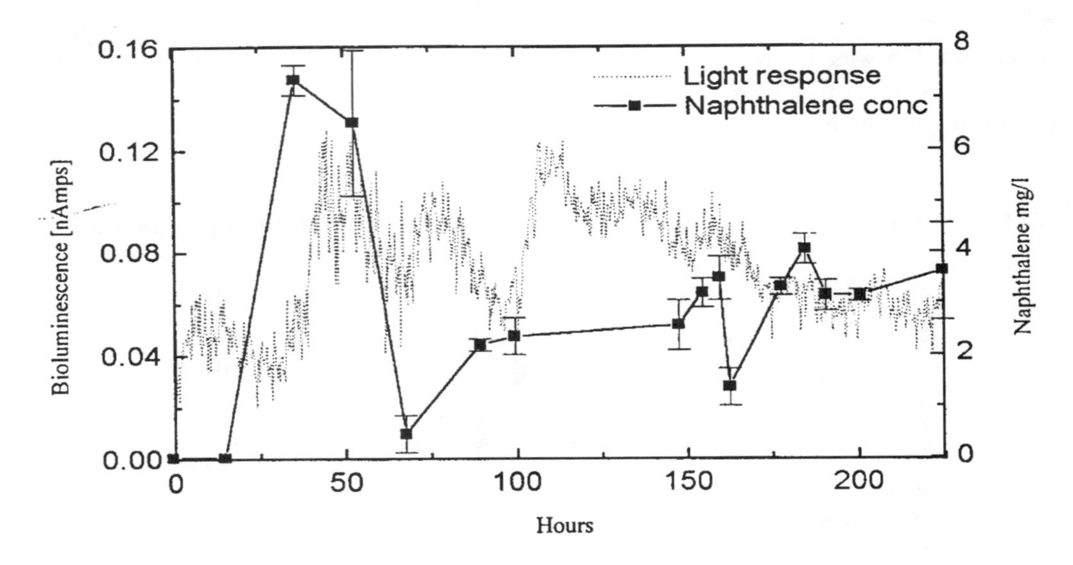

Source: Author.

Photodiode based detection of bioluminescent signals

With the intention of developing a cost-effective, more sensitive and easy-to-use optical detection system, we investigated the optoelectronics and designed an inexpensive optical detection device, using photodiode detector.

The photodiode (1 cm^2 surface area) preamplifier was fabricated. The photodiode was connected to a low-impedance preamplifier. The preamplifier was supplied power from two 9 volt batteries. The output from the preamplifier was fed into a computer. A glass flow chamber with immobilised HK44 cells was held in position above the photodiode in an electrically grounded metallic cabinet to provide total darkness. Naphthalene was fed to the beads at a constant rate (~1 ml min^{-1}). The naphthalene feed was sampled periodically to determine the concentration.

The bioluminescent response to the naphthalene concentration of the feed is shown in Figure 4. The signal gradually increased in response to naphthalene feed and fluctuated due to perturbation in the feed, revealing the light detection response and specificity of the photodiode. These results have led to the selection of low cost photodiodes for *in situ* analysis of HK44 in field released testing at Oak Ridge National Laboratory.

Figure 4. Photodiode based bioluminescence detection

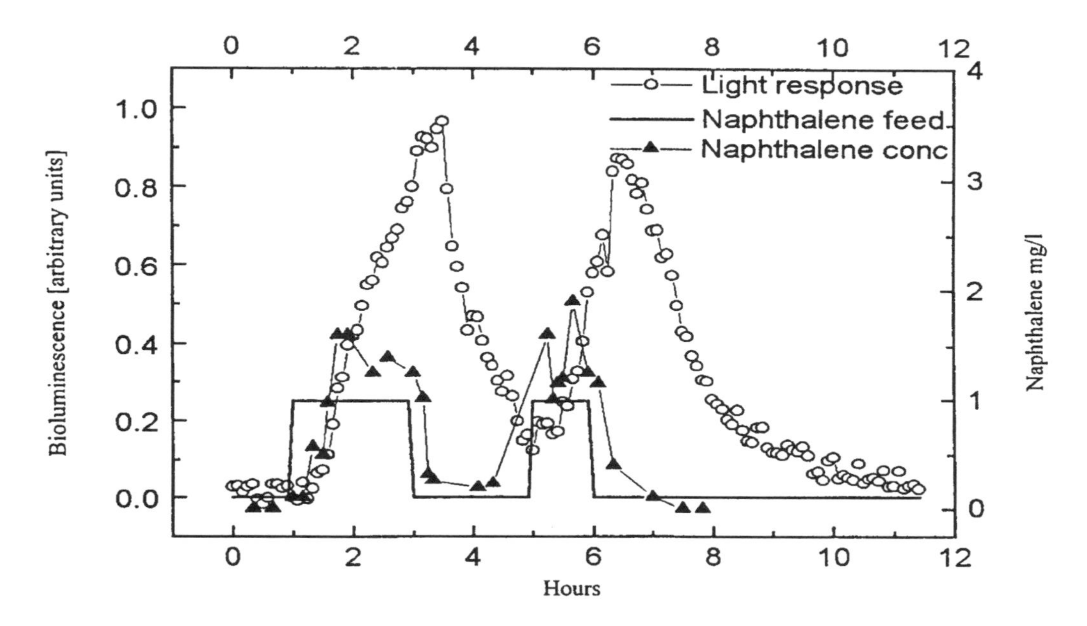

Source: Author.

Automated, multi-fibre light

To monitor light from multiple locations simultaneously, a prototype system of fibre optic probes is being constructed, bearing in mind the cost and sensitivity.

Figure 5, is a schematic diagram of a multi-fibre switch system. The distal ends of 25 plastic clad-silica fibres will be placed near the light source and the proxil (or output) ends mounted to a circular shaped bracket. A motorised rotational stage holds a large numerical aperture (3 mm core, NA = 0.47) liquid light guide such that it may be positioned to couple with any of the fibres mounted on the circular bracket. A photon-counting PMT, in a cooled housing, is used to measure the light collected by the light guide from the fibre probes. A computer containing a timer/counter board is used to record the PMT signal, and a motion control system is fitted to the same computer to control the movement of the rotational mount. The rotational stage, liquid light guide and PMT housing are contained in a light tight enclosure. The entire system cost about $ 25 000. Since up to 3 rotational mounts can be operated by the same motion control system, additional systems will be less expensive.

Figure 5. Multi-fibre switcher system for light monitoring

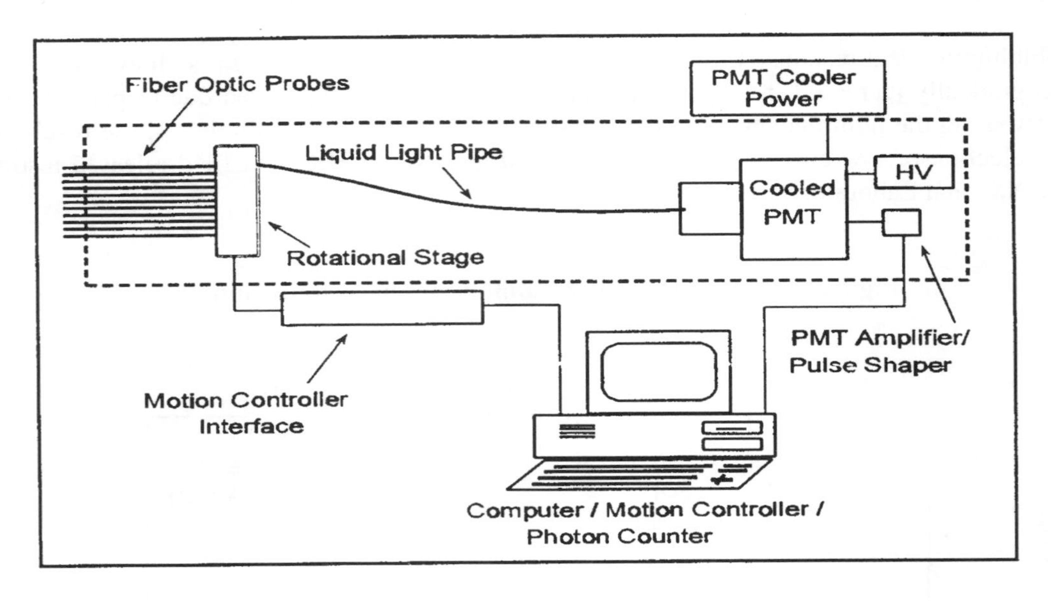

Source: Author.

Instrument operation

Computer control, via custom software, allows the fibres to be "read" sequentially, or in any desired order. First the rotational mount is moved to couple the liquid light pipe with the first fibreoptic probe. When the movement is complete, the counter board begins a counting cycle of a programmable period. The counting cycle is repeated a programmable number of times. The results of several counting cycles are averaged and the rotational stage is moved to couple the liquid light guide with the next fibre. The rotational stage requires less than 0.2s to move between two adjacent fibre probes. If light from each probe is sampled for 10 seconds, an overall duty cycle (fraction of total time spent measuring) of 98 per cent is achieved. The duty cycle for each probe is about 4 per cent. A maximum of 2 seconds is required to switch between any two fibre probes.

Significance

This system is unique from any commercially available fibre switching systems in its ability to handle very large core (1mm) fibres and in the number of fibres controlled be each switching device. A low-noise photon-counting PMT is used, so the system is capable of measuring less than 100 photons s^{-1} reaching the detector with a signal to noise ratio greater than three for a single one second counting period. Because of the limited numerical aperture and small size of the fibre probes, about 2-4 x 10^4 photon s^{-1} cm must be emitted in the soil for 100 photons s^{-1} to reach the detector. Longer counting periods will allow the measurement of lower light levels, however it is possible that a fibreoptic probe system will not have adequate sensitivity for some of the measurements needed for this study. A backup system involving a miniature PMT based sensors is also under development. These systems will allow maximum flexibility in the measurement of bioluminescence, while maintaining a reasonable cost.

Photon counting and single cell bioluminescence

The degree to which bioluminescence readings are truly quantitative has advanced rapidly due to concurrent advances in the light sensitivity of cameras. The development of intensified CCD cameras has yielded the spatial precision, the light-sensitivity, and the digital approach necessary for time-resolved low-light imaging of bioluminescence. Molecular biology has provided the tools to create reporter constructs which use bioluminescence as indicators of gene expression (Stewart and Williams, 1992). This co-development of molecular biology with bioluminescence-monitoring tools has resulted in rapid, sensitive methods of monitoring genetic activity in eukaryotic cells (e.g. Craig *et al.*, 1991; White *et al.*, 1995). Much previous work in prokaryotic systems has focused on light detection in naturally bioluminescent organisms (Masuko *et al.*, 1991*a*) and on detection of single cells by external enzymatically amplified luminescent reactions (Masuko *et al.*, 1991*b*). Photon-counting is a quantitative approach to bioluminescence at very low light levels; preliminary results from photon-counting of single bacterial cells that contain a bioluminescent reporter are present here for *Pseudomonas fluorescens* strain 5RL. This strain, a parent of HK44, harbours the bioluminescent reporter plasmid PUTK21 found in HK44 (King *et al.*, 1990). Strain 5RL was maintained (room temperature) on agar-solidified YEPSS (Burlage *et al.*, 1990) that contained 14 µg/ml tetracycline.

Photon-counting of single bacterial cells

Cells grown overnight in liquid YEPSS (YEPSS with 1/10 the standard concentration of peptone) were introduced into a silicon/glass microscopy flowcell (Palmer and Caldwell, in press). The flowcell was inverted to enhance attachment of the bacterial cells to the underside of the coverslip. After an attachment period of 15 minutes, the flowcell was mounted on an Axioplan microscope (Carl Zeiss Inc., Thornwood NY) and YEPSS was pumped through the cell (0.1 ml/min) using a low-volume peristaltic pump (Sarah, Manostat Inc, New York NY). Cells were visualised with a PlanApo 100x oil-immersion lens. Real-time photon-counting was performed at 20-minute accumulation intervals over 2 hours after 20 minutes of flow in the peptone-rich medium. The 20-minute equilibration period was necessary to wash out unattached cells and to allow for physiological upshift. Transmitted-light and photon-counting images were captured using the VIM 3/Argus 50 (Video Intensified Microscope/control software, Hamamatsu Photonic Systems, Bridgewater NJ). The microscope with attached camera was enclosed in a dark-box; photon-counting was performed at maximum gain.

Data indicate that bioluminescent reporter constructs can be used at the level of a single cell; such an approach can provide time-resolved, rapid-response data on gene expression. Data also demonstrate heterogeneity in light production within a clonal population. Much more work is necessary before this heterogeneity can be explained, however similar results (striking differences in light production between clonal cells) have been seen with the naturally bioluminescent bacterium *Vibrio harveyi*, and the absolute level of light production in the genetically engineered strain is several orders of magnitude lower than in *V. harveyi*. (Palmer and Caldwell, in press). Therefore, it seems premature to ascribe the light-production differences between individual cells in the genetically engineered strain to a simple molecular explanation such as plasmid copy-number. It is tempting to speculate that the differences reflect true phenotypic (physiological) variance within the clonal population. Future work will centre on the effects of inducer-molecule (salicylate, naphthalene) concentration on the absolute levels of light production and the time-course of light production, as well as on the effects of other requisite molecules (O_2, fatty aldehyde).

Tod-lux bioluminescence gene fusion for TCE bioremediation

Trichloroethylene (TCE) is a common and persistent groundwater contaminant that is a suspected carcinogen; therefore, there has been great interest in bioremediation of TCE. Several catabolic pathways have been shown to co-metabolise TCE, including oxidation by ammonia mono-oxygenase, aromatic mono and dioxygenases and methane and propane mono-oxygenases (Ensley, 1991). *Pseudomonas putida* F1 has been shown to co-metabolise TCE in the presence of toluene via the toluene dioxygenase pathway (Wackett *et al.*, 1988). A bioluminescent reporter was constructed from F1, and was designated *Pseudomonas putida* B2. A 2.75 kb fragment of DNA containing the *tod* promoter (courtesy of D. Gibson) was cloned in front of the promoterless *lux* genes of plasmid pUCD615 (courtesy of C. Kado) which was then mated into *Pseudomonas putida* F1 (Zylstra *et al.*, 1988). Figure 6 illustrates the construction of the reporter plasmid in strain B2.

Figure 6. Construction of *tod-lux* reporter plasmid

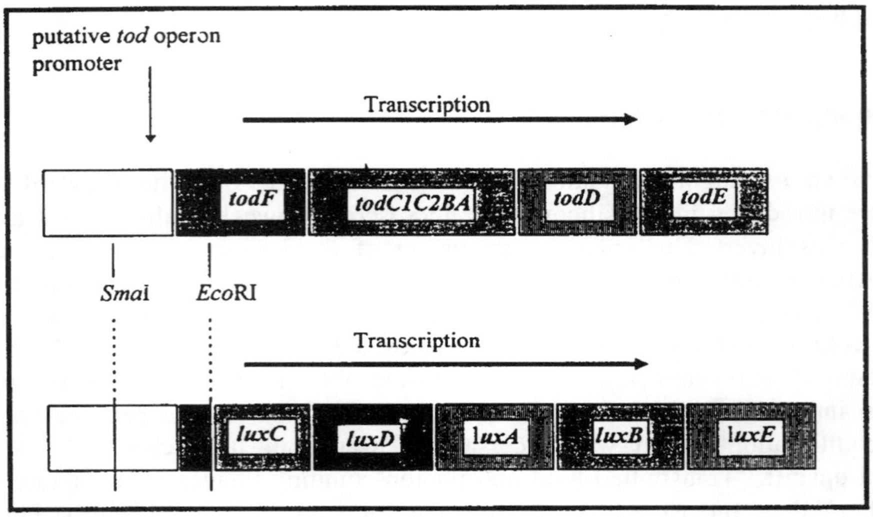

Source: Author.

Pseudomonas putida B2 was grown in a succinate media with 25 ppm kanamycin to a density of 0.9 OD_{600}, centrifuged, and resuspended in minimal media to a density of 0.4 OD. Batch experiments were performed in 32 ml glass vials with Mininert valves and 5 ml culture. After 20 hours TCE and toluene were added to the desired concentration (concentrations are expressed as if all the solvent was present in the liquid phase). Then 100 ppm toluene and 5 ppm TCE were added to triplicate vials and toluene, TCE and bioluminescence were modified. Vials were shaken at room temperature at 100 rpm on an orbital shaker.

Toluene and TCE concentrations were monitored using headspace injections of 25 *u*l to a Shimadzu gas chromatograph 14A equipped with an electron capture detector and a flame ionisation detector in series and a 30 m Supelco Vocol column. A constant oven temperature of 150°C was maintained. Bioluminescence was monitored with an Oriel photomultiplier detector system.

When 100 ppm toluene was added to 5 ml of culture, within 24 hours 100 per cent of the toluene was degraded, while 40 per cent of the initial 5 ppm TCE was degraded. Bioluminescence correlated to the presence of toluene and TCE co-metabolic activity as shown in Figure 7. In addition, the *tod-lux* bioreporter was found to be sensitive to 0.1 ppm - 100 ppm toluene within 35 minutes of exposure time and was qualitatively reproducible after multiple exposures and starvation (data not shown).

Figure 7. Toluene, TCE and bioluminescence in a batch study with *Pseudomonas putida* B2

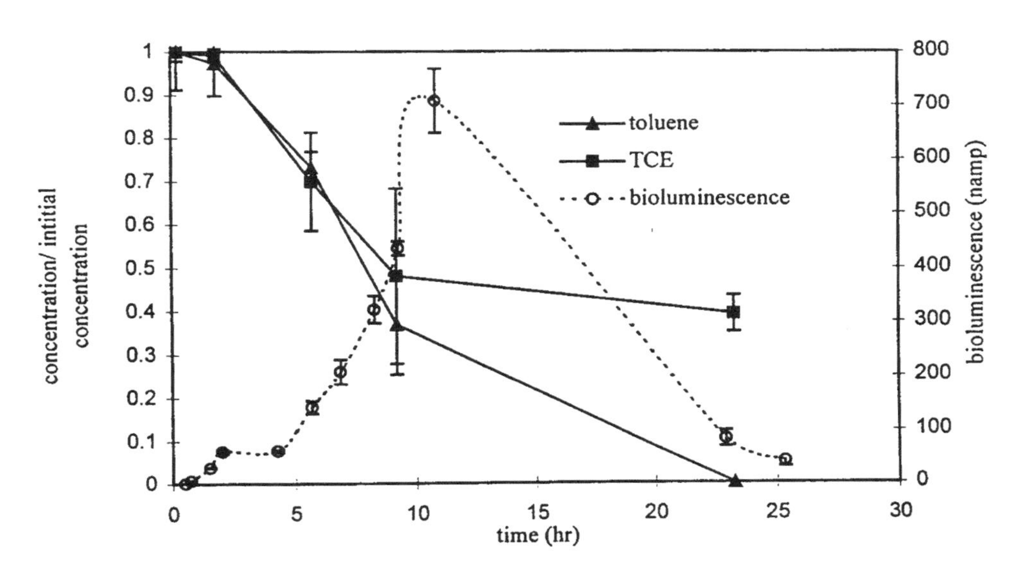

Source: Author.

Conclusions

Lux gene transcriptional fusions have been demonstrated for on-line sensing of hydrocarbon biodegradation in the case of naphthalene and toluene/TCE. Other *lux* gene fusions such as mercury (*merR*) and alginic acid exopolymer biosynthesis have been reported. All such fusions appear measurable by a variety of light sampling and quantitation methods.

Even though the bioluminescent reactions are complex and potentially effected by a number of environmental variables, whole cell biosensor organisms appear robust and function in complex environmental media.

While light detection is quantifiable at the single cell level, practical applications in soils and waste are likely dependent on bioluminescent cell concentrations in the range of 1 000 to 10 000 organisms per gram of sample. This is especially true for fibre optic interrogation of samples to detect light. It appears that robust and inexpensive photodiode light sensing methods may play a more immediate role in direct on-line sensing of microbial activity in soils, as a field deployable technology.

Questions, comments, and answers

Q: Is bioluminescence a metabolic burden on microbes? Is stability a problem - do other degrading strains out-compete this organism?

A: No, bioluminescence is not a great energy drainer for microbes and our cells have ample carbon. Stability is not a problem - the microbe is not in competition with other microbes in soil. It is robust and came from a real-life pollutant environment.

Q: The marker is good but we have not had it operating in Mycobacteria for PAH degradation. Have you been able to get it into gram positive strains?

A: No, there may be a vector or a gene expression problem. We pre-screen with a constitutive plasmid first to see if the aldehyde production system is adequate.

Q: Is there any delay between naphthalene consumption and reporter output because of the need for salicylate?

A: There is a 20-30 minute delay which corresponds to the induction time.

Q: Do the genes respond to TCE also?

A: No, that system has been put into a *Pseudomonas* and when it is turned on the toluene dioxygenase, which is chromosomally encoded, degrades the TCE. The system is limited by limiting the amount of regulatory proteins, but we hope to overcome this by making a chromosomal insertion so that we do not have multiple targets for the regulatory proteins.

Q: How near are you to validating and standardising this system and therefore qualifying it for international standards such as ISO 9 000?

A: The light detection system can be assembled quite quickly - it is standardised already. The organisms are available for research purposes world-wide. Patents have been applied for on the microbes. It is an exploitable technology and only needs to be approved for environmental use. This is currently being sought. Antibiotic resistance genes would be removed to improve the system.

REFERENCES

BURLAGE, R, G.S. SAYLER and F. LARIMER (1990), "Monitoring of naphthalene catabolism by bioluminescence with *nah-lux* transcriptional funsios", *Journal of Bacteriology* 172, pp. 4749-4757.

CRAIG, F.F., A.C. SIMMONDS, D. WATMORE, F. McCAPRA and M.R.H. WHITE (1991) "Membrane-permeable luciferin esters for assay of firefly luciferase in live intact cells", *Biochemistry Journal* 276, pp. 637-641.

ENSLEY, B.D. (1991), "Biochemical diversity of trichloroethylene metabolism", *Annual Reviews of Microbiology* 45, pp. 283-299.

HEITZER, A., O.F. WEBB, J.E. THONNARD and G.S. SAYLER (1992), "Specific and quantitative assessment of naphthalene and salicylate bioavailability using a Bioluminescent Catabolic Reporter Bacterium", *Applied and Environmental Microbiology* 58, No. 6, pp. 1839-1846.

HEITZER, A., K. MALACHOWSKY, J.E. THONNARD, P.R. BIENKOWSKI, D.C. WHITE and G.S. SAYLER (1994), "Optical biosensor for environmental on-line monitoring of naphthalene and salicylate bioavailability with an immobilized bioluminescent catabolic reporter bacterium", *Applied and Environmental Microbiology* 60, No. 5, pp. 1487-1494.

KING, J.M.H., P.M. DIGRAZIA, B. APPLEGATE, R. BURLAGE, J. SANSEVERINO, P. DUNBAR, F. LARIMER and G.S. SAYLER (1990), "Rapid, sensitive bioluminescent reporter technology for napthalene exposure and biodegradation", *Science* 249, pp. 778-781.

MASUKO, M., S. HOSOI and T. HAYAKAWA (1991*a*), "A novel method for detection and counting of single bacteria in a wide field using an ultra-high-sensitivity TV camera without a microscope", *FEMS Microbiology Letters* 81, pp. 287-290.

MASUKO, M., S. HOSOI and T. HAYAKAWA (1991*b*), "Rapid detection and counting of single bacteria in a wide field using a photon-counting TV camera", *FEMS Microbiology Letters* 83, pp. 231-238.

PALMER, R.J. Jr. and D.E. CALDWELL, "A flowcell for the study of plaque removal and regrowth", *J. Microbiological Methods*, (in press).

STEWART, G.A.S.B. and P. WILLIAMS (1992), "Review Article: *lux* genes and the applications of bacterial bioluminescence", *Journal of General Microbiology* 138, pp. 1289-1300.

WACKETT, L.P. and D.T. GIBSON (1988), "Degradation of trichloroethylene by toluene dioxygenase in whole-cell studies with *Pseudomonas putida* F1", *Applied and Environmental Microbiology* 54, No. 7, pp. 1703-1708.

WHITE, M.R.H., M. MASUKO, L. AMET, G. ELLIOT, M. BRADDOCK, A.J. KINGSMAN and S.M. KINGSMAN (1995), "Real-time analysis of the transcriptional regulation of HIV and hCMV promoters in single mammalian cells", *J. Cell Science* 108, pp. 441-455.

ZYLSTRA, G.J., W.R. McCOMBIE, D.T. GIBSON and B.A. FINETTE (1988), "Toluene degradation by *Pseudomonas putida* F1: Genetic organization of the *tod* operon", *Applied and Environmental Microbiology* 54, No. 6, pp. 1498-1503.

Parallel Sessions/B1 and B2

B1.2. INFORMATION TRANSFER AND DISSEMINATION

DISCUSSION STARTER ON INFORMATION TRANSFER AND DISSEMINATION

Question-and-Answer Session

Bernard Dixon
Bio/Technology, Ruislip, United Kindom

Questions, comments, and answers

Bioremediation is uniquely well-positioned to enjoy public support. It is green and not categorically new. While there is not total agreement on efficacy and cost, the need is there, especially in Central and Eastern Europe. It is a technology which can develop hand-in-hand with public trust and confidence.

There are four approaches:

1. Focus on the media which can be used for good or bad. Conduct workshops for journalists and explain the principles, as well as using case studies.

2. Prepare briefing papers with background information, not forgetting the "gatekeepers", the news editors.

3. Talk directly to the public via consensus conferences and public awareness sessions within scientific meetings.

4. Schools are important - young people are more and more environmentally conscious.

C: I can confirm that meetings with newspaper editors can give rise to useful articles. The Web can be used for circulating briefing documents.

C: The OECD Science Directorate has a home page on the Web and a summary of this workshop will perhaps go out on the Internet.

C: We have had some success talking with politicians' staffers. Also, there are organisations in the United States, the GE foundation being one of them, which work with journalism schools and disseminate case studies. However, a good story contains controversy - a 'nice' story about bioremediation will not go far.

A: Controversy is one ingredient but there are others. Bioremediation has counter-intuitive values which also make for "human interest".

Q: What about timeliness of disseminating the information?

A: Timeliness is very important. August is a low point because people are on holiday and this is therefore a good time to insert material; Mondays are good days because there is a lack of journalistic material at the beginning of the week. Immediacy and topicality are important considerations.

C: We have seen public debates hijacked and not allowed to proceed in the form envisaged.

A: There is always a danger of this but the answer is to use the same technique.

Q: Are there any negative actions to avoid concerning media coverage?

A: Biotechnology is as often hyped up in the media as it is put down. What not to do is do nothing!

COMMUNICATION TO THE PUBLIC AND TO GOVERNMENTS: THE RELEASE OF GENETICALLY-ENGINEERED MICRO-ORGANISMS

by

Ronald M. Atlas
Department of Biology, University of Louisville, Louisville, United States

Scientific literacy: Finding a common language for communication

One of the tasks facing the scientific community, government policy makers, and the public is finding a common language in which to communicate. Scientific communication as it occurs in the technical journals in which scientists publish the results of their works is aimed at communication with other scientists in the same or related fields. These journal publications are extremely precise and technical; they use terms that have specific meanings which are not understood by the public and which are viewed as scientific jargon by the lay public, some of whom may believe that the language is specifically designed to mislead. The semantics of these communications are filled with words whose nuances are understood almost exclusively by experts in the field. This is especially true in the field of molecular biology, the science that underpins biotechnology. In molecular biology specific nucleotide sequences are the "words" of the genetics language and publications are filled with details of nucleotide sequences and descriptors such as "-35 region promoter recognition sequences", "open reading frames", "coding sequences", "contigs", and so forth that have very precise scientific meaning but convey no useful information to nonscientists.

Additionally, communication among scientists tends to be extraordinarily narrow in scope, focusing on specific details, rather than placing the work into a more global context that would illicit any public interest or that could be used in formulating public policy. Only rarely do scientists try to place their studies in the broader context use in the less technical language that could be understood by the public and government policy makers. Most scientists do not have the capability or the interest in placing their studies into a context and a language that can be accurately conveyed to the public in terms that can be understood. In contrast, public communication focuses on broader issues such as environmental quality that are difficult to define in objective terms.

In order to begin to overcome the communication gap between the scientific and public communities, there is a clear need for educational programs to increase scientific literacy. Only when the public begins to understand the language (terms) used by scientists and the framework of scientific studies can accurate communication occur. In this regard scientists have a responsibility to devote time to the educational process beginning with conveying to students at all levels of education how science works. In particular, the public must begin to learn that the scientific method, which is the fundamental approach to scientific investigation, uses a process of hypothesis testing and statistical analyses to assess the confidence in that testing that is aimed at rejecting false hypotheses rather then proving true hypotheses. Thus, scientists reject ideas that are false and indirectly gain confidence in what is true, but they do not actually prove the truth. Hence there is always some uncertainty about the possibility that there may be another explanation that has yet to be explored. The public wants absolute assurance, for example, about the safety of a biotechnological product, which science cannot provide; science can show specific instances in which a product is safe but cannot test all possibilities and therefore cannot provide absolute answers. Rather, scientists make an assessment of the degree of uncertainty. The uncertainty, the inability to give answers in absolute terms, is often interpreted by the public as either ignorance or proof that what they believe was right in the first place.

The American Society of Microbiology (ASM) has begun a proactive project to increase scientific literacy among the public and public officials. The ASM has been trying to communicate an accurate image of micro-organisms, including those that may be employed for biotechnology. Recognising that micro-organisms are viewed by most people exclusively as agents of disease and death – most individuals view viruses, bacteria, fungi and other micro-organisms as germs that should be eliminated – the ASM hopes to educate the public about microbial diversity and the beneficial aspects of microbiology and biotechnology, as well as to form a better basis for public understanding of pathogenic (disease-causing) micro-organisms that will translate into support for appropriate public policies. The ASM intends to convey messages about ubiquity of micro-organisms in the environment, the fact that we are 90 per cent bacteria and only 10 per cent human on a cell count basis, and that without micro-organisms all life on Earth would cease within a few weeks. These basic facts are not well known by the public.

To overcome the view that all micro-organisms are evil germs, the ASM has sponsored luncheon roundtables in Washington D.C. with Congressional representatives and staff where various beneficial aspects of microbiology, including the benefits that are gained from biotechnology and bioremediaton, were described. These presentations include extensive periods for questions and answers. This format has proven to very valuable in establishing a dialog which permits communication between scientists and public officials and overcomes a priori fear of micro-organisms. Additionally, microbiologists from the ASM's Board of Education and Training regularly address local schools on the benefits and hazards of micro-organisms. The ASM also tries to provide the news media with experts who can help convey accurate science to the public in terms that can be understood. Reaching the public with accurate, complete and understandable scientific information via the news media is an important aspect of improving scientific literacy. One problem with the news media, is that in the editorial process scientific points are often made so simplistic as to be inaccurate. Unfortunately the alternative is often presentation of scientific points that are too technical and complicated for general understanding. Few science journalists are able to understand the scientific material well enough to provide definitions for the technical terms which are used so that the public can understand them and even fewer are able to establish a context for presentation that retains interest while presenting scientific details. There are very few journalists who can bridge the gap between the scientist and members of the general public

The ASM is in the process of undertaking its most extensive outreach program to enhance scientific literacy – the Hollywood-style production of six television segments that highlight the diversity of micro-organisms and the scientific process of investigation. The project has encountered many problems, starting with its multimillion dollar cost and, more importantly, with defining the message to be brought to the public. The Hollywood producers want to grab and ensure the viewers' attention by presenting only the dramatic elements in the scientific process, whereas the scientific advisers want to present the more mundane aspects of science that are at the core of the scientific discovery process. While the scientists want to convey information, for example, about the nature of a microbial genome, how DNA is sequenced, and how organisms are genetically engineered, knowledge and information that underpins molecular biology and biotechnology, the producers would rather focus on the outbreak of Ebola virus and the great containment precautions taken when working with biological containment level 4 micro-organisms. The outcome of the debate about the content and focus of the series will have a major impact on the effectiveness of the original goal of enhancing scientific literacy.

Bringing science to the debate about the safety of biotechnology

As already noted, the public has a very strong aversion to micro-organisms and finds it difficult to understand why anyone would want to release micro-organisms deliberately into the environment or otherwise create conditions in which we might be exposed to them. The populous is too concerned with controlling disease and those few micro-organisms that are human pathogens to focus on the majority of micro-organisms that are beneficial – those that make beer, wine, cheese, and antibiotics. It is with the conventional wisdom that micro-organisms are germs as a background that the technology of genetic engineering emerged. The public concept that micro-organisms were dangerous and should be contained at all cost was reinforced when scientists at the Asilomar Conference in February 1975 declared a moratorium on recombinant DNA research until such time as a fail safe bacterial strain could be made that could not escape the laboratory. The unknown aspects of recombinant DNA and the potential risks to human health and planet Earth were considered so great that only research on such a strain was considered safe. The view that scientists could and would create an Andromeda strain that would devastate humanity was fostered by the book by Michael Crichton and the widely seen film of that title. The scientific community did little or nothing to counter this image. Rather scientists sought to design and build containment systems like those used to safeguard the public against radiation at nuclear power plants.

When recombinant DNA technology was first developed, many scientists described it as very powerful, conveying an image to the public that biotechnology was analogous to nuclear power. The public understanding of powerful was like that of an atomic bomb. The scientists simply meant that they could do things by using recombinant DNA technology that they otherwise could not. Power meant that genes could be moved from one organism to another across species boundaries with great precision. The idea that genes always move among organisms, after all that is the basis of human sexual reproduction, was not conveyed to the public. The concept of power was oversimplified and misinterpreted. This early setting of the science has had a long-term impact on the public conception of biotechnology. The perceived similarities of biotechnology and nuclear radiation continues as a public focus and influences societal acceptance of environmental applications of biotechnology (see Munson 1995 for a discussion of public concerns).

Concern over the release of genetically engineered micro-organisms continues in part because of a lack of knowledge about the environment, in addition to a lack of knowledge about the genetically engineered organisms. Scientists can also contribute to the fear based upon lack of familiarity by publicly expressing ideas and concepts outside the narrow confines of their own highly specialised interests and research. This was exemplified by quotations attributed to Gwyn Jones, a British microbiologist: micro-organisms may hold the greatest scope for anticipated and unpleasant effects because we know absolutely nothing of their effects on the ecosystem [attributed to Gwyn Jones (Munson 1995)]. Munson differentiates between ecologists, who she feels have a more realistic view of the world and the potential impacts of genetically modified micro-organisms, and laboratory scientists (biotechnologists) who she feels are naive in not recognising the potential doomsday scenarios of releasing genetically engineered micro-organisms. She feels the ecologists view of unknown and therefore presumably dangerous adverse effects outweighs the scientific community view that genetically engineered micro-organisms are not fundamentally different than other genetically selected micro-organisms with regard to their potential for causing environmental harm.

A major feature of the debates in the United States was the progressive development of a well-organised, articulate and balanced response by the scientific community. A leading role was played by the American Society for Microbiology, but many other professional associations of biological and medical sciences joined with ASM in a broad alliance (Cantley 1995). The recommendations of the ASM favoured responsible uniform national oversight of scientists and institutions engaged in recombinant DNA research and that the degree of oversight should be based upon risk. Importantly, the ASM worked with the US Congress in providing objective scientific advice in an open public dialog that allowed the exchange of information and viewpoints between the scientific community and the political community representing the general public.

The Public and Scientific Affairs Board of the ASM has maintained a role of adviser to governments, the public, industry, and academia. While expressing clear and cogent views based on science, the ASM has avoided acting as a lobbying group. As a result of the dialogue between public officials and representatives of the scientific community, the US Congress have been willing to listen and to alter their views in the face of reason and scientific evidence. As a direct result unnecessary legislation was avoided. The scientific community similarly has heard the public concerns about recombinant DNA research and sought to answer those concerns in a responsible manner. There have been no accidents or harm as a result of recombinant DNA related activities.

One of the important activities undertaken by the ASM was to hold a conference at the request of the US Congress on the scientific issues surrounding the potential or deliberate release of genetically engineered micro-organisms (Halvorson 1986). The conference brought together scientists from many disciplines. At times the meeting was raucous, particularly when the discussions ran late into the night, as microbiologists, ecologists and scientists from other disciplines argued about whether genetically engineered micro-organisms were fundamentally different from other micro-organisms and what aspects of the environment might be adversely impacted. A proceedings volume was published by the ASM that conveyed to the scientific community the outcome of the debate and the specific aspects of concern, such as the potential for gene exchange. Many of the papers published in that volume sparked important research projects on the risks associated with the environmental release of genetically engineered micro-organisms. In order to meet the goal of communicating the sense of the meeting to the public and the US Congress, a science journalist was engaged to write a summary of the meeting. The separate lay summary written by Bernard Dixon was distributed to policy makers and science journals. This proved very important in communicating with the US Congress and was a very influential document in formulating US policy about genetically engineered micro-organisms.

The ASM Public and Scientific Affairs Board has continued to examine questions related to oversight of recombinant DNA research and to maintain a dialogue with regulators and public officials. The position that the ASM has brought to public forum is that regulatory oversight should be based upon risk analysis and should focus entirely on the characteristics of the organism (its genotype and phenotype), rather than the process by which that organism evolved or was created. This same position has been adopted by the broader scientific community as put forward by the United States National Academy of Sciences; as expressed by the United States National Academy, "the risks associated with introducing genetically engineered micro-organisms into the environment are the same in kind as those associated with the introduction of unmodified organisms and organisms modified by other genetic techniques".

Nevertheless, regulatory agencies in the United States have continued to focus on biotechnology products, that is genetically engineered micro-organisms. Some oversight groups have greatly altered their perspective as they have gained familiarity with recombinant organisms, gradually focusing on areas where the unknown raised questions as to safety and where there was significant potential for direct impact on humans. This certainly has been true of the recombinant advisory committee (RAC) of the National Institutes of Health which came into being following the Asilomar Conference as a means of co-ordinating federal oversight of recombinant DNA technology so as to ensure public safety. The RAC from its inception included both scientists and non-scientists (lawyers, ethicists, etc.) so that there was always lively debate that included great diversity of opinion. This was critical for the airing of public concerns and forcing the scientific community to deal with those concerns. Over the years the RAC provided guidance to local biosafety committees and brought to the national central debate only those issues of greatest concern. Some recombinant experiments, such as those involving gene transfers between organisms known to naturally exchange genes, were exempt from regulatory oversight based upon the logic that scientists would not bring about any genetic combinations that might not occur naturally. More and more the RAC focused on human gene therapy and left to other agencies, such as the US Environmental Protection Agency, the Food and Drug Administration and the Department of Agriculture, the task of regulatory oversight of genetically engineered micro-organisms, including those that might be released into the environment.

The EPA has struggled to develop an efficient process for the regulatory oversight of deliberate release of genetically engineered micro-organisms which would safeguard the environment and be based upon scientific principles and risk. Despite the advice of several scientific advisory councils, the EPA has yet to achieve approval of a regulatory framework for biotechnology products under the Toxic Substances Control Act. In part the problem is with deciding how to measure environmental quality so as to define what to protect and how to measure environmental impact. The problem of assessing environmental impact due to the release of genetically modified micro-organisms is in stark contrast to the established procedures for assessing the impact of micro-organisms used as vaccines to prevent disease. There is a well established process for the licensing of vaccines that uses accepted protocols for determining adverse side-effects on humans. Risk analyses are performed and the benefits that may be accrued are balanced against the risks. Because only a limited number of physiological responses of the human body need be considered, it has been possible to declare several recombinant micro-organisms safe for full scale direct injection into humans. The same cannot be said for the release of genetically engineered micro-organisms into the environment. The benefits are difficult to determine and the potential impacts are diverse.

There is no consensus as to whether every species should be protected and what balance of nature should be maintained. Scientists should be engaged in this debate which will formulate public policy regarding the future of environmental biotechnology. They must help define the consequences of releasing genetically engineered micro-organisms, for example to clean up toxic wastes, and also the ramifications of not employing recombinant DNA technology to improve environmental quality. They must remain engaged in the communication process both individually and through national organisations, such as the American Society for Microbiology.

Questions, comments, and answers

Q: The production of video materials pre-supposes that someone will show it. Would it be possible for an interactive video to be produced, where the intent of the movie is to educate people to the use of GMOs?

A: Concerning the education of young people, the Microbiological Society has taken the decision to let children first interact with microbes and not be frightened of them and to this end microscopes have been distributed to schools so that children can understand more about microbes before we attempt to teach them about the molecular level. In order to have our film seen we have gone to Hollywood to have it produced and it will be shown on a major network.

REFERENCES

CANTLEY, M. (1995), "The regulation of modern biotechnology: A historical and European perspective", in *Biotechnology*, H.J. Rehm and G. Reed (eds.), pp. 508-681, VCH Weinheim, Germany.

HALVORSON, H.O., D. PRAMER and M. ROGUL (eds.) (1986), *Engineered organisms in the environment: Scientific issues*, American Society for Microbiology, Washington, DC.

MUNSON, A. (1995), "Risk associated with and liability arising from, releases of genetically manipulated organisms into the environment", *Science and Public Policy* 22 (February), pp. 51-63.

OECD INITIATIVES ON INFORMATION TRANSFER TO GOVERNMENTS

by

D.C. Mahon
Commercial Chemicals Evaluation Branch, Toxics Pollution Prevention Directorate,
Environmental Protection Service, Environment Canada, Quebec, Canada

Information is the fuel of governments. Low grade information results in poor performance, occasional stalling and, on occasion, an inability to get up to speed. Governments use different brands of information depending on whether the object is policy development, development of regulations, economic structure or research initiatives, but the quality is still the most critical criterion in achieving the objective.

Where does the information come from? The sources are diverse, ranging from public sources, to industry, to the scientific community, to international inter-governmental fora, and to policy fora. Government must have confidence in the information it uses. There is a need for governments to be certain of the quality of the information on which they are basing their decisions. They must also be assured that the information is similar for other governments with whom they work in the international fora. This brings up the topic at hand, the OECD and the initiatives of the OECD in transferring information to governments.

In discussing OECD initiatives on information transfer to governments, it is critical to know who the target audience is within the governments. Governments are not homogeneous structures. There are, for example, Cabinets, departments, regulatory or promotional units, policy units, scientific or industrial groups, and public response groups. The same information can be generated in a variety of ways that best suit the needs of the target audience.

Several characteristics determine the usefulness of information to governments:

- source;
- quality;
- type;
- intent.

Governments' objectives with respect to the diffusion of biotechnology in the field of bioremediation may be broad in terms of policy, or singular in terms of possible regulation. Depending on the national objective, the scope of the policy may encompass both naturally occurring and genetically modified organisms; it may result in the development of a regulatory framework or guidance; it might address the entire range of activities, or only a final result such as commercialisation.

OECD initiatives on information transfer to governments' bioremediation technologies will be reviewed using these parameters to exemplify the value of the projects.

OECD initiatives in information transfer on bioremediation

The OECD mandate is to assist governments in economic development through co-operation and mutually acceptable activities. Part of this mandate is to ensure that Member governments obtain information that can enable them to reach informed decisions, with the underlying theme that if Member governments have the same information, they will reach the same conclusions and establish similar policies or programmes. Where similar policies are in place, a secondary objective for the OECD is to ensure that governments have the necessary information on which to design implementation strategies for these policies, that will result in equivalent legislation, thus reducing impediments to economic development or trade. In no other area has this been more clearly demonstrated than in the developing field of biotechnology in all its applications.

The reason for this can be seen in a timeline. Modern biotechnology is a scientific and industrial discipline that has progressed at an extremely rapid rate. Consider that the first recombinant organism was produced in 1973, and 20 years later we have hundreds of rDNA organisms close to commercialisation throughout the world. On a parallel track is the rapidly developing awareness of the global problem of pollution production, and the drive for either pollution prevention or remediation. These themes come together in this workshop.

In the OECD there are activities on biotechnology conducted under the auspices of two Committees: the Committee for Science and Technological Policy (CSTP), and the corresponding Directorate for Science, Technology and Industry (DSTI), the Directorate responsible for organising this workshop; and the Environment Policy Committee (EPOC), and the corresponding Environment Directorate. In discussing the theme of information transfer, the activities of both of these bodies will exemplify the role of the OECD in this important activity.

A third venue for information generation within the OECD, and one that may well become more important in this time of shrinking budgets and limited resources for government initiatives, is the use of multi-programme consultations.

In addition to the integral function of the OECD, information transfer to its Member governments, there are several parallel information transfer systems that may achieve increasing importance as the OECD reaches out to non-Member governments and organisations. These systems are interactive webs with such organisations as UNIDO, FAO, UNEP, and IUCN. As the larger international bodies such as the United Nations family of organisations become increasingly involved in economic and scientific activities, these functions will probably increase in importance. Through these mechanisms, the OECD may play a role as a major technical information source for the United Nations.

Committee for Scientific and Technological Policy

As identified above, government activity could be envisaged as a pyramid, with policy development at the apex. From the policies flow a series of requirements that translate into government or departmental actions such as implementations mechanisms, departmental programmes, such as research and development programmes, economic strategies, industrial incentive development, co-operative research, government-industry partnerships, and in some instances, regulatory programmes where necessary. Each has clearly defined requirements in terms of information.

The CSTP make available to governments information that addresses industrial and science policy. In developing science or industrial policy, it is useful for governments to have an appreciation of directions that other governments might be taking in establishing, for instance, science policy regarding developing technologies, and a complementary view of where industry (global) might be moving. It is appropriate, therefore, that the sources of the information comes from beyond the government participant. Similarly, in the science policy field, it is critical for national governments to have information on the views of the international science community, but provided in a format that is useful to non-scientists, and in the context of policy development.

Two examples of information transfer to governments from the CSTP programme, with somewhat different characteristics, exemplify this.

Biotechnology for a Clean Environment

This initiative was an attempt to summarise the state of the science and the potential for industrial applications of biotechnology to environmental protection and remediation. The document was developed through a series of consultations with government experts, industry experts, scientists, and economists. It is by definition a sweeping overview of the potential of the technology in a critical area, with specific recommendations on research and development policy.

The OECD publication, *Biotechnology for a Clean Environment: Prevention, Detection, Remediation* (OECD, 1994) is a clear example of information generated for the express purpose of informing governments of the current and projected applications of a technology, with a broad economic base but with possible gaps in the science base. The objective is to enable government to consider the need for developing or reviewing industrial policy with respect to emerging technologies. The report addresses many of the issues germane to the development of a national policy.

OECD CSTP workshops on specific topics

This is a series of projects under a single initiative, to provide information on the state of the art with respect to bioremediation applications. The first workshop, Tokyo 1994, established a broad overview of bioremediation, and addressed such issues as efficacy measurement, criteria for reliability, and the potential role of biotechnology in pollution prevention through such applications as the development of renewable materials and in-process treatments. Amsterdam is the second workshop in the series and is directed to air and soil remediation. The third in the series is planned for 1996 in Mexico, and addresses water pollution remediation.

By the criteria, the workshop fulfils the necessary objectives for providing governments with useful information.

OECD Programme:	CSTP
Initiative:	Tokyo workshop
Source:	Government experts/industry/academics
Target:	Government research policy
Intent:	Identify areas of research/co-operation
Result:	Proceedings of workshop Conclusions and recommendations

The 1994 Tokyo workshop on bioremediation as applied to air and soil, and the proposed Mexico workshop specifically addressing bioremediation applied to water systems are examples of expert groups providing state of the art information on science as it apples to bioremediation. These are not peer-reviewed science articles, but are individual scientist or industrialist specialists providing a perspective on what technology is available, possible applications of the technology, or gaps in science or policy that could result in impediments to technology diffusion. The information, both to participants and, in the resulting proceedings are available to governments within the OECD, and indeed to other interested governments. It provides a broad scale overview of directions in research and industrial application and often a balanced view of developing concerns.

Information generated at such venues as the Tokyo and Amsterdam workshops are of special interest to two groups of people in government, the industrial research specialist and the policy maker. It is through such fora that government can obtain a view of the direction in which industry and science are moving, for example, the degree to which the use of genetically modified versus naturally occurring micro-organisms are perceived as likely candidates for technology. Tokyo 1994 clearly identified an area where there was significant concern for both the industry, and potentially for government, the transferability of the technology from site to site. Directly related to this concern, a need to develop criteria for efficacy was also identified. In many Member countries efficacy is not a government concern, yet it clearly relates to the diffusion of the technology.

In developing policy, especially where this relates to international agreements or objectives such as mutual acceptance of data, or multilateral research agreements, there is a need for governments to have clear ideas of where the information comes from, what the credibility of the information is, and the acceptability of the information. Any international organisation has, by definition, a broader base of knowledge than a national source. This is not to imply that the international information will always form the basis of decisions, but that the information may provide for a more informed decision.

Similarly, the OECD publication, *Bioremediation: the Tokyo '94 Workshop* (OECD, 1995*a*), and the proceedings of this workshop will form a significant information resource for governments taxed with developing policies with respect to the use of the biological technologies in the environmental field.

Environment Policy Committee

A different activity is where representatives of governments, either civil servants or scientists meet to generate consensus positions or documents that are accepted for consideration by the governments of the OECD Member countries. These documents are at a higher level of international accountability since they are generated by government representatives and represent the position of the government of Member countries. For example, in both the CSTP programme on Safety Considerations for Biotechnology, and the Environmental Policy Committee work on Harmonization of Regulatory Oversight, consensus documents have been developed. This process results in information documents in which both the information and, where applicable, the recommendations to governments or conclusions are agreed to by the Member countries. These documents must be approved by higher level bodies within the OECD.

The Environment Directorate biotechnology programme is geared to harmonization of regulatory oversight. This programme provides a venue for government representatives to formulate approaches to regulatory harmonization. As such it impacts on both national and international activities. The thrust of the information generated in this venue is clearly different but complementary to that of the Directorate for Science and Technology. In the overall time scale, when policy has been developed and translated into legislation or regulation, it is immensely beneficial to all parties if Member countries know that their national regulatory/oversight structure reflects consensus on the issues of concern, and where possible, that the approach taken for resolution is the same.

As an example, the Ad Hoc Group on Harmonization of Regulatory Oversight has a similar series of workshops to the CSTP series, but addresses regulatory harmonization. The first such workshop was held in Brussels in 1993, and the second in Friburg (Switzerland) in 1994, at which government experts reviewed and evaluated the information that would be needed for an evaluation of organisms resulting from modern biotechnology, using as a targeted subset for evaluation, those organisms likely to be used in environmental applications such as bioremediation. The report of this workshop was a consensus document (OECD, 1995*b*) in which it was clearly identified that Member countries do require essentially the same information in performing an assessment. The successor group to the Ad Hoc Group, the Group of Experts on Regulatory Harmonization, is continuing this activity by further analysis of the information needs to clarify the depth of information required and the methods for generating the information.

A review of the information generated in Brussels, and contained in Environment Monograph 100, similar to that for the CSTP information generating activity reveals certain differences.

OECD Programme:	EPOC
Initiative:	Brussels workshop
Source:	Government regulatory experts
Target:	Government regulators
Intent:	Identify areas of commonality
Result:	Proceedings of workshop / Conclusions

The source of the information in Environmental Monograph 100 is government experts dealing with regulations, it is singular and directed to governments or departments which have regulatory authority for the substance. The target is quite clear, as is the intent. The consensus information is agreed to by all participants and therefore is directly useful to governments both in reviewing their own regulatory requirements and in considering national requirements, in the light of international activities.

Another example of a different type of information provided through the OECD to governments is the case of information-only documents, such as the documents currently under development in the EPOC programme in which a synopsis of available information on the basic biology of organisms known to be candidates for regulatory oversight is generated. The purpose of the document is for use in a regulatory setting. The value lies in several characteristics:

1. All Member governments have agreed to information documents that contain the same information.

2. The information has been approved by national and international experts to be accurate and appropriate.

3. Governments can act on this information knowing that the information is acceptable to other national agencies.

CSTP-EPOC: Joint consultation

A final example of information transfer to governments with a different intent was the recent Joint Consultation Meeting on Environmental Policy and Bioremediation/Bioprevention Technologies, held at the OECD in September, 1995. This was a consultation at which experts from the scientific side, the regulatory side and environmental policy/pollution prevention policy were asked to consider several questions relative to the diffusion of biological technologies in bioremediation, and bioprevention from a policy perspective. The thrust of the questions related to the impact policies, and in particular, the impact environment policies might have on the diffusion of bioremediation technologies.

The purpose of the consultation was to bring together government and industry representatives in science and policy development for an exchange of information and ideas about the diffusion of biotechnology.

JOINT CONSULTATION

Are there policies that affect diffusion of bioremediation technologies?

Do bioremediation technologies deserve special attention?

OECD Programme:	JOINT CSTP/EPOC
Initiative:	Joint Consultation on Biotechnology Pollution/Prevention
Source:	Government policy developers in pollution prevention; scientists; regulatory experts
Target:	Government policy developers
Intent:	Identify possible policy impediments to diffusion
Result:	Proceedings of Consultation / Conclusions

Conclusion

How do governments acquire and use the information generated in the various fora of the OECD? This depends in part on the scope of the information and the structure of the government as it relates to the topic under discussion. In Canada, policy is always the domain of the Government in Cabinet, i.e. the Ministers. A single Minister may have responsibility for an area, such as environmental protection, but government policy is established in Cabinet. Below this level, there may be Departmental policy, where the Minister has discretion on how to achieve a specific aim, for instance in establishing the efficacy or use of a specific application of a technology such as bioremediation. At an even lower level, scientists in a department will be responsible for providing information or advice to the Minister on particular aspects of a policy that may be implemented through regulation.

Each level requires a different mix of information, much of which can be generated from the OECD initiatives.

The OECD has several initiatives with the intent of providing information to governments on biotechnology and environmental applications. The initiatives are designed to meet the needs of specific areas of government from policy setting to specific scientific or regulatory concerns. The source, participants, in each activity reflect the intent of the project, from international consensus to state of the art snapshots of industry or science to recommendations directly on policy.

CONCLUSION

Criteria for information transfer to governments

Quality:	International science/government consensus
Source:	Clearly defined and preferably independent
Target:	What level of government
Intent:	Policy, state of art, specific application

Considering the general theme of information transfer to government in a manner that facilitates policy development, co-operation and harmonization, and the three-fold OECD initiatives of science and industrial policy, regulatory harmonization and appropriate information transfer to the effective level in government, it would appear that the system fulfils the needs.

REFERENCES

OECD (1994), *Biotechnology for a Clean Environment: Prevention, Detection and Remediation*, OECD, Paris.

OECD (1995*a*), *Bioremediation: the Tokyo '94 Workshop*, OECD, Paris.

OECD (1995*b*), *Analysis of Information Elements Used in the Assessment of Certain Products of Modern Biotechnology*, OECD Environment Monograph #100, OECD, Paris.

Parallel Sessions/B1 and B2

B.2 STANDARDISATION AND BEST PRACTICE: SOIL & AIR/OFF-GAS

STRATEGIES TOWARDS THE ABATEMENT OF AGRICULTURAL EMISSIONS TO AIR[1]

by

W. Day, V.R. Phillips, T.R. Cumby, C.H. Burton and A.G. Williams
Silsoe Research Institute, Silsoe, Bedford, United Kingdom

Introduction

Farming and the processing of farm products have been an essential part of human activity since pre-history and, as such, might be seen as part of the natural environment. However, the industrial revolution and worldwide population increase have brought pressures on farming systems to increase their production efficiency. The focus on productivity has moved farming away from being the management of the natural environment towards an industrial process, seeking optimal process performance. As the world now looks at its ability to manage the international and global environment, agriculture also comes under the spotlight, to ensure that the optimal process configurations being identified meet environmental constraints too.

Increased productivity has been achieved by the introduction of extra inputs or closer control over their application. Monoculture, mineral fertilisers and crop protection chemicals have obviously disturbed the balance of a range of natural processes, but predominantly in ways that have been seen as sustainable in the restricted context of the production cycle. Increasingly however, recognition of the impact that some of these disturbances can have on the wider environment, and on the long-term sustainability of farming systems as part of the global ecosystem, requires agriculture to address the question of acceptable optimisation of its production systems afresh.

An increase in the emissions of gases to the atmosphere has been one result of the intensification of crop and animal production. Some gases have been recognised as a consequence of farming systems for a long time – particularly those associated with the smells of housed livestock production. Among these, hydrogen sulphide has been known to have implications for human and animal safety in confined spaces. High concentrations of carbon dioxide that can accumulate in crop storage systems can also lead to safety problems. However, it is not these gases that are the prime concern of the international community at present. Concern focuses on two groups of emitted gases, those that can lead to acid rain and smog, and those that contribute to global warming. From the first group, ammonia is released from soils and from animal excreta, and there are also low level emissions of some of the reactive oxides of nitrogen. From the second group, carbon dioxide is, of course, absorbed by growing plants, so productive agricultural systems are important to efforts to limit the rise in carbon dioxide levels that result from the burning of fossil fuels. Methane, however, is emitted from both wetland farming systems, particularly rice paddies, and from anaerobic breakdown, of feed

[1] We acknowledge valuable discussions with colleagues, including Dr. Ian Davidson of MAFF, and the extensive editorial input of Mrs. Elizabeth Field.

during digestion by ruminant animals, and of animal excreta. Although soils can be a natural sink for methane, the net balance from agriculture is a considerable emission. Nitrous oxide is also a product of partial denitrification in soils and nitrogenous wastes.

This paper will lay out some of the key issues that need to be addressed in diminishing the gaseous emissions from agricultural systems. The diversity of agriculture is so great that it would not be sensible to catalogue the whole problem. Instead we will follow aspects of the problem through the use of some prime examples. Through the definition of the problems in these areas, the strategic questions and the scope for technical solutions will hopefully become apparent. Though the examples will predominantly relate to systems in the United Kingdom and Western Europe, the principles apply to much of the developed world.

Gaseous emissions from agriculture

The first step in defining the best approaches to abatement of emissions is to establish in detail the major contributions. This has been the focus of much national and international activity in recent years (van Ham *et al.*, 1994). Current studies in the United Kingdom, led by the Institute of Grassland and Environmental Research for ammonia (Pain *et al.,* 1995) and by Silsoe Research Institute, are compiling UK emission inventories of methane and other gases from animal agriculture. These complement other European and international ventures such as the UNICEF Workshop held at Culham Laboratory in the United Kingdom from 31 October – 2 November 1994 entitled "The potential for abatement of ammonia emissions from agriculture and the associated costs", including papers from eleven European countries, and the report of the IPCC Working Group 11, *Agricultural options for instigation of greenhouse gas emissions,* which will form Chapter 23 of the IPCC 95 Assessment, to be published early in 1996. To give this paper a context, Table 1 gives an indication of the contribution that agriculture makes to the UK emissions of methane and nitrous oxide. The figures for ammonia are less certain but agricultural emissions may represent 90 per cent of total UK emissions (Pain *et al.*, 1995).

Table 1. Indication of the contribution agriculture makes to the UK emissions of methane and nitrous oxide

Source	% contribution (1990)	
	Methane	Nitrous oxide
Nylon industry *		73
Agriculture	31	17
Landfill	40	
Fuel industries **	25	
Vehicles		7
Fuel combustion		3
Sewage	2	
End use	2	
TOTAL	100	100

Notes:

1. * Reducing rapidly following action by manufacturers, so that the percentage from agriculture may now be about 40 per cent to 50 per cent; ** Includes coal mining 15 per cent and gas distribution 8 per cent.

Sources: HM Government (1994), Williams (1994).

An agricultural problem: emissions from animal production

As animal production systems are major contributors to the gaseous emission story, they will be used as the theme on which questions of abatement will be developed. In essence, the animal production cycle involves the feeding of animals with crop products, the handling, storage and disposal of the excreta, and the disposal of that excreta on cropped land. The nitrogen in this cycle can be an important crop nutrient, though the term "animal wastes" that will be used frequently in this paper implies that these materials have not been greatly valued in intensive production systems.

Emissions from animal housing

Intensively farmed animals are housed in buildings with various systems for collecting and storing excreta. When the excreta are collected as slurry, the surfaces contaminated with wet waste can be significant sources of ammonia, and to a lesser extent nitrous oxide. Regular cleaning can be achieved in a slurry-based system by frequent flushing and/or scraping of surfaces. However, the quantities of waste material will increase with the application of water for regular washing, and this poses problems at the next stage of waste storage. Straw-based systems may lead to greater nitrous oxide losses because conditions are more aerobic. The major source of methane production at this stage in the cycle is ruminant animals themselves, in whose rumens the action of fermentative organisms causes the production of methane in considerable quantities.

Emissions during waste storage

It is increasingly necessary to store animal waste for a considerable period before it can be applied to the land. This is primarily because of the risk of water pollution if the wastes are spread onto land in the winter. At this time crop demand for nutrients is limited and rainfall is generally greater than the evaporative demand of the crop so that there is a considerable flux of water and solutes through the soil, leaching out nitrates and other chemicals. In storage, the production and emission of gases depends on the physical and chemical conditions in which the waste is kept. Enclosing the wastes, for example in a covered silo or tank, can substantially reduce the rate of loss of ammonia gas to the atmosphere (van der Kamp *et al.*, 1993), although this can also result in changes to the chemical state of the waste. The conditions in slurry stored in a tank are likely to be strongly anaerobic, leading to significant methane generation. Covering the tank may limit the rate of exchange of gas in the headspace with the atmosphere, but methane is only sparingly soluble and the anaerobic conditions will mean that methane production is maintained or enhanced. Headspace methane concentration will rise, so emissions are likely to be maintained unless the methane can be utilised as biogas.

Application to land

Although incineration and composting are options for waste disposal, most animal waste is applied to agricultural land. The land spreading can be adjusted to modify the gaseous loss processes. Surface spreading enhances the exchange process with the atmosphere and can therefore lead to rapid losses of any ammonia that is in the waste. However, the conditions in the waste are likely to be aerobic, so there will be minimal production of methane. The alternative is to incorporate the waste into the soil. For solid wastes this is only feasible if the land can be ploughed - i.e., incorporation is only possible for fallow land. For slurries, it is possible to inject the waste into the soil of a growing crop. This can be done at various depths; with greater machinery and energy requirements the greater the depth of injection. There is also some damage to the crop that will lead to a check to its growth. With less exposure to the atmosphere, the rate of loss of ammonia from injected slurry will be lower than from surface application (van der Kamp *et al.*, 1993). However, conditions in the waste are likely to be anaerobic, so that methane production may continue giving greater overall losses. Ammonia losses may also continue over a longer period if there are not alternative sinks.

The fate of nutrients in soil

The final stage in the cycle is the further breakdown of the organic matter in the waste when it is in the soil, and the uptake of some of the components, particularly nitrate and ammonium, by the growing crop. The ability of a crop to take up these nutrients is strongly dependent on the rate of growth of the crop, which determines the demand for nitrogenous compounds within the plant. There is growing evidence that the presence of active sinks for nutrients has a major effect on further breakdown and consequent gaseous loss that occurs from the soil.

If animal manures are not used as a source of nutrients for plant growth, artificial fertilisers must be, and they are used extensively in addition to manures in high productivity systems. Gaseous emissions from fertiliser use can be significant, and urea is particularly implicated as a source of nitrous oxide and ammonia losses.

Abatement: strategic approaches and problems

In establishing the most appropriate strategies for abating gaseous emissions from agriculture, it is necessary to consider the context in which agriculture operates. The aim is to harness natural biological production processes to produce food (and increasingly industrial raw materials) at low cost. The harnessing of biology is not a precise engineering process, in part because it operates in an unsteady, heterogeneous and often unpredictable environment. Soils have varying physical, chemical and biological properties, and the weather varies unpredictably in time and often quite markedly in space, so that the conditions experienced by a crop in one region can be very different from those just a few hundred miles away. It is therefore important that the principles behind the emissions processes are understood, so that decisions on abatement measures can be tested against the real context in which they will be applied.

The goal for an abatement strategy may be quite simple to formulate – to reduce the emissions of a particular pollutant by a certain percentage or return them to the levels of a previous time. Unfortunately, many of the pollutants that flow from agricultural production systems cannot be considered in isolation. Both ammonia and nitrous oxide are products of the nitrogen cycle. Methane is formed during anaerobic processes in organic biomass, but aerobic conditions, which will reduce the likelihood of methane production, usually lead to enhanced production of nitrous oxide. Highly aerobic conditions may favour the conversion of nitrogen into nitrates, but nitrates can move rapidly through the soil and hence are likely to contaminate water courses. Therefore, it is essential to look at the total systems in which these gaseous pollutants are contributing part of the pollutant burden, and define strategies that have a net benefit for the overall environment.

One aspect of a systems approach is to consider the mass balance of inputs, i.e. where do the extra agricultural inputs of nitrogen, etc., come from and where do they go? For example, taking the animal production enterprise as the system, inputs may be bought in as feed, or fertiliser if feed crops are grown. Some commodities will leave the farm as useful products (animals, cereals, possibly compost, etc.). The difference between the inputs and outputs of each element represents an accumulation (short-term) and ultimately a loss leading to pollution. Any strategy that is to reduce the pollution problem must enable the reduction of the imbalance, e.g. more efficient use of nutrients. This mass balance concept is increasingly being applied (e.g. in the Netherlands) to highlight the extent of the overall problem, particularly to the farmer, and then seek remedies.

In addition to considering the production cycle, it is important to consider the whole life cycle of products and pollutants when weighing up the strategies for abatement. Life cycle analysis methodologies provide help with this. They seek to establish a complete inventory of the impacts and outputs of all the component processes involved in producing a specific product. Figure 1 shows an outline of a typical agricultural system on which such an inventory might be based. The method allows different options to be compared, knowing the integrated effect of all these processes per unit of product. In relation to agricultural systems, this raises some important issues. Waste products from one system can be valuable inputs to another; thus manure as a fertiliser can substitute for chemical fertilisers and reduce emissions associated with chemical fertiliser production. Extensive systems produce less per unit land area. Their emissions must be assessed per unit of production, not per unit of land area, if we are to be sure that the net effect has been to reduce the pollutant burden. Intensive production systems, on the other hand, may lead to land being left idle (set aside), and then the role of that land as a source or sink of pollutants will need to be considered.

Figure 1. Outline of an agricultural system

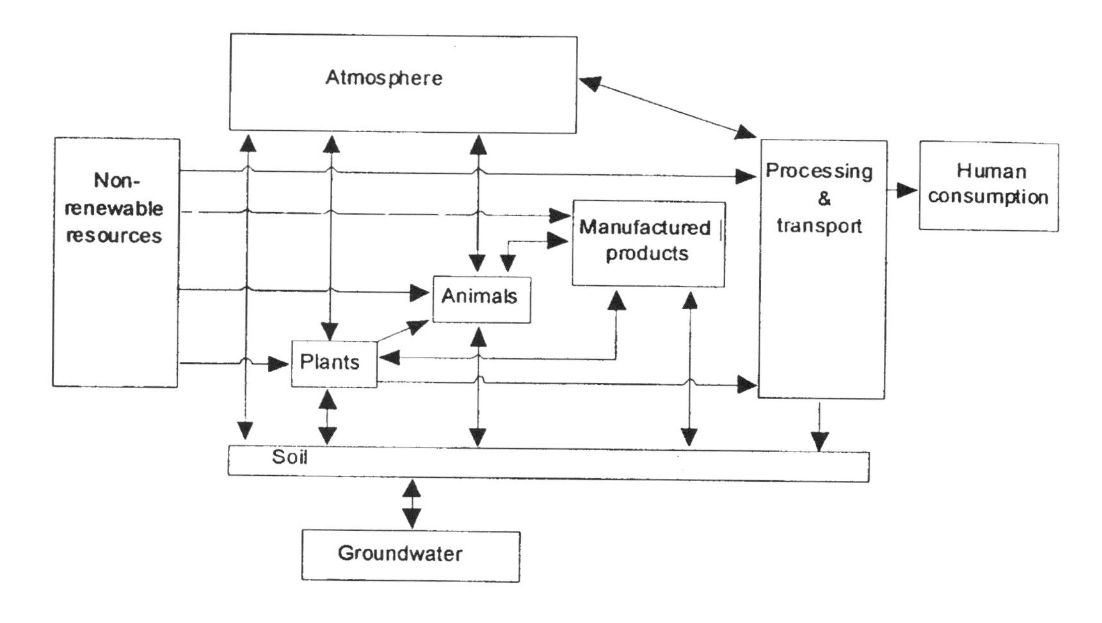

Source: Author.

Farming systems in Europe have developed in varied ways, even where the climate is similar. Thus in the Netherlands, Belgium and Denmark, and in some specific regions of other countries, very high densities of animal production have been established, often relying on the import of cheap feed materials from overseas to sustain the competitiveness of the system. As a result, there is not enough land available for disposal of the nutrient-rich waste materials. This demonstrates how necessary it is to consider broader systems than just the individual farm, or even the agricultural practices of a single country, if the problems of emissions from wastes are to be overcome in a sustainable way.

Lastly, in defining a strategy, it is important to bear in mind the economic context of the industry. There is no point in imposing abatement measures that make a production process unprofitable if, in consequence, the same commodity is then imported from a country in which there are no measures in place to minimise these emissions. It certainly does not do the global climate any good to transfer gaseous emissions from one part of the globe to another. Therefore, the strategy to be adopted must be internationally sustainable.

Technical solutions

The extensive and diverse nature of agricultural systems means that technical solutions are likely to be applicable only to a subset of operations. However, the principles can be considered through examples of treatments and techniques to reduce gaseous emissions that could be applied at various stages of the cycle of animal waste production, storage and disposal, as illustrated in Figure 2.

Figure 2. Emissions from housed livestock - some technical solutions

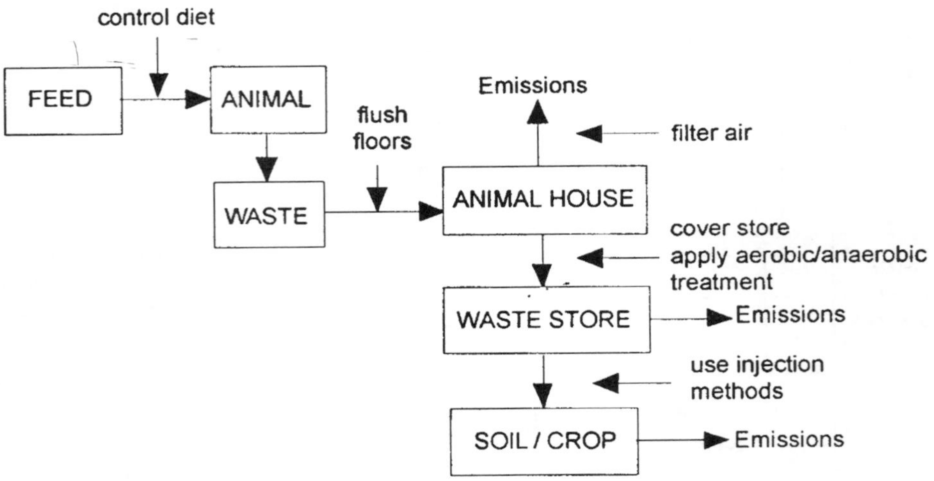

control diet
FEED
ANIMAL
WASTE
flush floors
ANIMAL HOUSE
Emissions
filter air
cover store
apply aerobic/anaerobic treatment
WASTE STORE
Emissions
use injection methods
SOIL / CROP
Emissions

Source: Author.

Solutions applicable at disposal

For most processes that give rise to gaseous emissions, end-of-pipe solutions must be considered, as they may offer immediate benefit with least disruption to the production process itself. However, for agriculture there is usually no obvious pipe end at which to apply a solution. For example, when wastes have been landspread, the emission source is so closely coupled with the atmosphere and so broad in extent, that it is obviously impossible to apply a scrubbing or similar abatement measure.

The use of injection techniques or soil incorporation is a close parallel to an end-of-pipe process in that the biological activity in the soil is being used as a filter on the emissions. In a number of European countries, injection has been made mandatory for the disposal of slurries. However, some caution is needed before this is seen as an effective solution everywhere. Firstly, the benefit from reductions in ammonia loss must be set against probable increases in methane losses. In wetter areas especially, surface-spread wastes may get rapidly washed into the soil, preventing excessive ammonia losses, whilst injected wastes will emit more methane in the anaerobic conditions. Secondly, injection techniques are not appropriate in areas where soil conditions, e.g. stonyness, make them impractical. Thirdly, the expected benefits may disappear under specific conditions: in cracking soils, injection followed by a dry period may mean that subsequent rainfall can cause nutrients to be rapidly flushed through into watercourses.

Waste treatment options before disposal

It is possible to modify the condition of the waste before it is landspread, for example by aerobic, anaerobic or acidification treatments. Aerobic and anaerobic treatment of slurries can significantly modify the condition of the waste, and hence influence the likely emissions during subsequent storage and disposal.

Anaerobic treatment can be a means of obtaining valuable energy from waste in the form of methane, provided the methane can be successfully contained and effectively used, and it has been extensively adopted in Denmark. It is not so attractive at the individual farm level, because a substantial throughput of wastes of dependable and high biomass content is required to justify the capital expenditure, and because some technical sophistication is necessary to manage the plant in good operational order 365 days in the year. It is also important to handle the digested waste carefully, as the presence of methanogenic bacteria can lead to continuing low levels of methane production, even when the waste has completed its useful digestion in the biogas plant itself. Thus the method will be attractive where a central processing approach to waste handling is acceptable, for example, where there is a large concentration of waste production in a region, and infrastructure can support the transport implications. It is also likely that access to non-agricultural sources of wastes will be necessary to maintain the efficient performance of the plant all the year round.

Aerobic treatment can be used to modify the condition of the nitrogen in wastes. By appropriate choice of the treatment regime, it is possible to transform much of the nitrogen into microbial biomass (short treatment time with biomass recovery and recycle of soluble pollutants), or to remove much of the nitrogen as nitrogen gas (long treatment times incorporating sequential aerated and anoxic periods). The former approach puts the nitrogen in a form that should make it available for crop uptake over a prolonged period, while the latter would be a possible option for farming systems in which there was too much nitrogenous waste for the land area available. In these circumstances, decreasing the nitrogen content of the waste could be seen as a pragmatic solution, although it would have to be weighed against the need to supply the nitrogen instead as mineral fertiliser. Aerobic treatment involves cost for the energy to aerate, with none of the benefits of energy production associated with anaerobic treatment. It is likely that its role will be limited to circumstances where other demands, like reducing the odour content of the waste and/or managing the conversion of ammoniacal nitrogen to other forms, make it a useful technique . It also has a role in some situations to reduce the risks of water pollution associated with dilute organic wastes by destroying BOD.

Acidification of wastes has been used to reduce ammonia emissions. Lowering the pH of a waste slurry shifts the dissociation of dissolved ammonia further towards the ionic ammonium form, thus decreasing the source strength of ammonia for volatilisation. As an abatement technique, its weakness is probably the increased nitrogen content of the waste, arising from the use of nitric acid for the acidification, and it may also lead to some subsequent nitrous oxide emission by denitrification of the nitrate (Wouters and Verboon, 1993). The safety of using concentrated acids must also be considered.

Abating emissions from stores and buildings

Covering waste storage systems has already been mentioned as a way of reducing ammonia losses. However, with this and other techniques, the changes made to emissions in this phase of the process may have a deleterious effect on the later phases. For example, if the process of covering leads to an increase in the pH of the waste, or to higher temperatures which accelerate the production of ammonia by the hydrolysis of urea and other organic nitrogen compounds, then losses from the land-spread waste could be greater. Store management is also important. In some countries it is common practice to mix stored slurries regularly in order to ensure ease of handling and land spreading when the store is emptied. However, the mixing process can substantially increase production and emission of gases.

The animal house itself is potentially a major source of both ammonia and methane. A scrubbing technique may be used to reduce emissions from housed livestock systems, particularly where the ventilation of the building is under mechanical control. The potential for biofiltration to treat the air from such buildings has been confirmed by Scholtens and Demmers (1991) and Pearson *et al.*, (1992). The key factors in the acceptance of this approach will be physical compatibility with the ventilation system, availability of cheap substrates for the biofilter, the ability to readily dispose of used substrate, and the total cost of the exercise to the farmer. To be able to interface biofilters with ventilation systems, the pressure resistance of the medium must be low. Phillips *et al.*, (1995) have suggested that wood chips may be appropriate, and with the increased interest in wood as a biomass fuel, they may both be readily available and disposable after use. However, the total cost of a biofilter system is in the order of £0.20 per bird produced in a broiler farm, which is similar to the profit of such enterprises. Uptake will therefore be limited, and only occur in countries where there is a legal requirement to limit emissions. The method would be more attractive if it were possible to show a direct economic benefit in production terms. This would be achieved if, instead of removing the ammonia from the exhaust air, it was removed within the animal house itself. There is evidence that the animals do not perform well in high concentrations of ammonia, and this may lead to direct farmer interest in ammonia abatement from such production systems (Hamilton *et al.*, (1993).

The option of biofiltration is not available in systems that rely upon natural ventilation to maintain an acceptable environment for the animals. Losses of ammonia from the wastes deposited in the building can be reduced by regular and efficient cleaning, for example in slurry systems by flushing the floors and gutters with liquid. The usefulness of this approach depends on the acceptability of some features that would optimise its effectiveness. Firstly, it has been shown that flushing works best on smooth concrete surfaces rather than rough ones. However, wet smooth surfaces are entirely unacceptable from the point of view of animal welfare. Slatted floor systems themselves have some welfare problems. Reducing the exercise area per animal can also contribute to reduced emissions in a similar way, by decreasing the area of emitting surfaces, but this approach may also be unacceptable for animal welfare reasons. Secondly, a regular flushing system requires the availability of appropriate water. It may be possible to use separated slurry, though this may limit the effectiveness of the cleaning process and may cause odour problems in the building (aerobic treatment of the flushing water may prevent this). Fresh water is unlikely to be an acceptable option because it will lead to further increase in the quantities of slurry for handling, storage and disposal.

Solid wastes must be treated differently. As an example, drying poultry waste as part of the housing system does lead to reductions in the emission from the house. Whether it reduces the emissions from the total cycle depends on the conditions and timing when it is land-spread. The combustion of broiler and turkey litter for energy production is already operational in some power stations in the United Kingdom. The emissions from combustion of wastes must of course be evaluated, though stack monitoring should be more straightforward than integrating emissions after field spreading.

Reducing emissions at source

Having examined solutions applicable to the end-of-pipe problem and then moved upstream, we are left with the source. There is an association between the nitrogen content of wastes and the diet of the animals, and modification to the diet may be a realistic option for the future. Protein is an important dietary requirement for rapid growth and meat quality, but the balance in amino acids that the animal really requires to satisfy its needs is not precisely met by any feed. Methods of modifying and tailoring diets to meet the true requirements of the animal are now showing that high productivity can be achieved with more efficient protein utilisation by the animals and consequently lower levels of nitrogen output in the wastes.

This is one example of how achieving precision in inputs may be a possible method of reducing losses to the environment. To complete the cycle, precision should also be part of the approach used out in the field for crop production. As pointed out earlier, the use of the term wastes reflects on the limited value ascribed to these materials. Yet part of the environmental problem is associated with the fact that they are rich in nitrogenous compounds, phosphorus, potassium and organic matter. In crop production, these chemical entities are seen as valuable contributors to productivity. What is the reason for this difference in perception? In part it reflects the uncertainty associated with the nutrient value of organic manures for crops. The nutrient content of animal manures can be very variable. It depends upon the animal rations, the waste handling methods (which can affect the degree of dilution of slurries), and the treatment and storage regimes. The difficulty in applying these materials uniformly, and the uncertainty about whether the required nutrients will be available to the crop, limit the confidence that the farmer will have in relying on animal manures.

Conclusions

There is much scientific information available about the extent and mechanisms of gaseous emissions from agricultural systems, exemplified here by this brief survey of some of the emissions from livestock production. There are a number of emission abatement strategies that might help to solve these problems. However, there is also a variety of constraints on the implementation of these strategies, and these are worth reviewing.

Agriculture is an industry in which the profit margins on the production process are tight, with ever increasing pressure from international competition and requirements to minimise subsidies or other safety nets. The changing position on regulatory constraints makes the industry wary of taking on board non-statutory modifications to its production processes that do not have a direct and proven benefit for profitability. The uptake of some of the technical measures will therefore depend on low cost or cost neutrality through parallel contribution to farm profitability, or on regulation.

Better utilisation of the nutrients in animal slurries and solid manures has the potential to replace some inorganic fertilisers, reducing costs and contributing to emission abatement. To achieve this, attention must be paid to the agronomic and technical constraints on maximising the value from the nutrients. Improved awareness of the need to manage livestock wastes more carefully will provide an incentive to adopt measures that reduce nutrient losses and emissions. Links between biological understanding and advances in the technologies for information gathering (e.g. sensing the nutrient content of manures) and decision support for the farmer (e.g. through IT methods) will be important here.

It is certainly possible that advances in the understanding of microbiological processes will give better ways of managing the microbial systems that are involved in gaseous emissions. Even now, engineering methods for managing aspects of the formation, emission or absorption process are available and, if they can be made cost effective, they can contribute to abatement.

At present it seems that the best way of reducing emissions in the short term will be the wider use of low emission application technologies. Practical evaluation of options in this area may provide a good vehicle for further advances based on concepts of better nutrient utilisation.

However, crucial to the delivery of these abatement techniques is the development of a confident and sustainable argument about the whole raft of pollution issues with which agriculture is concerned. This will call for systems studies that relate, as best possible given the variability of biological processes and the weather, to these various environmental impacts and which assess the overall benefit. Integration of these approaches with practical demonstration of effectiveness will give confidence that measures are worthwhile.

Questions, comments, and answers

Q: How important is human waste in comparison with animal waste since we spread some domestic sewage on land?

A: In terms of nitrogen cycle, it is a contributor but agricultural production systems produce much more nitrogenous material. If we look at land spread with animal waste then this dominates over sewage sludge at present.

Q: Is there any information on avoiding contamination of water and streams with pathogenic micro-organisms during spreading and is there any regulated system to check this in the United Kingdom?

A: Farming systems would say that pathogens are not their problem and, indeed, human waste is implicated. Avoidance of these problems will come from ascribing value to these materials - people will take more care. Drinking water regulations are usually applied.

Q: One has to be sure that decreasing emissions on one side does not increase emissions into another compartment. Can you compare routes to molecular nitrogen?

A: Looking at management, in order to turn wastes into nutrients we need to identify processes for optimum utilisation. More rapid utilisation by cropping systems decreases the pools that can be converted into N_2O or NH_3. The question is how to give the end-user confidence in the value of the product, bearing in mind the variation in nutrient quality.

REFERENCES

H. M. GOVERNMENT (1994), "Climate change – the UK programme", *United Kingdom's report under the Framework Convention on Climate Change,* Cmnd 2427, HMSO, London.

HAMILTON, T.D.C., J. ROE, F.G.R. TAYLOR, G. PEARSON and A. WEBSTER (1993), "Aerial pollution and exacerbating factors in 'atrophic' rhinitis of pigs", in *Proceedings of Fourth International Livestock Environment Symposium*, pp. 895-903, American Society of Agricultural Engineers, Warwick.

PAIN, B.F., T.J. VAN DER WEERDEN, B.J. CHAMBERS, V.R. PHILLIPS and S.C. JARVIS (1995), "A new inventory for ammonia emissions from UK agriculture", *Atmospheric Environment*, (in press).

PEARSON, C.C., V.R. PHILLIPS, G. GREEN and I.M. SCOTFORD (1992), "A minimum cost biofilter for reducing aerial emissions from a broiler chicken house", in *Biotechniques for air pollution abatement and odour control policies*, pp. 245-254, A.J. Dragt and J. van Ham (Eds.), Elsevier.

PHILLIPS V.R., I.M. SCOTFORD, R.P. WHITE. and R.L. HARTSHORN (1995), "Minimum cost biofilters for reducing odours and other emissions from livestock buildings: Part I Basic airflow aspects", *Journal of Agricultural Engineering Research* 62, pp. 203-214.

SCHOLTENS, R. and T.G.M. DEMMERS (1991), "Biofilters and air scrubbers in the Netherlands", in *Odour and ammonia emissions from livestock farming*, pp. 92-97, V.C. Nielsen, J.H. Voorburg and P. L'Hermite (Eds.), Elsevier.

VAN DER KAMP, A., P.P.H. KANT and A.J.H. VAN LENT (1993), "Farm economic and environmental consequences of low-emission farm systems on dairy farms", *Rapport 149*, Proefstation voor de Rundveehouderig, Schapenhouderig en Paardenhonderig (PR), Lelystaad, Netherlands.

VAN HAM, J., L.J.H. JANSSEN, R.J. SWART (Eds.) (1994), "Non-CO_2 greenhouse gases: why and how to control?", Proceedings of an International Symposium, Maastricht, The Netherlands, 13-15 December 1993.

WILLIAMS, A. (Ed.) (1994), "Methane emissions. Report of a Working Group appointed by the Watt Committee on Energy", *Watt Committee Report* No. 28, Watt Committee on Energy, London.

WOUTERS, A.P. and M.C. VERBOON (1993), "Handling of slurry in relation to the environment on dairy farms", in *Forward with grass into Europe*, British Grassland Society Occasional Symposium 27, pp. 85-96.

BEST PRACTICE IN SOIL BIOREMEDIATION

by

R.J.F. Bewley
Dames & Moore, Manchester, United Kingdom

"Best practice" in soil bioremediation requires the implementation of a series of protocols and procedures at various stages of the treatment process so as to fulfil project objectives in a safe and cost-effective manner. Clearly bioremediation is a site-specific process and what is appropriate for one particular remedial project may be unnecessary for another. However, there are a number of general principles of importance that need to be established during the site investigation phase, bioremediation feasibility assessment, the remedial design, bioremediation implementation, and verification and post-remediation monitoring.

This paper discusses the various issues which represent "best practice" at each of these stages.

Site investigation phase

Investigation of contaminated sites generally follows a phased approach which may include the following elements:

- desk study (historical review of site usage and operating procedures);
- initial investigation (targeting key on-site facilities which may be potential sources of contamination to assess nature and severity of contamination, assessing contaminant migration pathways and potentially vulnerable receptors);
- contaminant delineation investigation (assess the lateral and vertical extent of contamination).

A suitably thorough site investigation is a prerequisite for successful remediation. Some of the key issues concerning the potential applicability of bioremediation which should be identified at this stage of the process are provided in Table 1. The delineation investigation is particularly important in terms of identifying the extent of the contamination to avoid unnecessary remedial effort being applied to areas which do not require treatment, as well as identifying those areas of the site which are not amenable to bioremediation and where an alternative treatment process is required.

A risk-based approach is then applied to develop site-specific clean-up criteria; alternatively generic standards (such as the Dutch Intervention values) may be invoked. At this stage it is appropriate to undertake a review of remedial options to select the most effective strategy.

If bioremediation appears to be a viable option, then a feasibility assessment will be required to determine whether the clean-up targets can be met.

Table 1. Key issues to be evaluated during site investigation which will influence soil bioremediation strategy

Information from site investigation	Importance for bioremediation strategy
Present above-ground structures and operations on-site	Selection of *in situ* or *ex situ* treatment, availability of land for *ex situ* treatment, general operating constraints. Ongoing sources of contamination which may affect success of treatment process.
Nature of contamination (compounds involved, occurrence as LNAPLs, DNAPLs, etc.)	Suitability of bioremediation as a treatment strategy, e.g. if contamination is predominantly inorganic rather than organic, if particularly recalcitrant compounds are involved, or if contaminant present as immiscible product of low bioavailability (e.g. coal tar, free-phase mineral oil).
Depth and distribution of contamination, whether saturated as well as unsaturated zone is affected, geological and hydrogeological characterisation.	Selection of *in situ* or *ex situ* treatment, viability of process, e.g. presence of fine-grained material of low permeability will limit applicability. Requirement for groundwater treatment.

Source: Author.

Feasibility assessment

Although it is convenient to refer to this stage of the process as "feasibility assessment", the objectives of this phase should be not only to assess the ability of bioremediation to meet the remedial objectives, but also to gain information which will provide the key input parameters into the design of the remedial scheme. The scope of the feasibility study will be dependent upon the complexity of the site in question, but may involve further investigation of site parameters, as well as treatability testing (the latter typically involving microcosms of contaminated soil maintained under laboratory conditions).

Site chemical and microbiological factors

Characterisation of the key chemical properties of the soil environment which will influence soil microbial activity is a prerequisite for successful bioremediation, particularly where treatment of the saturated zone is concerned. Here there is a requirement for determination of redox potential, dissolved oxygen, presence of electron acceptors (nitrate, sulphate, ferrous iron) and nutrient status (nitrogen, phosphorous, potassium, etc.) to develop an appropriate strategy for provision of electron acceptors and nutrients. At the same time, other physicochemical factors which may influence microbial activity should be evaluated, e.g. pH, and the occurrence of inhibitors such as heavy metals. In practice, however, even if the latter are elevated they may be adsorbed onto cation exchange complexes and be rendered unavailable to the microbiota. Total organic carbon is a useful determinant for evaluating whether there may be problems of reduced contaminant availability and hence reduced biodegradation through sorption processes, or through the occurrence of a "competitive" substrate (i.e. a more readily available source of organic material than the contaminant(s) in question). Total viable counts provide a gross assessment of the microbial populations present but given their limitations, in terms of selecting only a small fraction (probably <10 per cent) of the soil microflora, such data should be treated with caution. A similar limitation applies to counts of contaminant-specific degrading organisms which may not reflect the activity of consortia of micro-organisms responsible for contaminant degradation. Where for example, it is proposed to undertake bioventing or vapour extraction, on-site measurements of soil gas composition, in terms of oxygen and carbon dioxide concentrations will provide a better index of microbial activity than population levels as measured by viable counts.

Site physical factors

Geological characterisation of the contaminated environment is of importance in determining the requirements for processing of soil for *ex situ* treatment (to achieve appropriate particle size distribution) and is essential for assessing the applicability of any form of *in situ* treatment.

For treatment of the saturated zone, hydrogeological characterisation will be necessary, involving pump tests/slug tests for assessing hydraulic conductivity, which should not be less than 10^{-4} cm/s for water re-circulation. Similarly, if treatment of the unsaturated zone by bioventing or soil vapour extraction is envisaged, intrinsic permeability should not be less than 10^{-9} cm^2. For assessing this parameter, laboratory tests such as grain size distribution and the triaxial test are suitable, although an on-site double O ring infiltrometer permeability test should be used whenever possible. (This test is not applicable, however, to highly permeable soils.) For systems involving bioventing, an *in situ* test such as described by Hinchee and Ong (1992) should be performed which consists of ventilating the contaminated soil of the unsaturated zone with air and periodically monitoring oxygen depletion and carbon dioxide evolution over time after the air supply is turned off. By undertaking such measurements across an appropriate number of monitoring points, an assessment can be made of the feasibility of bioventing as a treatment technology in terms of providing sufficient oxygen to the sub-surface, and input data for remedial design can also be obtained. Helium may also be mixed with the air injected into the soil gas monitoring points to ensure that the gas sampled is actually the gas injected so that any changes observed are not attributable to diffusion or leakage.

Treatability testing

Treatability testing using microcosms of contaminated soil is always advisable, even where the degradability of the contaminant is well-established, in order to evaluate any issues pertaining to the physicochemical properties of the soil in question, which may influence contaminant degradation. At the same time, microcosms allow for comparative testing of various nutrient supplements, designed to enhance indigenous microbial activity.

The application of nutrients, particularly for *in situ* projects needs to be evaluated in terms of concerns over potential eutrophication of adjacent receptors such as surface water or groundwater. Organic compounds, as slower release nutrients, may be suitable but may require a longer time-frame and in some cases may prove inhibitory to the degradation process as competitive substrates. Excessive concentrations of some inorganic nutrients are also inhibitory to microbial activity as concentrations may reach toxic levels within soil pore water. In view of this, and because of availability issues, generically derived "optimal" carbon:nitrogen ratios are often unreliable in terms of estimating nutrient requirements. Microcosms may therefore be established to assess the effects of different permutations of nutrients on contaminant degradation and other factors requiring optimisation such as pH adjustment, or the benefits of adding a specific inoculum can also be evaluated at this stage.

Figure 1. Reductions in total petroleum hydrocarbons (TPH) observed with various nutrient supplements applied to soil in laboratory microcosms and in the field (mean ± S.E.M.)

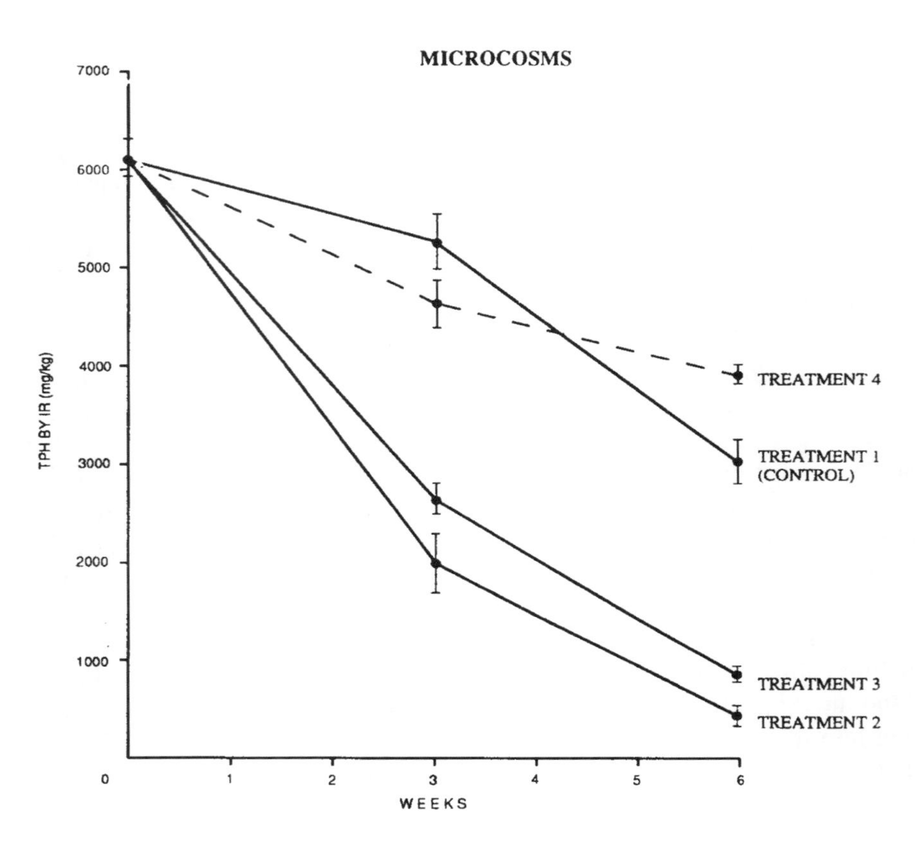

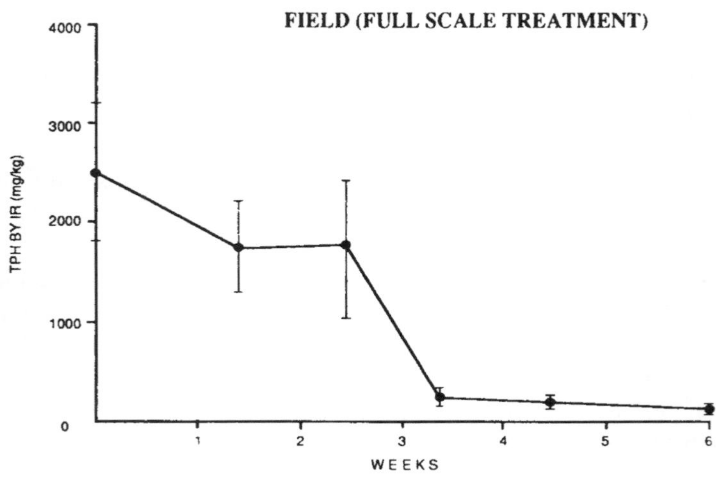

Source: Author.

An example of the application of microcosm testing is illustrated in Figure 1[1], where the effects of three treatments (plus a control) are compared, in terms of reducing total petroleum hydrocarbon (TPH) concentrations (Dames & Moore, unpublished data), in a soil contaminated with diesel oil. These comprised:

- Treatment 1: a control (no addition);

- Treatment 2: a mixture of organic and inorganic nutrients previously used in a full-scale bioremediation of a former lube oil site (Bewley *et al.*, 1995) and consisting of 0.075 per cent w/w "Purisol 100" (a liquid fertiliser manufactured by ICI Chance & Hunt, consisting of a mixture of ammonium phosphate and urea), together with 0.01 per cent w/w molasses, 0.1 per cent w/w food yeast powder, 0.0035 per cent w/w ammonium sulphate and 0.00878 per cent w/w potassium chloride. The concentrations added were based on what had previously been established as "optimal" levels through previous published (Bewley *et al.*, 1990) and unpublished studies, without regard to the specific physicochemical properties of the present soil. This treatment gave a C:N ratio of 70:1 and a C:N:P ratio of 460:6.5:1 based upon the total organic content of the soil (1.8 per cent), i.e. addition of 0.026 per cent w/w N and 0.004 per cent w/w P;

- Treatment 3: a mixture of urea (0.019 per cent w/w), ammonium nitrate (0.026 per cent w/w), potassium dihydrogen phosphate (0.036 per cent w/w), dipotassium phosphate (0.055 per cent w/w) and magnesium sulphate (0.002 per cent w/w), to give a C:N:P ratio of 100:1:1, i.e. additions of 0.018 per cent w/w N and 0.018 per cent w/w P;

- Treatment 4: the same mixture as for treatment 3, although with urea and ammonium nitrate concentrations increased to 0.096 per cent and 0.129 per cent w/w respectively to give a C:N ratio of 20:1 and a C:N:P ratio of 100:5:1, i.e. additions of 0.09 per cent w/w nitrogen and 0.018 per cent w/w phosphorus.

As illustrated in Figure 1, the higher concentrations of nutrients in treatment 4 appeared to have an inhibitory effect on biodegradation, in comparison with treatment 3, despite the former being closer to the "optimal" C:N ratios reported in theoretical studies (Bewley, 1996). A significant loss of TPH was also obtained in the unamended control (treatment 1), as a result of enhanced aeration of the soil through periodic agitation and also through physical loss. The "richer" nutrient treatment (treatment 2), consisting of organic supplements in addition to the mineral salts, resulted in the greatest reduction in TPH. However, the rate of degradation did not differ substantially from treatment 3, so that additional costs involved in supplying the organic supplements may not be justified. This example therefore illustrates the value of microcosm testing in demonstrating levels at which nutrient concentrations may become inhibitory and allowing an evaluation of cost-effectiveness of different nutrient combinations. Further evaluation of the treatments and subsequent modification of the nutrient suite may be desirable prior to full-scale application. In this particular instance, for example, when full-scale treatment was applied to soil in the field undergoing *ex situ* treatment, the actual treatment used consisted simply of one initial application of yeast (at 0.017 per cent w/w), and two applications of purisol 100, one at 0.083 per cent (w/w), the second 31 days later, at 0.067 per cent (w/w) to give a total contribution of approximately 0.03 per cent w/w nitrogen and 0.007 per cent w/w phosphorus. Quantitatively, this is similar to the application rate of treatment 2,

---

[1] The permission of BP Oil in allowing me to make use of the data for the microcosms and field testing illustrated in Figure 1 is gratefully acknowledged.

but using a simplified and less costly mixture of ingredients. As can be seen from Figure 1, this achieved the required reduction in TPH concentrations over the summer period.

It is important to realise the limitations of laboratory microcosms, the most important of which concern problems of attempting to represent heterogeneity on a small scale. In this particular example, the soil was a relatively homogeneous coarse-grained material. In other cases, however, particularly where "made ground" is concerned and where *ex situ* treatments involving material processing are to be applied, a small-scale field trial, even one involving a few cubic metres of material, has much to commend it. Such trials allow for a better understanding of the extent to which physical processing of the material (e.g. reducing particle size, homogenisation, addition of a bulking agent) needs to be implemented, although the higher degree of variability will necessitate a greater degree of sampling. For *in situ* treatment where there are likely to be any concerns over physical issues such as channelling, impermeable substrata (clay lenses) and general site heterogeneity, a field trial is essential.

Remedial design

Once the feasibility of bioremediation has been established, the various details from the on-site and laboratory testing can be applied to the remedial design. The effective delivery of oxygen (or alternative electron acceptor) to the zone of contamination is usually the key criterion to be addressed, and site heterogeneity issues are generally the most problematic. *In situ* bioremediation for the unsaturated zone clearly has key advantage over *ex situ* treatment in terms of minimising disturbance of the site, but is of limited application where heterogeneous environments are concerned and lenses of less permeable soils exist. Bioventing, i.e. forcing air through the unsaturated zone, obviates the requirement for treatment of off-gas as would be required in the case of soil vapour extraction, but is more prone to channelling effects, with the air stream following the easiest pathway and potentially avoiding contaminated zones within the bulk of the material. *Ex situ* treatment allows for physical issues relating to mass transfer of oxygen to the sites of microbial activity to be addressed through optimisation of soil structure, e.g. crushing and processing material prior to treatment or adding a bulking agent.

For the saturated zone a range of possibilities exist for *in situ* treatment of soil and groundwater. Apart from evaluating technical effectiveness, best practice requires that additional consideration should also be given to the bearing of various elements of the remedial design on the end-use of the treated material, and on the wider health, safety and environmental issues involved.

Any supplements added to the soil should not adversely affect the proposed end-use of the ground or pose a risk to potential receptors, for example:

– organic bulking agents, such as wood chips, mulch or bark in biopiles, should not be applied in sufficient quantity so as to render the soil geotechnically unsuitable for future development;

– additions of chemical supplements, particularly surfactants, should not adversely affect soil permeability. Similarly, the application of hydrogen peroxide should require attention be given to potential pore blockage, particularly as a result of precipitation of oxidation products;

- additions of chemical supplements should also not promote the accumulation of intermediates or other compounds which are of significant toxicity and mobility, e.g. arising as oxidation products;

- any microbial inocula used must be demonstrated to be non-pathogenic and should not result in any adverse effect on ecological systems. To date, there is little convincing evidence on a field scale that the application of specific microbial inocula is necessary for, or has significantly accelerated bioremediation processes in any examples of full-scale treatment reported. However, there may be instances, especially for more complex xenobiotics, where bioaugmentation may be appropriate. The risks arising from the release of genetically-modified micro-organisms (GMMOs) is outside the scope of this review. However, it should be noted that the introduction of a GMMO with a particular biodegradative ability may influence the activity of the indigenous microbiota through changes to the soil environment, which may suppress the activity of particular groups of organisms. The formation of the intermediate 2,4-DCP arising from 2,4-D degradation initiated by a GMMO, resulted in a decrease in soil fungi and in the rate of carbon dioxide evolution (Short *et al.*, 1991);

- additions of nutrients, electron acceptors (particularly nitrate) or other chemical supplements to the soil should not pose a significant risk to groundwater or surface water bodies either during the treatment process or following reinstatement (where *ex situ* treatment is involved).

To avoid potential impact of supplements on receptors, the remedial design should incorporate containment measures, for example:

- in the case of *ex situ* treatment, by including provision of an impermeable membrane to separate the soil treatment bed from the underlying soil, together with a leachate collection systems and facilities for re-circulation/treatment;

- where extraction systems are involved, appropriate facilities for treatment of off-gas;

- hydraulic containment, as appropriate for *in situ* schemes for the saturated zone (especially where nutrient additions or nitrate supplementation as electron acceptor is involved).

Bioremediation implementation and verification

Once the remedial design has been implemented as the full-scale treatment, a monitoring programme should be instituted with the following objectives:

- to assess loss of contaminant and achievement of clean-up targets;

- to assess physicochemical factors;

- to assess microbial activity;

- to demonstrate toxicity reduction achieved by bioremediation and/or lack of intermediate accumulation.

Monitoring and verification of contaminant loss

Monitoring loss of contaminant is essential for determining whether degradation is proceeding satisfactorily and when treatment is complete. The number of samples required will be dependent upon the variability of the contaminant concentrations within the material undergoing treatment. Where highly heterogeneous material is involved a far greater sampling intensity will be required in order to obtain an acceptable level of confidence. Unpublished studies undertaken by Dames & Moore for sampling treatment beds have used as a starting point the numbers of sampling points recommended in the British Standards draft document DD175 (1988) for the sampling of contaminated land. This recommends 15 sampling points for an area of 0.5ha and 25 for 1.0ha. Extrapolating this to a treatment bed of 0.2ha, ten samples were taken, each being a composite of five randomly selected sub-samples within one of ten equal sections of the treatment bed. The soil undergoing treatment had a significant amount of silty clay present and the contaminant involved was a heavy-end lube oil. The coefficient of variation for mean petroleum hydrocarbon concentrations in various batches of soil undergoing treatment varied from 15 per cent to 81 per cent initially with a mean of 35 per cent and from 15 per cent to 46 per cent following treatment (mean, 28 per cent), which was acceptable for verification that clean-up criteria had been attained.

Sampling requirements during monitoring will be subject to a number of considerations, including economic constraints. For *in situ* treatment using vapour extraction, periodic sampling of soil by boreholes may be expensive and gas monitoring for carbon dioxide, oxygen and volatile hydrocarbons may provide a suitable indication of the progress of treatment.

However, whatever method is employed, final verification of the treated soil requires that a statistically-defined treatment goal has been attained. In most cases the simplest means of carrying this out is to determine the upper 95 per cent confidence limit of the mean concentration of the contaminant of concern from the samples analysed and compare this against the target concentration. The site can then be said to have been cleaned-up to the target concentration with 95 per cent confidence.

In some instances, further indications of attainment of treatment goals will be appropriate, for example where "biostabilisation" is also an objective, e.g. the production of an asphaltene residue from the biodegradation of a mixture of petroleum hydrocarbons. In such cases, undertaking leach tests on treated soil to evaluate potential mobility of residual contamination will be a requirement of the verification process.

For the saturated zone, where soil and groundwater treatment have been undertaken, there should also be a requirement for post-remediation monitoring following completion of the active phase of remediation. This may be undertaken by groundwater sampling via a network of monitoring wells to demonstrate that there is no significant desorption of untreated contaminants from soil surfaces into the aqueous phase. Additionally, such monitoring allows for an indication of the extent of further attenuation following active clean-up.

Assessment of physicochemical factors

The primary objective of assessing physicochemical factors during treatment is to assess whether such conditions are optimal for microbial activity. The key soil characteristics for evaluation are nutrient status, pH and moisture content. All of these may require adjustment during the treatment process to ensure a satisfactory rate of biodegradation. Temperature monitoring will also be an important consideration where composting systems are involved, and for vapour extraction or bioventing systems, monitoring the soil gaseous composition (oxygen, carbon dioxide) will also be a critical parameter.

For "open" systems, particularly for *in situ* treatment, monitoring of nutrient status in adjacent receptors, will also be a key consideration to ensure that no migration is taking place which could adversely impact the latter. Although complete containment of soil during treatment may represent best practice, situations may arise where land-farming could represent the only practicable option. Under such circumstances, monitoring of indices of nutrient migration into adjacent receptors, e.g. BOD/COD in surface water bodies, will be a key requirement.

Assessment of microbial activity

The loss of contaminant will generally be the key indicator of the success of the process, although assessment of microbial activity during treatment will provide an important insight into the performance of the treatment being applied, this generally being performed by measurements of oxygen and carbon dioxide concentration from a series of soil gas monitoring points located across the treatment area. This will be essential for bioventing or soil vapour extraction systems to assess whether sufficient quantities of oxygen supplied are reaching the sites of microbial activity where contaminant degradation is required.

Should the release of GEMMOs be deemed as acceptable practice,, then molecular techniques such as the use of gene probes in nucleic acid hybridisation, DNA amplification by the polymerase chain reaction and the use of reporter genes may be appropriate for enumerating inoculated micro-organisms (Harker *et al.*, 1994), where clearly there will be a requirement for tracking and monitoring their release into the environment.

Reduction of toxicity/absence of intermediates

The potential production of intermediate compounds which may be of significant toxicity is sometimes raised as a concern during the application of bioremediation. Adoption of appropriate analytical protocols during the monitoring process provides a useful indication of the potential for intermediate production, e.g. periodic scans using GC/MS (gas chromatography/mass spectrometry) to determine the occurrence of "TICs" (Tentatively Identified Compounds) during the degradation process.

Alternatively, and particularly where aromatic hydrocarbons are involved, application of Microtox testing or mutageneity tests (such as the Ames test) may be used to compare the toxicity of extracts from the soil at various stages of the treatment process. This approach has been successfully demonstrated by Wang and Bartha (1990) and Wang *et al.* (1990), to compare changes in toxicity of petroleum hydrocarbon-contaminated soil undergoing active treatment compared to an untreated system over time. In general these demonstrated that any increases in toxicity following nutrient addition were only of a transitory nature and that toxicity declined in the treated systems to background levels.

The incorporation of such assays into field monitoring protocols would represent an important addition to existing procedures in providing assurance that overall improvement to soil quality were being achieved as well as reductions in contaminant mass alone.

Conclusions

A significant range of treatment possibilities for *ex situ* and *in situ* bioremediation exist and it is not possible to set out a definitive set of procedures for best practice to cover all such scenarios. However, this review has illustrated that the adoption of specific protocols at each stage of the remediation process will ensure that the treatment is conducted, not only in a cost-effective manner, but also one which is acceptable in terms of minimising risks to both human and environmental receptors.

Questions, comments, and answers

Q: Would you normally apply a similar level of best practice to non-biological clean-up and how do you make comparisons of cost effectiveness?

A: In terms of competing treatments, removal to landfill is the biggest competitor in the United Kingdom, although this may be described as moving the problem somewhere else. We need to build best practice protocols into any type of clean-up to verify statistically that there is a reduction in contamination in the remaining soil so that all treatments are on a level footing.

Q: After the site you describe was cleaned the owner may wish to sell it, but if a new owner can find a clean site he will not take the risk. What is the perception of the site by the new owner?

A: All results of monitoring were made available to the new owner. There is an underlying ultimate pollution which is presented on a risk assessment basis of exposure route. The site is going to be used for industrial purposes and the new owner would be reassured by the results which would compare favourably with sites of generally unknown quality.

Q: Do you include a sensitivity analysis in protocols; for example balancing manpower tariffs against nitrogen and phosphate addition?

A: This depends on each site but in this case it was not an issue because we were adding minimal nutrients. We did a cost evaluation and the trade-off factor was time.

REFERENCES

BEWLEY, R.J.F. (1996), "Field implementation of *in situ* bioremediation: key physicochemical and biological factors", in *Soil Biochemistry, Volume* 9, G. Stotzky and J.M. Bollag (eds.), pp. 473-541, Marcel Dekker Inc., New York.

BEWLEY, R.J.F., B. ELLIS and J.F. REES (1990), "Development of a microbial treatment for restoration of oil-contaminated soil", *Land degradation and rehabilitation* 2, pp. 1-11.

BEWLEY, R.J.F., G.H. WEBB and J.G. ALEXANDER (1995), "Bioremediation of soil and groundwater at a former lubricating oil terminal", *Society for General Microbiology Abstracts, 132nd Ordinary Meeting, University of Aberdeen 10-13 September 1995,* p. 92.

BRITISH STANDARDS INSTITUTION (1988), "Draft for development. Code of practice for the identification of potentially contaminated land and its investigation", DD175, BSI, London.

HARKER, A.R., Y. KIM and U. MATRUBUTHAM (1994), "Application of genetic engineering to the field of bioremediation", in *Remediation of hazardous waste contaminated soils*, D.L. Wise and D.J. Trantolo (eds.), pp. 77-96., Marcel Dekker Inc., New York.

HINCHEE, R.E. and S.K. ONG (1992), "A rapid *in situ* respiration test for measuring aerobic biodegradation rates of hydrocarbons in soil", *Journal of the Air & Waste Management Association* 42, pp. 1305-1312.

SHORT, K.A., J.D. DOYLE, R.J. KING, R.J. SEIDLER, G. STOTZKY and R.H. OLSEN (1991), "Effects of 2,4-dichlorophenol, a metabolite of a genetically-engineered bacterium, and 2,4-dichlorophenoxyacetate on some microorganism-mediated ecological processes in soil", *Applied and Environmental Microbiology* 57, pp. 412-418.

WANG, X., and R. BARTHA (1990), "Effects of bioremediation on residues, activity and toxicity in soil contaminated by fuel spills", *Soil Biology and Biochemistry* 22, pp. 501-505.

WANG, X., X. YU and R. BARTHA (1990), "Effect of bioremediation on polycyclic aromatic hydrocarbon residues in soil", *Environmental Science and Technology* 24, pp. 1086-1089.

ON THE MICROBIOLOGICAL CONTRIBUTION TO PRACTICAL SOLUTIONS IN BIOREMEDIATION[1]

by

Kornel L. Kovacs, Levente Bodrossy and Csaba Bagyinka
Institute of Biophysics, Biological Research Center, Hungarian Academy of Sciences
Szeged, Hungary

Katalin Perei and Bela Polyak
Institute for Biotechnology, Z. Bay Foundation, Szeged, Hungary

Introduction

Controlled microbial consortia

Environmental contamination usually consists of a mixture of pollutants and their partially degraded derivatives. Such an ill-defined chemical mixture will eventually lead to the formation of an ecosystem of microbes. The individual member species cannot survive in a toxic and hostile environment. Effective bioremediation technologies should therefore invoke a mixture of micro-organisms forming synergistic consortia. Any realistic bioremediation concept is based on the recognition that it is the *concerted action* of various species which may bring about the desired clean-up effect.

A significant drawback for remediation microbiology is that there is a very limited basic microbiology knowledge on co-operative effects and interactions among micro-organisms. This lack of information derives from the first and most important prerequisite in a classical microbiology experiment: the purity of the strains. The requirement of working with isolated individual strains is fully justified by the rules of fundamental microbiology, yet this strategy automatically excludes any observation of the potential beneficial effects arising from interactions among various strains. Therefore, no such effects can be taken into account when biodegradation of a complex hazardous chemical mixture is considered. The simplistic view of assuming the mere summation of individual microbial contributions in a microbial consortia is inherently included in most approaches. One consequence is the widespread belief, that given certain minimum nutrient conditions, the microbial activity is not rate-limiting in biodegradation systems.

From an applied environmental biotechnology point of view, this lack of fundamental microbiology knowledge results in poorly understood system characteristics and inferior or

---

[1] The experimental work discussed in this paper was supported by several grants from the OTKA, OMFB, PHARE Accord programmes, and from the US-Hungarian Joint Fund. The authors are thankful for the financial assistance from these agencies.

unpredictable performance. A solution for this dilemma employed in several laboratories, including ours, is the assembly of *controlled mixed cultures*, in which pure cultures of bacteria are deliberately mixed in order to enhance bioconversion/biodegradation yields. To further support the microbiological interaction, one can invoke *immobilisation* techniques which compile high concentration of biomass confined into a small space. This increases both the active biomass concentration and the concerted action among various species. In addition, immobilisation may prevent escaping of the cells from the reaction volume (Caplan, 1993; Liu and Suflita, 1993).

Selection pressure by rare substrates

A general presupposition in developing biotechnological remediation systems is based on the notion that a natural microbiological activity exists in practically any environment which is capable of degrading even the most obnoxious chemicals, albeit at a slow rate. It follows from this principle that micro-organisms, specially adopted to decompose a certain type of contamination, will be most abundant at the site of contamination. The more incompatible a given chemical is to common forms of life, the stronger is the *selection pressure* to give preference to the growth of micro-organisms that are capable of metabolising that compound (Blackburn and Hafker, 1993, Lowe *et al.*, 1993).

From our practice, evidence supports the significance of bacterial interactions as an important additional dimension in applied microbiology, e.g. in the utilisation of bacterial hydrogen metabolism for environmental biotechnology applications. These advantages are demonstrated in the case of biogas production and denitrification. Other examples show the importance of selective pressure in developing laboratory scale microbiological answers to pressing bioremediation needs.

Results and discussion

Utilisation of hydrogen metabolism in biotechnological applications

Hydrogen evolution by intact bacterial cells is frequently observed in nature. In microbial ecosystems, the role of these micro-organisms is the creation and maintenance of an anaerobic, reductive environment, as well as supplying the universal reducing agent, molecular hydrogen. Gaseous hydrogen is usually not released from the natural ecosystems unless there is an excess of reductive power which needs to be disposed of in order to ensure the optimal metabolic and growth equilibrium in the population. H_2 generated *in vivo* by hydrogen-forming bacteria is utilised by hydrogen-consuming members of the microbiological community. Hydrogen is transferred to the recipient micro-organism very effectively by inter-species hydrogen transfer. The molecular details of this process are not fully understood, but its significance in safeguarding the optimum performance of the entire ecosystem and the delicate regulatory mechanisms should be appreciated. In the mixed population presented here, bacterial systems exploit the advantages of inter-species hydrogen transfer.

Biogas production from wastes

Biological methane production is carried out in three stages, performed by separate groups of micro-organisms. Complex organic materials are first hydrolysed and fermented by facultative and anaerobic micro-organisms into fatty acids. The fatty acids are then oxidised to produce H_2 and organic acids (primarily acetate and propionate), processes termed dehydrogenation and acetogenesis, respectively. The last stage is methanogenesis. Some methanogens can combine the hydrogen directly with carbon dioxide to form methane. Others split acetate into carbon dioxide and methane (Hall *et al.*, 1992).

Among the significant recent advances in understanding the ecology of anaerobic biodegradation of organic wastes is the recognition of the close syntropic relationship among the three distinct microbe populations and the importance of H_2 in process control (Benstead *et al.*, 1990). The regulatory role of hydrogen levels and inter-species hydrogen transfer optimise the concerted action of the entire population. The concentration of either acetate or hydrogen, or both together, can be reduced sufficiently to provide a favourable free-energy change for propionate oxidation.

During anaerobic biodegradation, hydrogen concentration is reduced to a much lower level than that of acetate. The acetate concentration in an anaerobic digester tends to range between 10^{-4} and 10^{-1} M; H_2 ranges between 10^{-8} and 10^{-5} M, or about four orders of magnitude less. In addition, the hydrogen partial pressure can change rapidly, perhaps varying by one order of magnitude or more within a few minutes. This is related to its rapid turnover rate. The energy available to the acetate-using methanogens is independent of hydrogen partial pressure, whereas that of the hydrogen-producing and hydrogen-consuming species is very much a function of it.

The low concentrations of free hydrogen within the digester has implications for the rate of H_2 turnover in the system. In a typical system the turnover rate of the hydrogen pool is 6.8×10^6/day or about 80/s. The inter-species H_2 transport is therefore a very rapid, delicate, and rate-limiting step. For processes on this micro-scale, diffusion is the main mechanism for hydrogen transport between species. The calculations clearly indicate that the optimal distance between interacting hydrogen-producing and consuming species is about 10 micrometers under practical conditions. This distance is equal to about 10 bacterial widths (Kovacs and Polyak, 1991). Clearly, the bacteria must be close together, otherwise inter-species hydrogen transfer will be the rate-limiting step in the overall process. This is the case in an ideal, completely mixed, or continuous stirred tank reactor where the individual bacterial species are dispersed uniformly throughout the system, and the distribution of reactants, intermediates, products of biotransformation, and bacterial species is homogeneous throughout the reactor.

New process design, such as the biofilm reactors (VanLoosdrecht and Heijnen, 1993), encourage different species to live in proximity to one another, and the result is a much higher rate of conversion per unit volume of reactor, although not yet optimal. Biofilms also permit a diversity of environments to develop in proximity to one another so that suitable conditions for the degradation of each substrate and intermediate can exist somewhere within the biofilm.

We have shown that under these circumstances, addition of hydrogen-producers to the system and thereby shifting the population balance, brings about advantageous effects for the entire methanogenic cascade. The decomposition rate of the organic substrate, which was animal manure in our first experiments, increases, and both the acetogenic and methanogenic activities are remarkably amplified. In laboratory experiments, some 2.6-fold intensification of biogas productivity has been routinely observed, and the same results were obtained in a 100 litres digester scale-up experiment (Figure 1.).

Figure 1. Cumulative biogas production of a 100-litres mesophilic anaerobic up-flow blanket fermentor

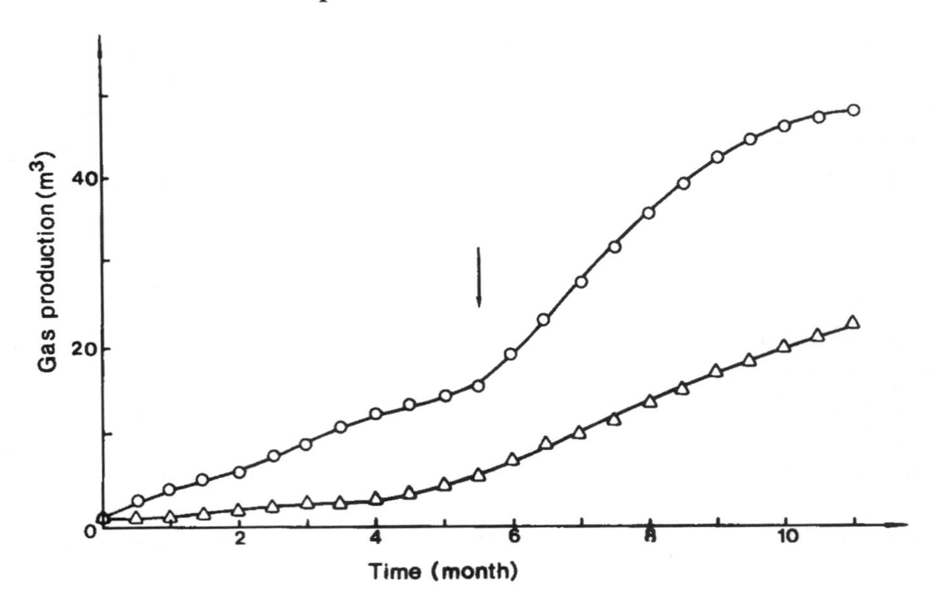

Notes: Feed material was pig manure (average dry weight 8 per cent), 8-10 litres/day.

Δ = control

O = inoculated with H2 producing bacteria (5 per cent fermentor volume × 33e) at the start of the experiment and at the point indicated by ↓ .

Source: Author.

In an ongoing field experiment, the intensification of biogas formation from solid municipal wastes is attempted. Municipal solid wastes are usually difficult to break down and the biodegradation process is particularly slow in landfill type of "bioreactors". Nevertheless, an economic biogas-producing system has been installed in the municipal waste landfill site of the city of Szeged, Hungary. The facility is operated by the City Council and receives about 200 000 m^3 of wastes annually, mostly solid household garbage. The biogas has been collected through an underground pipeline system, compressed, and fed into a natural gas-based heating centre that serves a residential area. Biogas production reached 700 000 m^3 by 1990. Since the system is simple and inexpensive, and the biogas collection and utilisation is solved in an ingenious and fortuitous way, the installation operates smoothly and efficiently.

Proper management of the bacterial population is expected to facilitate the start-up of the fermentation. In order to reduce the costs of this treatment supplemented bacteria are grown in diluted industrial waste-water. Spraying the microbes onto the surface of the garbage layer before covering the layer with dirt accelerates the formation of biofilms, the highly effective microbial centres of fermentative degradation. In addition to an increased biogas production, the faster decomposition rate is expected to shorten the long lifetime of the landfill site and the land will be ready for development within 10-15 years, instead of the usual 20-30 years needed for conventional treatment.

Denitrification

Nitrate contamination of natural waters is gradually increasing world-wide, in particular among the industrialised nations. Several methods have been considered for removal of nitrate from water. Ion exchange is currently the only effective technique used in full-scale treatment facilities to remove nitrate from drinking water. Continuous regeneration of the ion exchange resin is possible and plants to clean groundwater containing 20-30 mg/l of nitrate-nitrogen can produce lower than 2 mg/l nitrate levels. Several ion exchange systems are in operation world-wide with mixed results as far as system performance and economy is considered.

Biological denitrification has been studied for the purification of both waste-water and drinking water. The technique applied today consists of the use of denitrifying micro-organisms in a filter bed to reduce the nitrate to nitrogen gas. The biological filter is followed by a conventional filter to remove the carry-over bacterial growth. The heterotroph denitrifying organisms require an organic energy source. An organic substance therefore must be provided for treating drinking water because groundwater supplies are essentially free of organic material. In spite of being able to truly eliminate nitrate contamination, biological denitrification has not been enthusiastically received by water utilities yet, primarily because of the potential to introduce contaminations of microbial origin into the purified water.

The system developed in our laboratories includes two novel features, that, to our knowledge, have not been applied in denitrification technologies before. We intend to exploit the benefits of inter-species hydrogen transfer and that of our novel immobilisation technology. Inter-species hydrogen transfer is essential to supply the necessary hydrogen for nitrate reduction by denitrifying micro-organisms. To accomplish this task, we employ well-defined mixtures of bacteria. One species in this artificial microbiological ecosystem is responsible for hydrogen production from added organic substrate (certain industrial waste-waters, sugars, or cellulose can serve as organic substrate). Because of the close spatial proximity, hydrogen is effectively transferred from the helper micro-organisms to the denitrifying bacteria immobilised together with the hydrogen producer species (Kovacs and Polyak, 1991).

Inter-species hydrogen transfer brings about synergistic effects for the denitrifying bacteria since it is a very efficient way to administer hydrogen for their function. From the operational safety point of view, *in situ* generation and utilisation of hydrogen is evidently superior to bubbling the system with highly explosive hydrogen gas. Immobilisation of the participating bacteria in beads of high physical resistance provides better performance because of dramatically increased bacterial population density and high flow rate. Moreover, with our unique immobilisation technique, essentially sterile fermentation conditions can be maintained which allows a strict control of the microbiological ecology within the immobilised system.

The process in the simplest form contains an ion exchange column producing potable water while another ion exchange column is regenerated through a biological denitrification reaction. During ion exchange resin regeneration a brine solution containing nitrate in high concentration is generated. Suitably engineered bacterial population is able to convert nitrate to nitrogen gas. In this way, the regenerant can be used again after it has been subjected to denitrification, and thus salt requirement and brine production are minimised. In laboratory experiments a reduction of 90 per cent in brine volume has been routinely achieved.

Biological regeneration of the brine solution from the nitrate-loaded anion-exchange resin can be achieved in the bioreactor-containing immobilised micro-organisms. In the system developed in our laboratory, a continuous flow circuit is used, containing the ion exchange column which has to be regenerated and a bioreactor accommodating the biological material. The regenerant is recirculated in the system through the ion exchange column and the denitrification bioreactor, which converts nitrate into nitrogen gas.

The best hydrogen producer helper bacterium strains convert the organic material found in dilute waste-waters of the food processing industry (sugars and proteins). They are selected because this way a treatment of polluting waste-waters can be linked to the elimination of polluting nitrate contamination. The ability to bring about denitrification is characteristic of a wide variety of common bacteria, including the genera *Pseudomonas*, *Achromobacter*, and *Bacillus*. In our hands, such a biological denitrification system can handle 1 000-2 000 mg nitrate/l (100 mmol $NaNO_3$/l) at a conversion rate of 10 mg nitrate/g bead/hr. Nitrite was not detectable in the effluent of our experimental bioreactor, even after several weeks of continuous operation (Kovacs and Polyak, 1991).

Degradation of hazardous chemicals by natural isolates

Halogenated hydrocarbons

Methanotrophic micro-organisms oxidise methane in four steps, producing methanol, formaldehyde, formate intermediers and eventually degrade methane to carbon dioxide and water. It is possible to separate the pathway into four steps in the cell free extract or after partial purification of the various enzymes. The key enzyme is a metalloenzyme, methane monooxygenase (MMO) which catalyses the oxidation of methane to methanol.

Methanotrophs have great biotechnological potential in other, environmentally crucial bioconversion assignments as well. MMO catalyses the incorporation of oxygen into a broad range of substrates including alkanes, alkenes, aromatic hydrocarbons, alicyclic hydrocarbons, halogenated alkanes, and halogenated aromatic hydrocarbons (Murrell and Dalton, 1992).

Environmental contamination by chlorinated solvents, a model compound of them is trichloroethylene (TCE), presents serious health problems. The widespread use of halogenated hydrocarbons and concomitant careless handling, storage, and disposal, coupled with their outstanding chemical stability, put halogenated compounds among the most frequently detected groundwater and soil contaminants in many industrialised nations. The prevailing nature of drinking and well water contamination cause concern because halogenated organics can be both directly toxic and carcinogenic (Ensley, 1991; Bouwer and Zehnder, 1993).

It has been known since 1985 that methanotroph micro-organisms may be involved in aerobic TCE metabolism. Haloalkane oxidation is accompanied with intra-molecular hydrogen or hydride migration similar to reactions observed with cytochrome P-450 and the iron cluster of MMO generates an activated oxygen species with reactivity similar to the hem of cytochrome P-450. As a result of a comparative survey, it has been suggested that any microbial oxygenase can catalyse the oxidation of TCE, although substrate specificity and reaction rates vary (Murrell and Dalton, 1992).

The data available to date concerning aerobic oxidation of chlorinated hydrocarbons contain a considerable promise for the eventual biological treatment of these groundwater contaminants. An oxidation rate of 150 nmol/min × mg protein measured with whole cells of *Methylosinus trichosporium* OB3b expressing MMO means that a suspension of this bacterium at 1 mg/ml protein would eliminate 20 mg/litre (20 ppm) TCE, a fairly high contamination level, within 1 minute. This rate is very appealing for practical applications. The ideal degradation rate, however, contrasts with the persistence of TCE in the environment and emphasises the difficulty of applying laboratory degradation rates to the field conditions. At least two problems emerge from these studies. On the one hand, in order to achieve a steady and fast degradation of halogenated hydrocarbon, enough reducing power reserves should be provided to the cells to sustain their redox balance disturbed by the oxidation of halogenated compounds. On the other hand, MMO active site is apparently inactivated by the reactive metabolites generated by the oxidation process itself (Tsien *et al.*, 1989).

Sulphanilic acid

Sulphanilic acid is a typical representative of aromatic sulphonated amines widely used and manufactured as an intermediate in the production of azo dyes, plant protectives and pharmaceuticals. Its degradation is slow and incomplete by most biological systems because of the sulphonic acid group as a xenobiotic structural element. In addition, being an extremely strongly charged anion, penetration into intact bacteria is restricted. As a consequence, these compounds are recalcitrant in aerobic waste-water treatment plants. Sulphanilic acid is the intermediary for the synthesis of various sulphonamide drugs, noted for their strong bactericide effect. Their physiological effect is based on the capability to inhibit nucleotide biosynthesis.

Figure 2. Degradation of sulfanilic acid (- ● -) and increase of cell number (- ○ -) in a batch experiment with a culture isolated after enrichment on sulphanilic acid

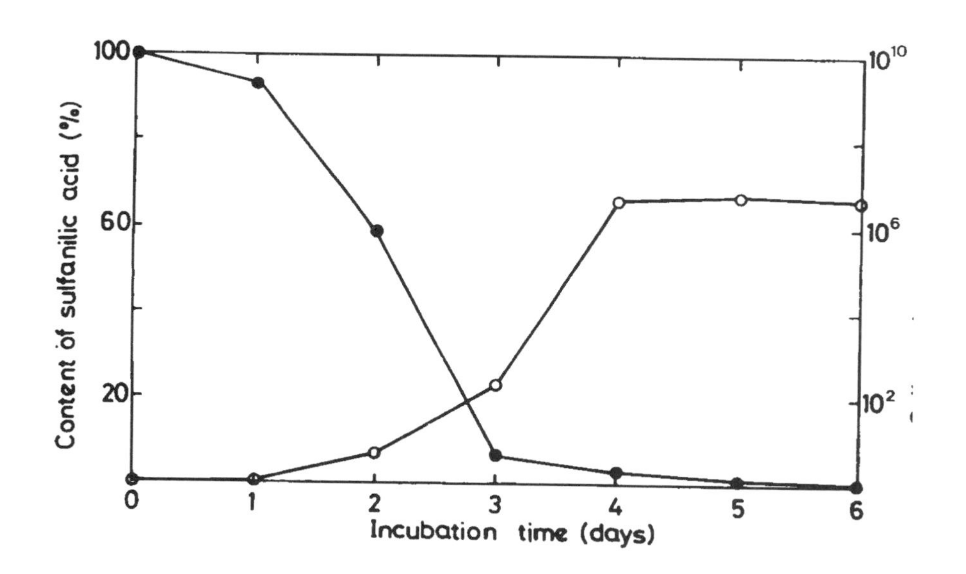

Source: Author.

Biodegradation of sulphonated aromatic compounds has been studied for about 40 years (Feigel and Knackmuss, 1988; Feigel and Knackmuss, 1991), but the only effective reactions with published experimental support involve dioxygenases or monooxygenases acting on the carbon atom carrying the sulphonate moiety (Feigel and Knackmuss, 1988; Locher *et al.*, 1991). By employing oxygenases natural microbiological systems, capable of degrading halogenated hydrocarbons and sulphonated aromatics, utilise strategies similar to other bacterial activities in the environment, e.g. decomposition of aromatic or aliphatic hydrocarbons. The common element of this strategy includes a first attack on the chemically inert target molecule by enzymatic oxygenation. Oxygenated derivatives become less stable and amenable to further chemical and/or biological degradation into products that are easy to metabolise (Locher *et al.*, 1991; Goszczynski *et al.*, 1994).

A microbiological degradation method has been developed to supplement the waste-water treatment system of Nitrokémia Ltd., a major Hungarian chemical plant producing sulphanilic acid in large quantities. For the biological sulphanilic acid eliminating reactor bacterial strains, capable of effectively degrading sulphanilic acid, have been isolated and immobilised. The optimised system appeared to function effectively at laboratory scale. Industrial experiments are in progress (Figure 2).

Questions, comments, and answers

Q: You describe thermophilic methanotrophs that degrade chlorinated aromatics - are these co-metabolic processes and couldn't you find organisms that grow on these materials?

A: No, they are not co-metabolic and require no extras. They were obtained from a local source hot water spring saturated with methane. We have done only preliminary characterisation, but the organisms can grow at temperatures some 10 degrees higher than known methanotrophs.

REFERENCES

BENSTEAD, J., D.B. ARCHER, D. LLOYD, (1990), "Role of hydrogen in the growth of mutualistic methanogenic cultures", in *Microbiology and Biochemistry of Strict Anaerobes Involved in Inter-species Transfer*, J.P. Belaich *et al.* (eds.), pp. 161-171, Plenum Press, New York-London.

BLACKBURN, J.W. and W.R. HAFKER (1993), "The impact of biochemistry, bioavailability and bioactivity on the selection of bioremediation techniques", *TIBTECH* 11, pp. 328-333.

BOUWER, E.J. and A.J.B. ZEHNDER (1993), "Bioremediation of organic compounds - putting microbial metabolism to work", *TIBTECH* 11, pp. 360-367.

CAPLAN, J.A. (1993), "The world-wide bioremediation industry: prospects for profit", *TIBTECH* 11, pp. 320-323.

COMMISSION OF THE EUROPEAN COMMUNITIES (1992), *Treatment and use of sewage sludge and liquid agricultural wastes*, J.E. Hall *et al.* (eds.), ECSC-EEC-EAEC, Brussels, ISBN 92-826-4142-2.

ENSLEY, B.D. (1991), "Biochemical diversity of trichloroethylene metabolism", *Annual Review of Microbiology* 45, pp. 283-299.

FEIGEL, B.J. and H.J. KNACKMUSS (1988), "Bacterial catabolism of sulfanilic acid via catechol-4-sulfonic acid", *FEMS Microbiology Letters* 55, pp. 113-118.

FEIGEL, B.J. and H.J. KNACKMUSS (1991), "Degradation of sulfanilic acid by a syntropic culture: Sulfanilic acid degradation to maleylacetic acid by *Pseudomonas paileronii* and *Agrobacterium radiobacter* mixed culture", *Industrial Waste Disposal* M1, p. 125.

GOSZCYNSKY, S., A. PASZCYNSKI, M.B. PASTI-GRINGSBY, R.L. CRAWFORD and D.L. CRAWFORD (1994), "New pathway for degradation of sulfonated azo dyes by microbial peroxidases of *Phanerochaeta chrysosporium* and *Streptomyces chromofuscus*", *Journal of Bacteriology* 176, pp. 1339-1347.

KOVACS, K.L. and B. POLYAK (1991), "Hydrogenase reactions and utilisation of hydrogen in biogas production and microbiological denitrification systems", Proceedings of the 4th IGT Symposium, Chapter 5, pp. 1-16, Colorado Springs.

LIU, S. and J.M. SUFLITA (1993), "Ecology and evolution of microbial populations for bioremediation", *TIBTECH* 11, pp. 344-352.

LOCHER, H.H., T. LEIRSINGER and A.M. COOK (1991), "4-toluene sulfonate methyl-monooxygenase from *Comamonas testosteroni* T-2: purification and some properties of the oxygenase component", *Journal of Bacteriology* 173, pp. 3741-3748.

LOWE, S.E., M.K. JAIN and J.G. ZEIKUS (1993), "Biology, ecology, and biotechnological applications of anaerobic bacteria adapted to environmental stresses in temperature, pH, salinity, or substrates", *Microbiological Review* 57, pp. 451-509.

MURRELL, J.C. and H. DALTON (1992), *Methane and methanol utilizers*, Plenum Press, New York.

TSIEN, H.C., G.A. BRUSSEAU, R.S. BRUSSEAU, R.S. HANSON and L. WACKETT (1989), "Biodegradation of trichloroethylene by *Methylosinus trichosporium* OB3b", *Applied and Environmental Microbiology* 55, pp. 2960-2964.

VANLOOSDRECHT, M.C.M. and S.J. HEIJNEN (1993), "Biofilm bioreactors for waste-water treatment", *TIBTECH* 11, pp. 117-121.

MONITORING MICROBES WITH BIODEGRADATIVE PROPERTIES IN THE ENVIRONMENT

by

J.L. Ramos
Department of Organic Chemistry, School of Sciences, University of Granada, Campus Fuentenueva, Granada, Spain; and CSIC Research Scientist

J.J. Rodríguez-Herva and C. Ramos
CSIC-Department of Biochemistry and Molecular and Cellular Biology of Plants, Granada, Spain

Life on this planet is based on the continuous cycling of elements such as carbon, phosphorous, nitrogen and others, between organic and inorganic states. Most organic compounds are degraded by soil micro-organisms. Bacteria, for example, rapidly mineralise many substances and play important roles in the carbon cycle and in maintaining ecosystems in balance. In recent years, large quantities of synthetic organic chemicals have been released into the biosphere. Although many of these are rapidly biodegraded, others are catabolized only slowly or incompletely, if at all. Toxic chemicals in this latter group constitute a source of environmental pollution. Although soil and water micro-organisms have considerable evolutionary potential, and under appropriate selective conditions evolve pathways able to degrade many synthetic compounds, such conditions do not always arise in nature, and the evolution of effective pathways for certain compounds may thus be relatively slow.

Genetic engineering is a way to accelerate the evolution of desired metabolic pathways (see Ramos *et al.*, 1995 for a review). Two general strategies can be envisaged for the experimental evolution of new catabolic activities, namely, the restructuring of an existing pathway, or the assembly of a new route through the functional combination of appropriate sections of different pathways. An existing pathway can be restructured by substituting enzymes or other proteins through the recruitment of new genes. This has been successfully used to evolve derivatives of *Pseudomonas* sp. B13 that can degrade an extended range of chlorinated aromatics. Earlier experiments involved the recruitment of single enzymes exhibiting broad substrate specificities (Reineke and Knackmuss, 1979). In other cases, expansion involved the modification of the profile of effectors for regulators and the substrate specificity of certain enzymes (Ramos *et al.*, 1987). The assembly of different catabolic blocks from different microbes is another approach which was successfully used to generate a "new" pathway for the simultaneous elimination of alkyl- and chloroaromatics (Rojo *et al.*, 1987).

Bioremediation is an area of growing interest for the decontamination of polluted areas. Such treatments involve, among other strategies, the release of large numbers of wild type or genetically modified micro-organisms (GMMs) that decontaminate polluted soils, waters, and sediments, to enhance waste degradation. However, concerns about the potential risks associated with the deliberate release of large numbers of wild type or GMMs have been raised because their immediate and long-term behaviour in the environment are not fully predictable (Ramos *et al.*, 1995). Soil, water, underground water, and other habitats represent sinks for released pollutants and GMMs, either directly as a result of waste disposal or inoculation (in the case of microbes), or indirectly through environmental transport. Therefore, rapid, reliable, and efficient techniques for detecting and identifying pollutants and microbes are necessary, and these techniques must be much more sensitive in detecting pollutants and micro-organisms in these habitats than what has been previously available with traditional technology. These techniques include antibodies, bioluminescence markers and PCR probes.

Immunological methods

Polyclonal or monoclonal antibodies (mAbs) provide a highly sensitive and specific detection technique in comparison with traditional methods of microbial detection. Antibodies can be used to identify specific gene products or even intact micro-organisms that express an appropriate antigenic determinant on their cell surface. Fluorescent-labelled antibodies, commonly used in medical microbiology and pathology, have been used in combination with immuno-epifluorescence microscopy to detect micro-organisms in natural samples.

Pseudomonas putida 2440, a restriction-negative strain which was derived from the natural bacteria *P. putida* mt-2, has been used as a host for recombinant DNA. We produced mAbs against the surface determinants of this microbe in order to trace it after it was released into the environment. The mAbs were tested against (1) several strains of *P. putida*; (2) different bacteria belonging to the pseudomonad group; and (3) against different bacterial phyla, including cyanobacteria, flavobacteria, gram-positive eubacteria, and purple bacteria. In all, about 30 different species of eubacteria were tested. Two mAbs were highly specific against *P. putida* and were called 7.3B and 7.4D (Ramos-González *et al.*, 1992). The target antigens recognised by mAbs 7.3B and 7.4D were found to be lipopolysaccharides (LPS) and a protein with an apparent molecular mass of 40 kDa, respectively.

Once a specific tool is available, it is of interest to determine its detection sensitivity. To establish the minimum number of *P. putida* 2440 bacteria that could be detected by the specific mAb, a dot blot ELISA was developed. Serial dilutions of a culture of *P. putida* 2440 grown on liquid medium were filtered through a nitrocellulose membrane so that the number of bacteria per well ranged from 10^5 CFU to less than 1 CFU per well. The immunoassay revealed that spots containing as few as 10^2 CFU were consistently positive.

The potential use of these mAbs as tracking reagents is based on the expression and stable maintenance of the antigen under different growth conditions. Cells of *P. putida* 2440 grown on rich and minimal medium, regardless of the growth phase, were recognised by both mAbs (Ramos-González *et al.*, 1992).

The mAbs that recognised *P. putida* 2440 were successfully used to track the bacterium in aquatic ecosystems. The mAb 7.3B was used to track *P. putida* 2440 after release in mesocosms installed in a eutrophic lake. The *P. putida* cells were easily distinguishable from autochthonous bacteria by their rod shape and FITC-immunofluorescence staining. One week after introduction into the lake water mesocosms, FITC-stained *P. putida* cells were also detected but they appeared as small coccoid cells, often in large agglomerations. Therefore, mAb 7.3B was a reliable tool for tracking *P. putida* bacteria, since it showed no unspecific binding to autochthonous bacteria.

In order to specifically track cells bearing recombinant DNA (rDNA), we concentrated on the identification of the antigen and the gene encoding for the target protein. We generated Tn*5* mutants of the wild type strain that were no longer recognised by mAb 7.4D. Then the mutations were cloned and the wild type gene identified and sequenced. The gene encoded for a cell outer membrane protein which, by protein homology, was identified as the peptidoglycan-associated lipoprotein (PAL or OprL) (Rodríguez-Herva *et al.*, 1996).

Our interest focused on the peptidoglycan-associated lipoprotein, which is particularly resistant to solubilization in SDS (Hancock *et al.*, 1990). This protein is modified at the cysteine amino terminal by a lipid moiety that is integrated into the outer membrane. The insertion of a new sequence into the amino-terminus of PAL therefore seemed a way of presenting a "new" epitope on the cell surface for antigen binding. We accordingly introduced single restriction sites within the *pal* gene so that a new antigen based on an animal virus sequence was introduced. mAbs against the new antigen were obtained, which allowed us to distinguish individual cells initially bearing or lacking rDNA.

Bioluminescence

Bioluminescence-based marker technology is also a promising approach for tracking GMMs. These marker systems involve the introduction of genes for light emission, originally cloned from naturally luminescent marine bacteria (with the *lux* marker) or from firefly luciferase (*luc* marker) (see Stewart and Williams, 1992, for a review). Techniques of genetic engineering, make it possible to exploit the phenomenon of bioluminescence for several rather different purposes, as they provide a real-time non-invasive reporter for measuring gene expression, a sensitive marker for bacterial detection, and an indicator of cellular viability.

Luminescence-marked organisms can be detected with several techniques. Many *lux*-marked strains that emit light from luminescent colonies can be detected with the unaided eye in a darkened room. Photography increases the sensitivity of detection of marked colonies; luminometry enables quantification of the light emitted; and, CCD image-enhanced microscopy makes it possible to detect and quantify single photons from which an image may be constructed.

Luminescence makes it possible to detect and localise marked organisms *in situ*. Bioluminecence measured with a luminometer is an effective aid in the detection in soil of *lux*-marked *Pseudomonas* strains (Flemming *et al.*, 1994) and *Xanthomonas campestris* (Shaw *et al.*, 1992).

Phytobioremediation makes use of the secretion of nutrients by plants that stimulate bacterial activities in the rhizosphere, and thus enhances the elimination of toxic compounds. In this context, *P. putida* -- a microbe with a wide range of catabolic activities -- is able to efficiently colonise the rhizosphere of a number of plants, including corn (maize), tomato, barley, and spinach. *lux*-marked *P. putida* strains have been shown to colonise corn roots, preferentially at the root tips, where they form micro-colonies as revealed by scanning electron microscopy and light detection by CCD camera.

lux has also been used as an environmental reporter gene to detect naphthalene and mercury in soil (King *et al.*, 1990). Aromatic hydrocarbons have been detected by constructing a *lux* gene fusion to the *meta*-pathway promoter from the TOL plasmid of *Pseudomonas*. The presence of an aromatic compound at low concentrations induces *lux* expression, and resulted in a detectable bioluminescent phenotype.

PCR and identification of bacteria in the environment

DNA probes in combination with PCR technology have been used to detect specific bacteria, i.e. human pathogens in food and micro-organisms from the natural environment (Amann *et al.*, 1990; Bej *et al.*, 1991). The molecular monitoring of microbial ecosystems has also recently received increasing attention as a suitable method to examine the fate of introduced and indigenous micro-organisms after deliberate release into the natural environment. However, the application of molecular techniques to studies of microbial diversity and heterogeneity in natural environments is limited because of the lack of probes that are highly specific for each type of indigenous bacteria.

The detection of specific micro-organisms is highly dependent on the specificity of the probes of the PCR primer sets. Many kinds of cloned chromosomal fragments or oligonucleotides have been used as probes. Among them, the nucleotide sequences of the 16S rRNA and 23S rRNA have most frequently been used. The hybridisation of rRNA-targeted probes and PCR amplification of the rRNA gene (rDNA) with specific primers have been shown to be useful for characterising and detecting specific populations of complex microbial communities. However, rRNA evolves so slowly that the specificity of probes based on rRNA sequences may not always be high enough to distinguish closely related strains.

The desired features of target genes for such DNA probes are: (1) They do not transmit horizontally, as plasmid DNA does; (2) Their rate of molecular evolution is higher than that of 16S rRNA; and (3) They are distributed universally among bacterial species. Yamamoto and Harayama (1995) selected the *gyrB* gene, which encodes the subunit B protein of DNA gyrase (topoisomerase type II) as the target of highly specific probes. DNA gyrase, which regulates the supercoiling of double-stranded DNA, is necessary for DNA replication, and the enzyme is distributed universally among bacterial species (Ochman and Wilson, 1987). Yamamoto and Harayama developed a set of PCR primers which allowed *gyrB* genes to be amplified from a large spectrum of bacteria, and nucleotide sequences of the amplified *gyrB* fragments to be obtained rapidly. The method provides a rapid and convenient system for the identification of bacteria, taxonomic analysis, and monitoring of bacteria in natural environments.

Conclusions

All the techniques discussed above are effective as environmental tests, because they provide rapid, reliable, specific and efficient ways to detect microbes in the environment. However, no single technique can provide all the information required for risk assessment, and the combined use of traditional and molecular techniques is essential. For monitoring pollutants, in addition to classical chemical techniques, monoclonal antibodies and bioluminescence provide powerful tools that in many instances make it possible not only to detect pollutants, but also to determine their bioavailability.

Questions, comments, and answers

Q: Did you get a strain-specific probe for *Pseudomanas putida*, and how stable is the epitope in the environment? You can make species-specific probes with 16S-RNA; if not, can you switch to 23S-RNA?

A: Lipopolysaccharides are expressed constitutively and are always located on the cell surface, and therefore constitute a good target for monitoring. Surface proteins involved in the maintenance of cell morphology are essential for survival and are also good targets. There are many complex communities in the environment, and thus, we cannot distinguish species with just r-RNA probes. You need a variety of probes; then the 23S-rRNA and the *gyr*B good alternatives.

Q: How do you perform the PCR, and do you have different sets of oligos for each strain? How laborious is it?

A: Protocols for PCR amplification of DNA are widely known. Once you amplify DNA from an unknown microbe, you can sequence, compare it with a DNA databank, and establish a set of oligos specific for your strain.

REFERENCES

AMANN, R.I., L. KRUMHOLZ and D.A. STAHL (1990), "Fluorescent-oligonucleotide probing of whole cells for determinative phylogenetic, and environmental studies in microbiology", *J. Bacteriol.* 177, pp. 762-770.

BEJ, A.K., M.H. MAHBUBANI, J.L. DICESARE and R.M. ATLAS (1991), "Polymerase chain reaction-gene probe detection of microorganisms by using filter-concentrated samples", *Appl. Environ. Microbiol.* 57, pp. 3529-3534.

FLEMMING, C.A., H. LEE and J.T. TREVORS (1994), "Bioluminiscent most-probable-number to enumerate *lux*-marked *Pseudomonas aeruginosa* UG2Lr in soil", *Appl. Environ. Microbiol.* 60, pp. 3458-3461.

HANCOCK, R.E.W., R. SIEHNEL and N. MARTIN (1990), "Outer membrane proteins of *Pseudomonas*", *Mol. Microbiol.* 4, pp. 1069-1075.

KING, J.M.H., P.M. DIGRAZIA, B. APPLEGATE, R. BURLAGE, J. SANSEVERINO, P. DUNBAR, F. LARIMER and G.S. SAYLER (1990), "Rapid, sensitive bioluminescent reporter technology for naphtalene exposure and biodegradation", *Science* 249, pp. 778-781.

OCHMAN, H. and A.C. WILSON (1987), "Evolution bacteria: evidence for a universal substitution rate in cellular genomes", *J. Mol. Evol.* 26, pp. 74-86.

RAMOS, J.L., A. WASSERFALLEN, K. ROSE and K.N. TIMMIS (1987), "Redesigning metabolic routes: manipulation of TOL plasmid pathway for catabolism of alkylbenzoates", *Science* 235, pp. 593-596.

RAMOS, J.L., E. DÍAZ, D. DOWLING, V. DE LORENZO, S. MOLIN, F. O'GARA, C. RAMOS and K.N. TIMMIS (1994), "Behavior of designed bacteria for biodegradation", *Bio/Technology* 12, pp. 1349-1358.

RAMOS, J.L., P. ANDERSSON, L.B. JENSEN, C. RAMOS, M.C. RONCHEL, E. DÍAZ, K.N. TIMMIS and S. MOLIN (1995), "Suicide microbes on the loose", *Bio/Technology* 3, pp. 35-37.

RAMOS-GONZÁLEZ, M.I., F. RUÍZ-CABELLO, I. BRETTAR, F. GARRIDO and J.L. RAMOS (1992), "Tracking genetically engineered bacteria: monoclonal antibodies against surface determinants of the soil bacterium *Pseudomonas putida* 2440", *J. Bacteriol.* 174, pp. 2978-2985.

REINEKE, W. and H.-J. KNACKMUSS (1979), "Construction of haloaromatic-utilizing bacteria", *Nature* 277, pp. 385-386.

RODRIGUEZ-HERVA, J.J., M.I. RAMOS-GONZALEZ and J.L. RAMOS (1996), "The *Pseudomonas putida* peptidoglycan-associated outer membrane lipoprotein is involved in maintenance of the integrity of the cell envelope", *J. Bacteriol.* 178, in press.

ROJO, F., D. PIEPER, K.H. ENGESSER, H.-J. KNACKMUSS, and K.N. TIMMIS (1987), "Assemblage of ortho cleavage routes for degradation of chloro- and methylaromatics", *Science* 238, pp. 1395-1398.

SHAW, J.J., F. DANE, D. GEIGER, and J.W. KLOEPPER (1992), "Use of bioluminescence for detection of genetically engineered microorganisms released into the environment", *Appl. Environ. Microbiol.* 58, pp. 267-273.

STEWART, G.S.A. and P. WILLIAMS (1992), "*lux* gene and the application of bacterial bioluminescence", *J. Gen. Microbiol.* 138, pp. 1289-1300.

YAMAMOTO, S. and S. HARAYAMA (1995), "PCR application and direct sequencing of *gyr*B genes with universal primers and their application to the detection and taxonomic analysis of *Pseudomonas putida* strain", *Appl. Environ. Microbiol.* 61, pp. 1104-1109.

MEASURING AND MONITORING IN PRACTICE: MICROBIOLOGICAL AND ECO-TOXICOLOGICAL ASPECTS OF SOIL BIOREMEDIATION

by

Anton Hartmann
GSF-Research Center for Environment and Health GmbH, Institute of Soil Ecology
Oberschleissheim, Germany

Introduction

In contrast to the physico-chemical cleaning processes, the detailed parameters of microbiological cleaning technologies should be considered very carefully. The feasibility of microbiological soil remediation needs to be tested via a set of basic experiments using samples from the contaminated site. Proper measures for optimising the degradative activity in a given plot of soil are essential for successful remediation. The autochthonous microbial population as well as the level of general activity and specific metabolic and genetic potentials of the microflora needs to be determined and monitored. In some cases, the effectivity and speed of decontamination can be greatly improved by adding microbial inocula with high metabolic potential. In the course of bioremediation processes, polar metabolites are usually produced, which might have increased toxicity for humans and/or environmental biota. Therefore, eco-toxicological tests need to be performed to evaluate successful remediation. The intended use of a plot of soil after remediation determines which eco-toxicological tests need to be performed in order to prove successful cleaning.

Testing the feasibility of microbial soil remediation

Before any type of clean-up can be carried out (*in situ* or on/off site), the microflora of the polluted site has to be characterised and tested for their degradative abilities on the pollutants in a given plot of contaminated soil. Sampling has to be carried out differently in various types of clean-up processes. In on-site remediations, at least three representative samples from the prospecting place need to be taken. In the case of planned *in situ*-remediation, at least three samples of undisturbed contaminated soil layers from three different locations are necessary. From inner cores of the soil samples, micro-organisms are separated and quantified by determining the viable counts on nutrient agar plates (e.g. R_2A-Agar). However, it has become obvious in recent years that only a very limited number (0.1-10 per cent) of soil microbiota is responding to cultivation techniques. Therefore plate counts can only provide a very limited insight into the actual residing soil microbial community.

To determine the actual and potential activity of the microflora, the respiratory activity of the soil samples is measured. All biological test methods determining oxygen consumption or carbon dioxide production are suitable. Since CO_2-concentration is influenced not only by the respiratory activity of microbes, but also by carbonates in the soil, a definite determination of the activity can only be accomplished by measuring the O_2-consumption. The water content of soil needs to be standardized at 50 per cent of the maximum water retaining capacity, because it has a large influence on microbial activity. The potential microbial activity is measured after adding 1 per cent (w/w) glucose as easily degradable carbon and energy source to a given plot of soil. Respiration of glucose is detected by measuring O_2-consumption in a closed system (respirometer). If the consumed oxygen is not continuously replenished by the measuring device, biological activity should not be limited by the oxygen supply. The actual microbial activity is measured without the addition of glucose. The incubation time depends on the microbial activity and should not exceed seven days. The microbiological testing of sites is supposed to give a first statement about the impairment of soil micro-organisms by the pollutant -- apart from physical and chemical tests of the risk potential of the site. In order to avoid additional expenses, these tests will have to be integrated into the initial assessment. The results of the tests provide evidence of an inhibitory effect of the contamination on the micro-organisms, consequently limiting microbial remediation, if less than 1 000 colony-forming units per gram of soil are determined (Kreysa and Wiesner, 1992). The actual respiratory activity should exceed 0.4 mg O_2/100 g soil per day and the potential respiratory activity should be higher than 4 mg O_2/100 g soil per day (Kreysa and Wiesner, 1992). Under strongly inhibitory conditions, added microbial inocula specialised on certain degradative ability may be inhibited. The nature of inhibition (e.g. heavy metal contamination) needs to be determined, which can be circumvented by adding adsorbing or chelating substances.

The microbial potential for bioremediation can be evaluated in more depth using biomarker techniques and molecular biotechnological approaches which do not need cultivation of the organisms for characterisation. The fatty acid profile of phospholipids extracted from soil allow an analysis of the microbial biodiversity present in a given plot of soil (Zelles and Bai, 1992). This approach can also be used to determine the population structure of soil microflora after bioremediation. The BIOLOG-approach, which uses a series of fluorigenic enzyme reactions, could provide information regarding functional diversity. The presence of particular biodegradative genes can be determined using specific gene probes, i.e. primers, PCR amplification and DNA hybridization techniques (Applegate *et al.*, 1995). As reported recently, the frequency of the naphtalene dioxygenase gene, *nah A*, was found to have a strong correlation with the presence of contaminant concentration in various Manufactured Gas Plant soils (Sanseverino *et al.*, 1993). Further development and application of specific gene probes could improve the predictability and reliability of bioremediation efforts. Using conventional microbiological methods, the determination of aerobic and facultatively anaerobic copiotrophic micro-organisms is possible, which are supposed to represent the most active soil microbes. However, the genetic potential for degradation of xenobiotics may very well reside in soil micro-organisms with special nutrient demands.

To generally evaluate possible pollutant biodegradation in contaminated soils, several experimental designs are used. Actual biodegradation can be tested under slurry conditions with stirring, where a soil suspension is incubated for an appropriate length of time in a fairly spacious closed vessel. Alternatively, most soil samples (50 per cent WRC) can be incubated in a closed vessel. It is important that volatile pollutants (e.g. motor-fuels, chlorinated hydrocarbons or aromatic compounds) require closed test systems. Pollutants with low volatility like polycyclic aromatic compounds or polychlorinated biphenyls do not require these precautions. The results of the degradation experiments show microbial restorability, if (i) the pollutant concentration could be reduced by 70 per cent in the experiments *and* (ii) the pollutant concentration in the control assays is at least 50 per cent higher than in the incubated samples. If the control assays show over 50 per cent pollutant reduction, further experiments detecting chemical and physical effects are required (Kreysa and Wiesner, 1992).

Approaches to optimise biodegradation

Bioremediation activity can be improved by adding particular microbes with high degradation activity to contaminated soil. These microbes could have been isolated from the same or similar contaminated sites and need to show good persistence in the contaminated soil. Microbes with low demands for nutrients or additives, good growth vigour in inexpensive media, as well as high activity at relatively low temperatures (15-20 °C) are superior to highly sophisticated laboratory strains. In addition, the cultivation conditions of the inocula determines the degree of cleaning. The residual level of contaminants can be lowered by adapting the inoculum strain to low pollutant concentrations. Furthermore, it is advantageous -- or even necessary in some cases -- to work with microbial consortia to introduce the full capacity of biodegradation to the polluted soil. Alternatively, the efforts to engineer highly active bacteria with complete degradation pathways are very constructive and promising for improving the predictability and reliability of the bioremediation approach. The inoculation trials should be -- at least in the experimental stage -- accomplished by monitoring the inoculant bacteria and their activity during biodegradation. Serological (Schloter *et al.*, 1995) as well as molecular biological tools are available to monitor the fate of the introduced micro-organisms and their degradative genes (Applegate *et al.*, 1995). The success of the bioremediation process would be more predictable if one used strain-specific monoclonal antibodies against highly potentially degrading micro-organisms (GEMS or natural isolates) to monitor these populations.

The success of biological clean-up trials is dependant on a welladapted laboratory optimisation program which takes into account the special situation of soil contaminant complex. Usually, the soil pH, aeration, water content and nutrient availability are the key parameters which may limit optimal degradation activities. Among the nutrients, the input of carbon substrates stimulates the biological activity in general, but also phosphate or sulphate or even iron or molybdenum may limit the optimal development of biodegradative activity of microbes in particular cases.

Investigations to monitor pollutant biodegradation

Degradation processes in soils are complex, because soil compounds are capable of "humificating" pollutants. This means that the pollutants or their metabolites are integrated into the organic matrix of the soil. Additionally, intensive adsorption and desorption processes on the surfaces of clay-humus-complexes make pollutants, or its metabolites, inaccessible to analysis, simulating a degradation process. On the other hand, the interaction between soil and contaminants can reduce the toxicity of the pollutants, thereby facilitating the bioremediation process. By being involved in the sorption processes, the contaminants will become partially inaccessible to direct degradation by most micro-organisms. However, with the use of exo-enzymes, some micro-organisms can attack even sorption complexes of pollutants. Consequently, several different types of biodegradation can be observed in microbiologically active soil samples: (i) complete mineralization to carbon dioxide; (ii) incomplete degradation (transformation); and (iii) binding of the pollutants or their metabolites to the soil matrix (humification). Which of these reactions will dominate the degradation process is determined by the composition of the soil and by the types of pollutants. The degradation of some pollutants, such as heavy oil, polycyclic and other aromatic compounds, should therefore be tested under different soil conditions in order to determine the optimum conditions of bioremediation.

In biodegradation processes, usually polar metabolites are produced, which might increase the toxicity for humans and the environment. However, these compounds are further degraded or eliminated from the system by adsorption or by covalent binding to the matrix, which could be monitored by chemical analysis and eco-toxicological tests. Polycyclic aromatic hydrocarbon (PAH) and aromatic compounds are of particular concern, when soil contaminated by accidental diesel oil spill is bioremediated after optimisation of microbial degradation activity by pH control, nutrient balance, aeration and mixing. During the bioremediation process, mutagenicity and toxicity assays were performed (Wang *et al.*, 1990) using methylene chloride extracts from these soils. The Ames mutagenicity assay and the Microtox-test, which assess acute toxicity by measuring reduction in light emission by a *Phosphobacterium phosphoreum,* were conducted. From a low zero-time value, PAH mutagenicity in the Ames test increased sharply in the first weeks of bioremediation. Obviously, the PAH compounds were hydroxylated and converted to active mutagens as a consequence of microbial metabolism. As further metabolism proceeded, mutagenicity declined at 12 weeks to almost background level and became equal to background by week 20. Concomitantly, EC_{50} values in the Microtox-test declined after a sharp rise in the first weeks to approach background by week 12. This coincided with chemical PAH-analyses of soil extracts. It was concluded that bioremediation treatment of PAH-contaminated soils does not leave behind objectionable PAH residues.

Eco-toxicological assessment strategy

In connection with the final aim and future use of the bioremediated soils -- either for multi-functional purposes or for landfills -- the final degree of cleaning needs to be discussed. Usually, in mineral oil spills, the first cleaning is about 70-80 per cent, while only in the second phase with increased effort, further reduction of pollutants is possible. As has been exemplified above, eco-toxicological monitoring is important to assess the degree of cleaning. Biological test systems are appropriate to detect adverse effects of chemicals of a different nature which could escape chemical analysis. Table 1 presents criteria for choosing biological test systems in relation to the planned use of the treated soils (Kreysa and Wiesner, 1995). For treated soils, which will be used under closed cover without excess of water, no biological tests are necessary. Soils used as landfills in non-covered, but commercially used, land should be examined in aquatic ecotox-tests (bioluminescence, algae and daphnia tox-tests) to prove that no further contamination of neighbouring ecosystems can occur through the water path. For bioremediated soils to be used in parks and recreation areas, eco-toxicological tests to prove the successful reconstitution of the habitate function of soil for plants, soil fauna and microbes are recommended. In this case, tests for plants and soil habitate function of soil microbes are recommended. The terrestrical ecotox-tests using plants, earthworm and soil microflora should be performed in a qualitative way to detect any risk potential. If the soils are going to be used in recreation areas, a human toxicological risk evaluation is also necessary. Finally, in the case of future horticultural or agricultural usage of bioremediated soils, the complete set of biological ecotox-tests needs to be applied (Table 1).

The water extracts should be prepared with a ratio of 1 to 2 (soil and water) according to DIN 38414. As test micro-organisms for the bioluminescence test, *Vibrio fischeri/Phosphobacterium phosphoricum* NRRL-B 11177 can be used with these soil extracts according to DIN 38412 - parts 34 and 341. As algal test, inhibition of biomass production of *Scenedesmus subspicatus* is assessed by measuring the chlorophyll fluorescence at 685 nm (DIN 38312 - part 33). As representative of sensitive aquatic animals, the effects of soil eluates on the mobility of *Daphnia magna* can be quantified (DIN 38412 - part 30). In terrestrical ecotox-tests the inhibition of plant growth is determined using *Avena sativa*, *Brassica rapa*, *Lepidium sativum* or *Phaseolus aureus* as test plants according to the ISO/DIS 11269-2 guideline. To test the undisturbed habitate function of soils for biocoenosis, the inhibition of substrate induced microbial respiration (ISO/DIN 14240-1) as well as the inhibition of the potential nitrification (ammonium oxidase activity) is recommended (Kreysa and Wiesner, 1995). Finally, the mortality test of the earthworm *Eisenia fetida* (ISO 11268-1) can also provide eco-toxicological relevant information on residual toxicity in soils. It is important, that the intrinsic toxicity of soils is measured using soil biota. In these test systems the bioavailability of toxic compounds is also assessed, providing information on the actual toxicity status of the soil. Clearly, there is further need for ecologically-relevant, easily performable, and sensitive, terrestrical eco-toxicological evaluation measures.

Table 1. Criteria for choosing biological test systems in relation to the planned use of the treated soils

Soil functions	Binding function	Habitate function	
	Water path	Plants	Soil biocoenosis
Biological test systems	Aquatic tests	Terrestrical tests	
	Bioluminescence (bact.) Algae Daphnia	Higher plants	Microflora: nitrification respiration earth worm
Soil usage	Examination of biological effects		
Soil under closed cover	no	no	no
Not covered, but commercially used land	yes	no	no
Cover of dumps	yes	yes[1]	no
Parks and recreation areas	yes	yes[1]	yes[1]
Land for horticultural and agricultural use	yes	yes	yes

Notes:

1. Only qualitative testing is required.

Source: Kreysa and Wiesner, 1995.

Questions, comments, and answers

Q: A problem of chemical analysis is that variation in results can be so high that they are not reliable. Is this true also of ecotoxicity tests? Is there the same variability?

A: If there is high heterogeneity in the samples, this is also true for ecotoxicity tests. However, the advantage of biological tests is that they give an indication of overall effects rather than specific causes. If there is an indication of some toxicity, you must go back to the detailed chemical analysis.

Q: What about acceptability of the use of GMOs? You seemed to say that the visibility of the organism will facilitate its acceptance, but De Lorenzo made the opposite point; remove all traces of engineering.

A: We should not discriminate between engineered and efficient wild-type organisms. We should treat both as potential inoculants, and make sure they do not have special risks. We also should provide specific tools to monitor these micro-organisms.

REFERENCES

APPLEGATE, B.M., U. MATRUBATHAM, J. SANSEVERINO and G.S. SAYLER (1995), "Biodegradation genes as marker genes in microbial ecosystems", in *Molecular Microbial Ecology Manua,* A. Akkermans and J.D. van Elsas (eds.), pp. 1-14, Kluwer Academic Publisher, Netherlands.

KÄMPFER, P., M. STEIOF and W. DOTT (1991), "Microbiological characterisation of a fuel-oil contaminated site including numerical identification of heterotrophic water and soil bacteria", *Microbial Ecology* 21, pp. 227-251.

KING, E.F. (1984), *Toxicity Screening Using Bacterial Systems,* D.L. Liu and B.J. Dutka (eds.), pp. 175-194, Marcel Dekker, New York.

KREYSA, G. and J. WEISNER, eds., (1992), "Laboratory Methods for the Evaluation of Biological Soil Clean-up Processes", Second Report of the interdisciplinary working group "Environmental Biotechnology - Soils", headed by J. Klein and W. Dott, DECHEMA E.V., Frankfurt, ISBN 3-926959-38-X.

KREYSA, G. and J. WEISNER, eds., (1995), "Bioassays for Soils", Fourth Report of the interdisciplinary working group "Environmental Biotechnology - Soils", headed by J. Klein and W. Dott, DECHEMA E.V.,Frankfurt, ISBN 3-926959-52-5.

SANSEVERINO, J., C. WERNER, J. FLEMING, B. APPLEGATE, J.M.H. KING and G.S. SAYLER (1993), "Molecular diagnostics of polycyclic aromatic hydrocarbon biodegradation in manufactured gas plant soils", *Biodegradation* 4, pp. 625-648.

SCHLOTER, M.A.B. and A. HARTMANN (1995), "The use of immunological methods to detect and identify bacteria in the environment", *Biotechnology Advances* 13, pp. 75-90.

WANG, X., X. YU and R. BARTHA (1990), "Effect of bioremediation on polycyclic aromatic hydrocarbon residues in soil", *Environmental Science Technology* 24, pp. 1086-1089.

ZELLES, L. and Q. BAI (1992), "Fractionation of fatty acids derived from soil lipids by solid phase extraction and their quantitative analysis by GC-MS", *Soil Biology and Biochemistry* 25, pp. 495-507.

AIR POLLUTION: BIOTREATMENT STRATEGIES AND THEIR ENVIRONMENTAL IMPACT[1]

by

Enrique J. Marroquín
Grupo Cydsa, S.A. de C.V., Garza García, Mexico

Victor M. Morales
Celulosa y Derivados, S.A. de C.V., Monterrey, Mexico

Sergio Revah
Dept. of Chemical Eng., Universidad Autónoma Metropolitana- Iztapalapa, Mexico City, Mexico

Introduction

Industries long established in what decades ago seemed remote rural regions of Mexico, are now surrounded by communities in cities such as Mexico City, Guadalajara and Monterrey. Indeed, the most important contribution from the industrial sector of Mexico comes from facilities located within or near these crowded urban areas. This unsustainable development has created severe problems such as pollution, overpopulation, and corruption in these cities.

Government, industry and community awareness regarding this situation began to develop in the mid 80s when unfortunate industry-related tragedies occurred in Mexico (e.g. the tragic explosions of a propane distribution facility located within the densely populated area of San Juanico, north of Mexico City, which killed several hundred people and injured more than 1 000, and levelled several blocks of houses surrounding the facility; and the disastrous hydrocarbon-related sewage explosions in Guadalajara, which caused the death of more than 200 people and drove a trench several kilometres long and several meters deep).

Furthermore, community awareness towards the impacts caused by past industrial-related activities, as well as those being caused by current ones, has surpassed their need for living close to their jobs, and prefer that industries relocate rather than be exposed to any hazards. This awareness has triggered a tougher enforcement by environmental authorities, which was practically non-existent a decade ago.

The situation in the city of Monterrey is no different than that of the other industrialised cities in Mexico. The oldest, and most respected industries of Monterrey (e.g. glass, cement, steel, chemical, paper, and breweries), from which the industrialisation of the city sprouted around the middle of this

1 The authors would like to thank Ms. Lourdes Ugarte for her assistance in the preparation of this report, specifically for her insights that contributed to finishing it as planned.

century, and have given Monterrey the status as the second biggest contributor in the industrial sector of the country, lie within densely populated areas throughout the city. Even though some of the industries have relocated towards less-crowded suburbs, or even to farther locations, the majority of them cannot afford to do so.

Due to these latter issues, some industries in Monterrey have had to shut down operations or, for those who could support the costs, relocated. Nonetheless, other industries have also worked pro-actively by devising solution strategies to reduce their environmental impact. The strategies to reduce and control emissions of contaminants include 'commodity' solutions, (e.g. reduction of fly ash and sulphur dioxide emissions from power generation plants by switching energy sources from fuel-oil to natural gas); and 'tailor-made' innovative solutions, such as use of the biotreatment alternative, as is the case of Cydsa Corp.

Although both strategies are equally effective in terms of mitigating impacts, we will concentrate in this paper on those that include 'tailor-made' sustainable biotreatment solutions for abatement of reduced sulphur gaseous emissions (i.e. hydrogen sulphide and carbon disulphide), as applied by Cydsa in two of its Monterrey facilities. It is important to note that this article is not intended to review nor discuss in detail any technical or scientific aspects related to gaseous biotreatment applications, but rather as a narrative of our experiences in the application of a particular biotreatment application at our facilities. The reader is encouraged to consult the reference section for technical reports and papers on this matter.

The situation

We will concentrate on a situation caused by a particular set of malodorous emissions generated at a group of facilities, in the City of Monterrey (900 kilometres north of Mexico City), within an industrial complex which is located in the midst of a thriving community. The perimeter of the complex is surrounded by houses, some of them having the complex's battery limits walls as their backyard walls. The chemical compounds that cause the foul smelling odours are hydrogen sulphide and carbon disulphide, generated by the production of rayon fibre and cellophane film, both using the viscose process.

It is important to note that these plants were erected 20 years or so before any homes were built at or near their perimeters. In other words, the foul smells had always existed, but, as the population in the area increased, more and more people moved into the area surrounding the complex. Nevertheless, the plants' operators felt that it was their right to continue operations as they were there first. On the other hand, the community felt that the industry was harming them on purpose, and were blind to the fact that it was the thriving industrial activity of this, and other complexes within the area, that had attracted them to the area. This plight created tension and anger between the community and the complex for more than a decade; and, as there were really no regulations stipulating the prevention of emissions, government authorities were not willing to attend to the situation beyond issuing reprimands to the plants' administrators.

Whilst the tension reached its height by the mid 80s, the top management of the company decided to begin two ventures.

The first venture was the creation of a community awareness program, in which, among other things, people who lived near the complex would be informed of the solutions being sought for eradicating the odours. The company would heed the community complaints, regardless of whether or not they were related to the operation of the plants, and when possible, address them in the most effective way. Moreover, during this time, the (environmental) law enforcement agencies did not issue reprimands anymore. A regulation was now in place that allowed them to legally act upon polluters. They were also considering stricter penalties, including fines and incarceration.

The second venture was a thorough search for a solution that would eliminate all the odours generated by these two plants. The result of this search was that none of the available abatement alternatives were fit for the current (economic) status of the plants; they were too expensive to implement. The stakes were too high -- 1 500 jobs, and 25 per cent of the complex's revenues were generated by these plants, notwithstanding the fact that, if moved or shut down, it would trigger a domino effect on the whole complex operation that would eventually lead to the shut down of the whole operation, and maybe of the whole corporation. Nevertheless, it was decided to explore the use of micro-organisms for this purpose. After all, both contaminants contained sulphur and were, theoretically speaking, easily degradable -- via the sulphur cycle -- by naturally occurring bacteria (Figure 1).

Figure 1. The naturally occurring sulphur cycle

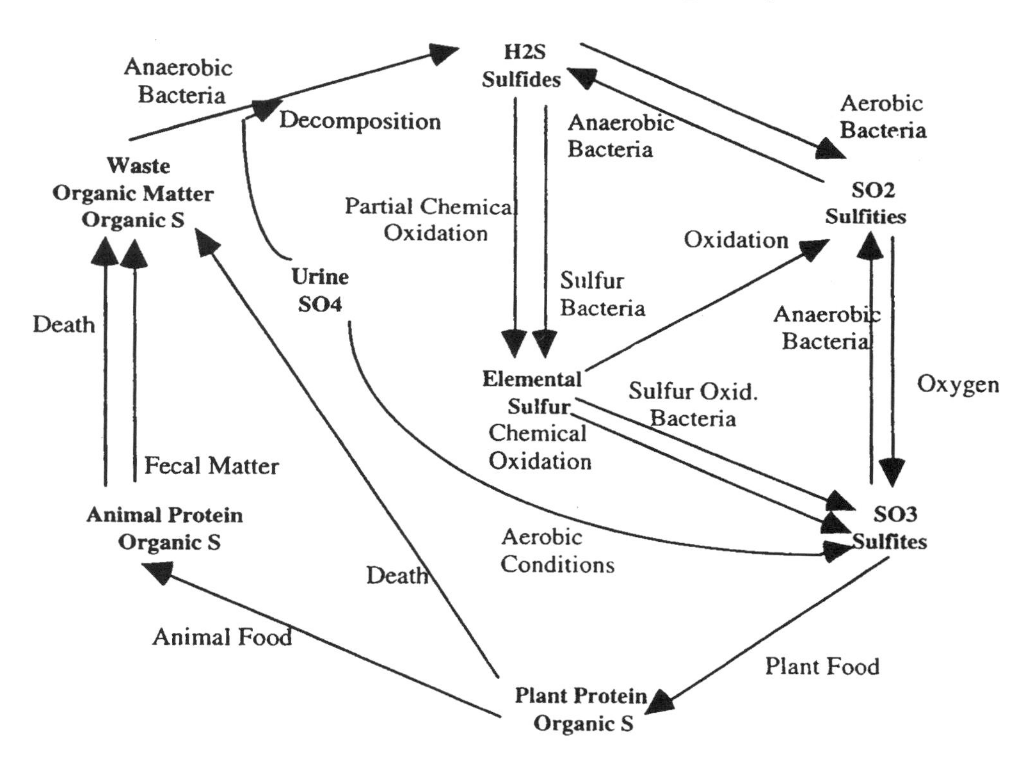

Source: Authors.

We will now focus on the strategy followed by our team for devising the biotreatment solution that is now in operation at these plants.

The treatment strategy

Selection of the treatment alternative

A team of four people evaluated five different treatment alternatives: chemical scrubbing; carbon adsorption; catalytic and thermal incineration; chemical and photochemical oxidation; and biological treatment. The following criteria was defined for the decision-making process:

- low capital cost throughout all the scale-up process;

- low maintenance cost solution;

- processes easy to design, build, operate and control -- keeping it as simple as possible;

- past experience with similar processes -- both operational and for design -- had to be considered; We wanted to use as many internal resources as possible.

Furthermore, two more criteria were defined as part of a long-term vision that Cydsa had for this development:

- The solution had to be sustainable (even though back in the late 80s this was not an immediate concern for the authorities, nor for that matter, in the public opinion).

- The solution had to be devised within a domain of a potential Intellectual Property business asset.

The latter two criteria were defined to eliminate the possibility of designing a short-term, end-of-pipe solution ,i.e. an aspirin for a migraine.

Based on this criteria, we decided to choose the biological treatment alternative. Briefly put, biological reactors, as known by us from other treatment applications (i.e. waste water treatment plants), were not complicated nor expensive to build; maintenance was low as the consortia of microbes would do most of the work, and there would be no major moving parts other than pumps and fans. Moreover, the company had had at least 30 years experience with biological processes -- in the waste water treatment domain.

In terms of sustainability, we knew that the product of the microbial degradation of sulphur compounds would be basically composed of organic matter (i.e. dead cells) and sulphur compounds (i.e. elemental sulphur and soluble sulphate salts), and that it could be used as an agricultural additive. Furthermore, the process could be patented, if needed, thus complying with the two latter criteria.

The only criteria that was not strong about this alternative, was the ease of operation and process automation. Biological processes can be controlled, but there was some uncertainty as to the relatively few documented process control schemes for these processes. Nevertheless, this opportunity was also addressed by the project team with an objective to design a process control loop which meant identifying key control parameters.

Micro-organisms

The choice of the biotreatment process was based on a group of micro-organisms that were known to degrade sulphides. For example, there were anaerobic photosynthetic bacteria, *Chlorobium thiosulfatum*, which is known to convert sulphides into elemental sulphur (Cork and Ma, 1982). Other reports showed that *Thiobacillus ferrooxidans* could transform H_2S into elemental sulphur (Schäfer-Treffefeldt *et al.*, 1984). In any case, we decided that our process was to be a simple one, so our search for a strain or consortium of bacteria meant that we needed to consider the following:

- The micro-organisms had to be aerobic.
- The carbon source had to be taken from both the CO_2 and the CS_2 present at the gaseous emissions (i.e. chemo-autotrophic micro-organism).
- The micro-organisms had to be easy to obtain, i.e. naturally occurring and not genetically altered.
- The micro-organisms had to derive their energy from the reduced sulphur compounds (i.e. H_2S and CS_2).

The micro-organisms belonging to the family Thiobacteriaceae and to the genus *Thiobacillus* were the ones that were to be used for our purpose, as we knew from published work that several *Thiobacillus* sp. strains had been used by others to degrade hydrogen sulphide (Buisman *et al.,* 1990; Sublette, 1987; Sublette and Sylvester, 1987).

Laboratory studies

The laboratory studies were conducted utilising micro-organisms obtained from different sources including activated sludge from an UASB-type pilot plant and mud from different sulphurous water sources.

The kinetic studies were realised with an enriched culture of sulphur-degrading bacteria in 5-litre stirred tank fermentors equipped with temperature and pH control, and inoculation was made from an Erlenmeyer flask, and in a 125 litre pilot tower constructed at the university site, Universidad Autónoma Metropolitana - Iztapalapa Campus. The details of these studies are described in "Biotechnological Process for the Treatment of H_2S and CS_2 from a Waste Gas from a Cellophane Plant" (Morales *et al.*, 1992).

Pilot plant

The pilot plant studies were conducted at a tower fermentor (7.25 m^3 packed section), constructed on the site of a Cellophane film producing facility in Monterrey, Mexico. This pilot plant was fed with the exhaust gas of the plant's production line, containing H_2S (0-1 000 ppmv) and CS_2 (0-400 ppmv). Elemental sulphur and sulphates were derived from the microbial oxidation of these sulphides. Removal efficiencies of 95 per cent for both compounds were attained after 10 weeks of operation (Morales *et al.*, 1992; Revah *et al.*, 1995).

Process

The pilot plant tests showed that the process could be scaled up to a size both technically and economically feasible. The process, named Biocyd®, which is thoroughly described in "Biological Process for the Elimination of Sulphur Compounds Present in Gas Mixtures" (Torres *et al.*, 1993) consists of a reactor filled with a high surface support that allows the immobilisation of the adapted sulphide-utilising population while having low pressure drops. Culture medium or water are continuously re-circulated from a balance tank in counter current or crossed flow with the gas flow. The pH is controlled through the addition of either NaOH or HCl. Part of the culture medium and water are re-circulated to the reactor, the rest is sent to separation process in which the remaining biomass, elemental sulphur and dissolved sulphates are recovered. Insofar as usage of these semi-solid materials is concerned (i.e. sustainability of the process), applications are currently being studied for agricultural uses; the results so far have been favourable.

Full-scale units

Several types of equipment based on these principles are now in operation; some have been used for pilot studies, while others are meant to treat the emissions of the above-mentioned plants. A list of different types of equipment is included in Table 1.

Table 1. Equipment used for gaseous H_2S and CS_2 removal from exhaust gas

Equipment	Packing volume (m^3)	Flow in thousands of (m^3/hr)	Source	Inlet gas concentration (ppmv)	Year of operation
BIOCYD®[1] I (pilot)	7.25	1.8-3.0	Cellophane Film Plant	H_2S: 250-500 CS_2: 200-900	1991
BIOCYD®-Rayon (pilot)	21.8	15-18	Rayon Fibre Plant	H_2S: 20-150 CS_2: 20-150	1992
BIOCYD®-Sponge (pilot)	4.4	1.5-2.1	Sponge Plant	H_2S: 40-150 CS_2: 50-500	1993
BIOCYD® II (full-scale)	51.3	52.0	Cellophane Film Plant	H_2S: 250-500 CS_2: 200-900	1995
BIOCYD®-Sponge II (pilot)	13.6 × 2 (reactors)	4.0	Sponge Plant	H_2S: approx. 0 CS_2: 0-1 100	1996 (est.)

Notes:

1. BIOCYD is a registered name of Grupo Cydsa, S.A. de C.V.

Source: Authors.

As can be seen from Table 1, the biotreatment strategy has been implemented thoroughly at three different facilities, two of which were mentioned above (i.e. Cellophane Film and Rayon Fibre manufacturing) and a third one, which is for a different application, (cellulose) sponge manufacturing. The reason for this being that cellulose sponge is produced via the viscose process; therefore, the plants' emissions, though not the same, are very similar, hence the interest in trying the applicability of the technology on this 'new' process.

Validation and documentation

As mentioned above, the biotreatment strategy once developed and devised for addressing our own problems, became interesting for other plants, with slightly different variations in their process stack emissions.

For this reason, and due to the fact that at the very conception of this project we set as a goal to obtain a patent, or some kind of intellectual property protection which would transform our investment cost in R&D into an asset, we started a thorough validation and documentation process within Cydsa's own Experimental Center for Applied Environmental Biotechnology -- a spin-off from the initial project, and necessary for conducting these activities, amongst others.

The strategy was to organise all the processes that conformed with the Biocyd® technology -- from research and development to marketing and customer service -- within the domain of a Quality Assurance and Quality Control philosophy; and, using as a reference, the ISO 9000 guidelines for documentation.

In the course of two years, we presented five technical and economical proposals to customers who were interested in installing a pilot unit at their plants in the United States, Italy, and China; and by the end of 1995, we had sold two pilot projects in the United States. The Italian project has advanced and will probably crystallise at the end of 1996.

The latter issues have forced us to fast track our Quality Assurance/Quality Control and ISO 9000 documentation program.

Our process validation strategy consists of two elements:

1) implementation of a statistical process control system on all Biocyd® reactors;

2) standardization by means of a Quality Control system based on the ISO 9000 standard.

We will discuss both elements of our strategy briefly.

1) Implementation of a statistical process control system on all Biocyd® reactors

It can be said that a process is stable when no exceptional variations occur; thus, the process is reproducible and predictable. Statistical process control (SPC) is a rational means to accumulate process performance knowledge that allows one to better understand how to modify process variables that will maintain results within an expected span. Furthermore, SPC leads to a stable process operation with reproducible and predictable results (Deming, 1982).

The following activities have been implemented by us towards the implementation of this SPC system:

- *Evaluation of all monitoring systems*: Activity that includes an evaluation program of the reproducibility, accuracy, calibration and cleanness of all measuring instruments.

- *Certification of analytical methods*: Activity that contemplates the evaluation of sampling techniques, verification of sampling points, standardization of the analytical methods, etc.

- *Identification of critical variables that influence the process' performance for achievement of desired targets (i.e. removal efficiency, removal rate):* In this activity, brainstorming sessions are conducted to generate cause-effect and Pareto diagrams to identify critical issues of the process performance.

- *Implementation of X-R control charts*: We have made a hypothesis that our process' variations can be described by a normal distribution curve, and that it can be defined by use of a mean and standard deviation. Almost all of the control diagrams utilised by us to describe the behaviour of the Biocyd® process use control limits calculated from the process data gathered at 3 *sigma* (i.e. 99.73 per cent of the measured variables are within limits).

Our work in this area is still unfinished, but we strongly believe that the use of control diagrams will allow us to reduce the variability of each and every one of the control parameters of our process, and that we will be able to predict and adjust the process towards stability every time there is a cause of instability.

2) Standardization by means of a Quality Control system based on the ISO 9000 standard

The second element is characterised by a thorough documentation process of all activities that conform with the Biocyd® technology, using as a reference the ISO 9000 standard. We have documented both administrative activities (e.g. marketing, customer service, procurement) and operation-related activities. We will concentrate more on these next. Our Quality Control system is based on the ISO 9000 Quality Pyramid depicted in Figure 2.

Figure 2. Documentation of Biocyd® process Quality Control system based on the ISO 900 Quality Pyramid

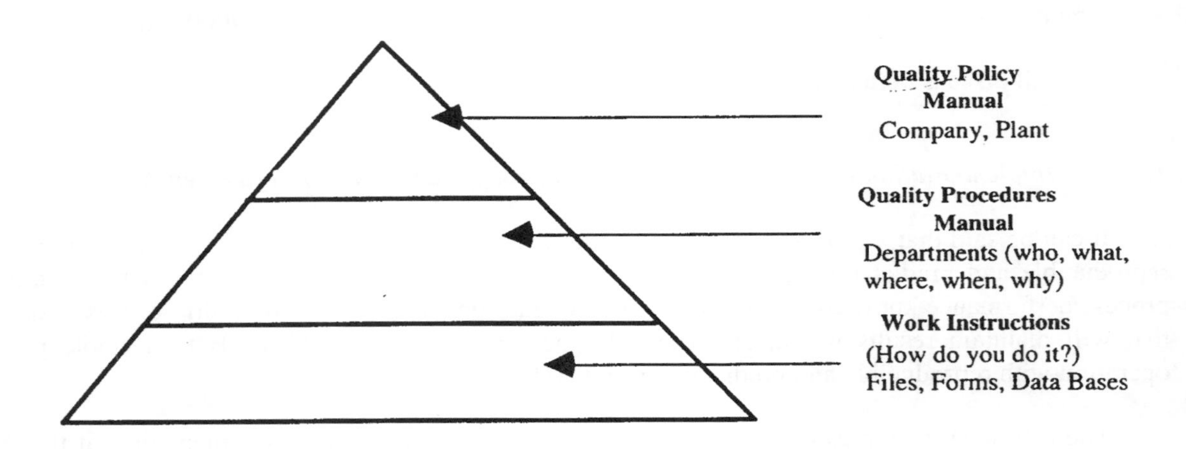

Source: Deming, 1982.

The lower level of the pyramid depicted in Figure 2 represents the third level documents that include all the detail process operational procedures, operation of equipment, nutrient and inocculum preparation, etc. It also includes all the formats that need to be filled when conducting start-up and shutdown procedures, analytical tests, logs, etc. These procedures, coupled with a robust operator training program, assures that all activities performed will be reproduced within the specified domain.

As for the other two levels, we have not started the documentation process yet; we expect to finish the whole documentation process by the end of 1996. When this operation is finished, we will decide if it is feasible to undergo the protocols for certification under the ISO 9001 standard, depending on the demand for the Biocyd® technology by customers whom require such certification.

Conclusions

The use of a biotreatment alternative for abatement of hydrogen sulphide and carbon disulphide emissions from a Rayon Fibre and Cellophane film plant has proven to yield good results and to become a competitive option vis-à-vis other control technologies.

However, the biotreatment alternative for air pollution control applications can be considered novel, and the public and government agencies are concerned with the outcome of using these technologies; in some countries, such as the United States, potential users of biotreatment alternatives prefer to use the old, but proven, alternatives such as carbon adsorption, caustic scrubbing, and thermal oxidation, rather than exploring the outcome that the former might bring. Government agencies in North America, on the other hand, apply the same validation criteria for biotreatment alternatives as they do for other conventional treatment technologies, which is making them less attractive for users who wish to comply with environmental standards.

The latter has pushed us to continue our efforts for validation and documentation of our technology in order to make it more appealing to not only quality-conscious clients, but to government agencies as well. This effort is consuming valuable resources which may otherwise be used for developing more know-how and continue to distinguish our technology from others that are beginning to appear in other parts of the world. Nevertheless, we are convinced that it is the only way to stand out and compete in the global market, and will indeed prepare the road for further work in this area for the future.

Questions, comments, and answers

Q: What is the load in terms of mg per m^3, and what was the efficiency?

A: We only have basic information here but can share the details with you later -- the removal efficiency depends on load and has been 98-99 per cent for H_2S, and 70-80 per cent for CS_2. CS_2 is less soluble and is harder for it to get to the biofilm.

Q: I'd like to know more about the lifetime of the biofilter, how frequently you have to replace it, and what you do with it afterwards.

A: In the original, which started up in 1991, we haven't yet changed the packing material. The tower is sealed and it would be very difficult! We inoculate on a bimonthly or monthly basis with fresh inoculum.

Q: Sulphur emerges as sulphate and elemental sulphur. How do you upgrade and dispose of this?

A: The emission we treat are mainly air, and so it is difficult to stop the process at the level of elemental sulphur. We are, however, using the bioliquor on soils lacking in sulphate.

REFERENCES

BUISMAN, C.J.N., P. YSPEERT, B.G. GERAATS and A.G. LETTING (1990), "Optimization of sulphur production in a biotechnological sulphide-removing reactor", *Biotechnology and Bioengineering* 35, pp. 50-56.

CORK, D.J. and S. MA (1982), "Acid-Gas Bioconversion Favors Sulphur Production", *Biotechnology Bioengineering Symposium* 12, pp; 285-290.

DEMING, W.E. (1982), *Quality, Productivity and Competitive Position, or Out of the Crisis*, Massachusetts Institute of Technology Center for Advanced Engineering Study, Cambridge.

MORALES, V., A. HINOJOSA, F. PAEZ, M. GONZÁLEZ, O. GONZÁLEZ, T. VIVEROS and S. REVAH (1992), "Biotechnological Process for the Treatment of H_2S and CS_2 from a Waste Gas from a Cellophane Plant", in Ninth International Biotechnology Symposium, August 1992, Washington D.C.

REVAH, S., A. HINOJOSA and V. MORALES (1995); "Air biodesulfurization in process plants", in *Bioremediation: the Tokyo '94 Workshop,* pp. 569-576, OECD, Paris.

SHÄFER-TREFFEFELDT, W., R. ENGEL and U. ONKEN (1984), "Removal of Hydrogen Sulphide from Exhaust Gas by Microbial Oxidation", in Proceedings, 3rd European Congress of Biotechnology, Munich, DECHEMA, pp. 3.123-3.128.

SUBLETTE, K. (1987); "Aerobic Oxidation of Hydrogen Sulphide by *Thiobacillus denitrificans*", *Biotechnology Bioengineering* 29, pp. 690-695.

SUBLETTE, K. and N. SYLVESTER (1987), "Oxidation of Hydrogen Sulphide by *Thiobacillus denitrificans*: Desulfurization of Natural Gas", *Biotechnology Bioengineering* 29, pp. 249-257.

TORRES, M., S. REVAH, F. PAEZ, A. HINOJOSA and V. MORALES (1993), "Biological Process for the Elimination of Sulphur Compounds Present in Gas Mixtures", U.S. Patent 5 236 677.

CONSEQUENCES OF BIOAVAILABILITY FOR CLEAN-UP AND ENVIRONMENTAL RISKS

by

Tom N.P. Bosma
Swiss Federal Institute of Environmental Science & Technology (EAWAG),
Kastanienbaum, Switzerland

Introduction

Organic pollutants in soil can be removed by means of biological treatment. A major problem in the application of these treatments is the efficiency of biodegradation in soil. The bulk of the pollution can often be removed but certain residual amounts remain unaltered. In addition, biodegradation rates are often much slower than expected on the basis of laboratory trials. Prerequisites for biotransformation are that the overall reaction is thermodynamically favorable and that micro-organisms possess biodegradative capacities, possibly after acquisition via gene transfer (Van der Meer *et al.*, 1992). The kinetics of microbial growth are not always sufficient to explain the slow biological removal rates in soil and the occurrence of residual amounts after bioremediation. Chemicals seem to be available for microbial degradation only when they are dissolved in water (Harms and Bosma, 1996). As a consequence, solid compounds have to dissolve (Volkering *et al.*, 1993) and sorbed substrates have to desorb to be available to micro-organisms (Rijnaarts *et al.*, 1990). Hence chemicals, particularly those with a low aqueous solubility or a high tendency to sorb to solid matter, may escape from microbial breakdown during the bioremediation process. In this contribution we will discuss i) the role of sorption and dissolution kinetics as the limiting factor for bioremediation, ii) ways to enhance the effectiveness of bioremediation and iii) the risk associated with the strongly sorbed, almost unavailable pollution that remains after removing the accessible part of the pollution.

Sorption

Sorption can be defined as the partitioning of a compound between a solid (e.g. soil particles) and a liquid phase. The adsorption isotherm of hydrophobic organic pollutants in soil is linear at low concentrations, which means that sorption is linearly correlated with the concentration in the liquid phase (Karickhoff *et al.*, 1979):

$$S = K_p C$$

S is the amount of compound adsorbed on the solid phase, K_p the partition coefficient and C the concentration of the compound in the liquid phase. Sorption of hydrophobic compounds in soil can be viewed as a partitioning process between the aqueous and organic phases (Schwarzenbach and

Westall, 1981). Equilibrium partition coefficients for sorption of hydrophobic organics to soil organic matter (K_{oc}) can be estimated from 1-octanol/water partition coefficients or water solubilities, using linear free energy relationships (Karickhoff *et al.*, 1979; Schwarzenbach and Westall, 1981). K_{oc}-values can be converted to K_p-values when the f_{oc} of a soil is known, by using the relationship:

$$K_p = f_{oc}K_{oc}$$

The partitioning of hydrophobic organic contaminants between organic matter and water is a physical process that reaches site equilibrium within milliseconds or, at most, seconds (Weber *et al.*, 1991). Sorption rates of organic pollutants found in soil and sediment are generally slower, reaching apparent equilibrium within a day, as a result of the existence of a diffusion layer between the bulk fluid and the sorption site (Weber *et al.,* 1991). However, several researchers have observed that, after contact times in the order magnitude of years, aggressive extraction methods are needed to remove organic chemicals sorbed to soil (Karickhoff, 1980; Steinberg *et al.,* 1987). A two-site model can be used to account for this apparent irreversible sorption (Karickhoff, 1980). One site accounts for a rapid equilibration and the second, defined by a small rate constant, accounts for the apparent strong binding after long contact times. Later, a sorption-retarded radial diffusion model was successfully applied to explain the very slow sorption of hydrophobic chemicals (Wu and Gschwend, 1986).

The restriction of the bioavailability by slow diffusion inside of natural and artificial aggregates has been observed in a number of studies (Rijnaarts *et al.*, 1990; Mihelcic and Luthy, 1991). A sorption-retarded radial diffusion model using one effective diffusivity parameter was shown to explain the effect of aggregate size on the α-hexachlorocyclohexane (α-HCH) desorption and bioconversion rate (Rijnaarts *et al.*, 1990). In many instances intraparticle diffusion was found to be a very slow process which may not reach equilibrium within years (Rijnaarts *et al.*, 1990; Ball and Roberts, 1991). The longer certain compounds are in contact with soil, the more resistant they become to desorption and degradation. In soils this has been found with chlorophenol (Salkinoja-Salonen *et al.*, 1989), TCE (Pavlostathis and Mathavan, 1992), 1,2-Dibromoethane (Steinberg *et al.,* 1987), hexachlorobenzene (Beurskens *et al.*, 1993) nitrophenol, and phenanthrene (Hatzinger and Alexander, 1995). Steinberg *et al.* (1987) attributed this so-called contaminant aging to progressive entrapment in microscopic pores. Using 3-chlorodibenzofuran that was sorbed by teflon particles, contaminant aging could be simulated in relatively short time scales (Harms and Zehnder, 1995).

Dissolution

Since hydrophobic pollutants are generally not equally dispersed in soil, they tend to stick together at high concentrations, forming particles or droplets. Since micro-organisms can only access dissolved species, the dissolution kinetics limit biotransformation (Luthy *et al.,* 1994; Volkering *et al.*, 1993). The dissolution from a solid or non-polar liquid to the aqueous phase can be described as:

$$J = KA(C_{max}-C)$$

J is the dissolution rate, K the mass transfer coefficient governing dissolution, A the solid/water contact surface, and C_{max} the maximum water solubility. When the uptake of pollutant by the degrading micro-organisms is limited by the dissolution rate, C will approach zero and the dissolution rate approaches its maximum value:

$$J_{max} = KAC_{max}$$

When dissolution is limiting the growth of the micro-organisms, the biomass will increase linearly with a rate that is controlled by Y and C_{max} (Volkering *et al.*, 1993). So, in case of a low water solubility of a pollutant or a small contact surface A, bioremediation of soil is limited by the dissolution rate. This limitation can be overcome by the addition of solvents that enhance the solubility of pollutants. Furthermore, a dispersion of the chemical resulting in a higher contact area, would also increase biotransformation rates.

Enhancement of bioremediation

In the late eighties, Zehnder and his co-workers undertook a case study to assess the feasibility of bioremediation of a soil contaminated with α-HCH since the early sixties (Bachmann *et al.,* 1988; Huntjens *et al.*, 1988; Doelman *et al.*, 1990; Rijnaarts *et al.*, 1990). The results of this study are summarised in Figure 1. The pollution had been persistent during the more than 20 years of its presence in the field (upper line). Biodegradation was stimulated after the soil was dug out and put into lysimeters. However, the biotransformation in the lysimeters could not be stimulated by adding nutrients or oxygen indicating that nutrient availability was not limiting. Also, no biodegradation was observed at concentrations below 150 mg/kg (Doelman *et al.*, 1990). In contrast, treatments involving a rigorous mixing of the soil breaking up the larger soil particles resulted in an almost complete degradation of the α-HCH within two weeks (Rijnaarts *et al.*, 1990). The biodegradation in these soil slurries was still desorption limited though. A pure culture isolated from the polluted soil completely mineralised the same amount of α-HCH within one day (Huntjens *et al.*, 1988). These results clearly indicate that a rapid remediation of polluted soil requires a pulverisation of soil particles to liberate the contaminants trapped inside, thus increasing the biotransformation rates and decreasing the potential residual concentrations (Harms and Bosma, 1996).

There is an ongoing discussion about the effectiveness of surfactants for bioremediation purposes. Enhancement of biodegradation is usually ascribed to the surfactant-mediated solubilisation or emulsification of sorbed, crystalline or separate phase pollutants. Detrimental effects were attributed to the inhibition of the micro-organisms by the surfactant, the prevention of bacterial adhesion to the substrate source, or the preferred utilisation of the surfactants. It may be argued that the effectiveness of surfactants to enhance the degradation of hydrophobic pollutants is mainly due to the homogenisation of the compounds. Surfactants do not increase the exclusively bioavailable aqueous concentration of a compound, but introduce the so-called micellar pseudophase (Liu *et al.,* 1995), which accumulates the chemical and is mobile. Degradation is therefore promoted by the micelle-mediated homogenisation of the chemical, rather than by increased available substrate concentrations.

Figure 1. Biodegradation of α-HCH in the field (λ), in lysimeters (σ), and in laboratory slurries (τ)

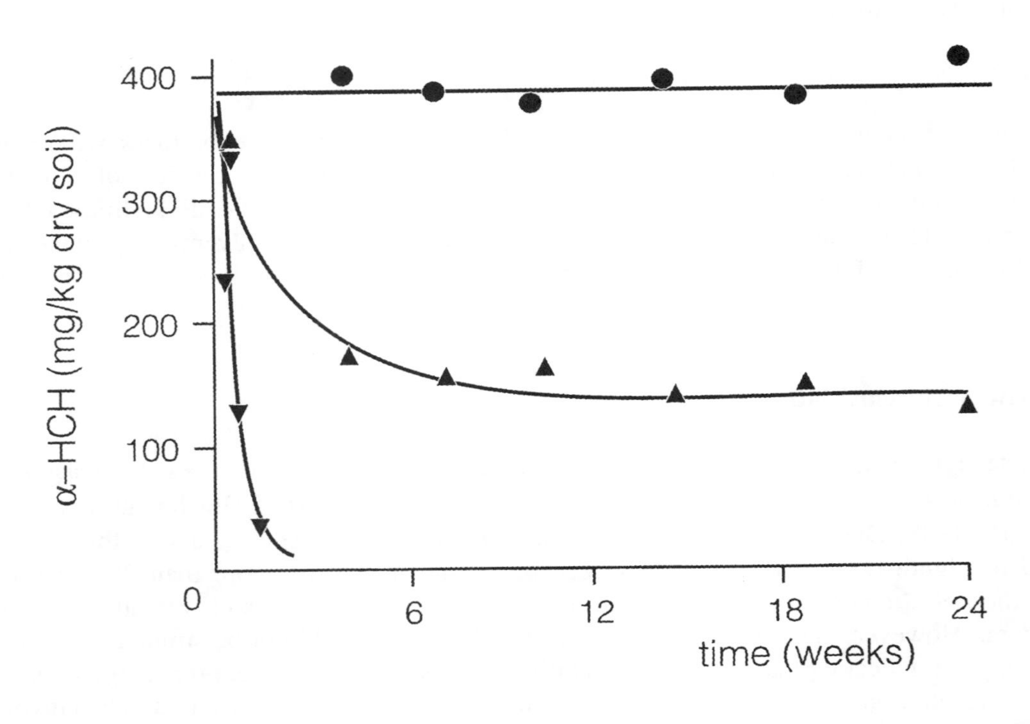

Source: Adapted from Bosma and Harms, 1996; data from Doelman *et al.*, 1990 and Rijnaarts *et al*, 1990.

Environmental risk of residuals

Considerable residual concentrations will always remain after bioremediation of polluted soil, due to strong sorption and incorporation in organic matter, unless special measures are taken to mobilise the pollutants. It is imaginable to remove the mobile fraction of pollutant in a relatively short time via a biological treatment during *in situ* remediation. During this phase, both mechanical homogenisation and addition of surfactants may be effective means to stimulate the biodegradation rate. However, the effectiveness of homogenization is minor in later stages of *in situ* bioremediation. The residual pollutant is trapped inside soil aggregates which cannot be broken up anymore by mechanical mixing and diffusion within the microporous system, retarded by sorption, is the limiting factor for biodegradation. It should be sufficient to monitor and control pollutants that are slowly leaching from the soil in this phase and to stimulate biotransformation by the addition of nutrients as soon as some critical level is reached. Bioremediation can be stopped as soon as the risk associated with the soil pollution is below acceptable limits.

The interest in the so-called intrinsic remediation of polluted sites is increasing in recent years. Intrinsic remediation is the attenuation of pollutants by, mostly extremely slow, natural degradation processes in the (sub)soil (for an overview see Hinchee *et al.*, 1995). It results in the establishment of a steady state pollutant plume that is stable in space and time. Anaerobic conversion processes appear to dominate the biodegradation during intrinsic remediation. The occurrence of intrinsic remediation is decisive for the acceptance of the risk associated with residual organic components at a polluted site. The successful application of intrinsic remediation involves an exposure and risk assessment that considers the location of potential receptors in relation to the extent of the contaminant plume (Rifai *et al.*, 1995).

Final remark

New pollutions have to be treated biologically as soon as possible to achieve optimal results because long contact times between pollutants and soil have a negative effect on the expected result of bioremediation. However, priority should be given to prevention of new pollutions.

Questions, comments, and answers

Q: What in actual practice is the equivalent of taking out samples and stirring to break up the particles? Can you do this *in situ*?

A: Land farming is in part an homogenising process. It is just not possible *in situ*.

Q: You say that the contaminants are occluded and that break-up allows access, but we also see diffusion into organic (humic) polymers. Do you see this?

A: In this case the HCH was deposited with calcium carbonate and this lead to a build-up of aggregates.

C: So it is not in a soil particle but rather a mineral particle?

Q: Are you not advising that we should give up *in situ* treatment because we have to wait for ever for the compounds to come out? I am in favour of fixing pollutants where they are and using enzymes to enter pores.

A: You need a good risk assessment, particularly of the sensitivity of the bound residue.

Q: Is the use of micro-organisms which excrete biopolymers and surfactants a good idea?

A: These are mainly effective for the mobile part of the contaminants. If pollutants are in very small micropores, then the surfactants have to diffuse to the contaminant. I see surfactants as effective in the first stages of remediation.

C: However, it is easier for small molecules to diffuse in than enzymes.

REFERENCES

BACHMANN, A., W. DE BRUIN, J.C. JUMELET, H.H.M. RIJNAARTS and A.J.B. ZEHNDER (1988), "Aerobic biomineralization of alpha-hexachlorocyclohexane in contaminated soil", Appl. Environ. Microbiol. 54, pp. 548-554.

BALL, W.P. and P.V. ROBERTS (1991), "Long-term sorption of halogenated organic chemicals by aquifer material. 2. Intraparticle Diffusion", Environ. Sci. Technol. 25, pp. 1237-1249.

BEURSKENS, J.E.M., C.G.C. DEKKER, J. JONKHOFF and L. POMPSTRA (1993), "Microbial dechlorination of hexachlorobenzene in a sedimentation area of the rhine river", Biogeochem. 19, pp. 61-81.

DOELMAN, P.L., H. HAANSTRA, H. LOONEN and A. VOS (1990), "Decomposition of alpha- and beta-hexachlorocyclohexane in soil under field conditions in a temporate climate", Soil Biol. Biochem. 22, pp. 629-639.

HARMS, H. and T.N.P. BOSMA (1996), "Mass transfer limitation of microbial growth and pollutant degradation", J. Ind. Microbiol., (in press).

HARMS, H. and A.J.B. ZEHNDER (1995), "Bioavailability of sorbed 3-chlorodibenzofuran", Appl. Environ. Microbiol. 61, pp. 27-33.

HATZINGER, P.B. and M. ALEXANDER (1995), "Effect of aging of chemicals in soil on their biodegradability and extractability", Environ. Sci. Technol. 29, pp. 537-545.

HINCHEE, R.E., J.T. WILSON and D.C. DOWNEY (1995), *Intrinsic bioremediation*, Battelle Press, Columbus, Richland.

HUNTJENS, J.L.M., W. BROUWER, K. GROBBEN, O. JANSMA, F. SCHEFFER and A.J.B. ZEHNDER (1988), "Biodegradation of alpha-hexachlorocyclohexane by a bacterium isolated from polluted soil", in K. Wolf, W.J. van den Brink, F.J. Colon (eds.), *Contaminated Soil* '88, pp. 733-737, Kluwer Academic Publishers, Dordrecht, NL.

KARICKHOFF, S.W. (1980), "Sorption kinetics of hydrophobic pollutants in natural sediments", in R.A. Baker (ed.), *Contaminants and sediments*, p. 193-205, Ann Arbor Science, Ann Arbor, MI.

KARICKHOFF, S.W., D.S. BROWN and T.A. SCOTT (1979), "Sorption of hydrophobic pollutants on natural sediments", Water Res. 13, pp. 241-248.

LIU, Z, A.M. JACOBSON and R.G. LUTHY (1995), "Biodegradation of naphthalene in aqueous nonionic surfactant systems", Appl. Environ. Microbiol. 61, pp. 145-151.

LUTHY, R.G., D.A. DZOMBAK, C.A. PETERS, S.B. ROY, A. RAMASWAMI, D.V. NAKLES and B.R. NOTT (1994), "Remediating tar-contaminated soil at manufactured gas plant sites", Environ. Sci. Technol. 28, pp. 266A-277A.

MIHELCIC, J.R. and R.G. LUTHY (1991), "Sorption and microbial degradation of naphthalene in soil-water suspensions under denitrification conditions", Environ. Sci. Technol. 25, pp. 169-177.

PAVLOSTATHIS, S.G. and G.N. MATHAVAN (1992) "Desorption kinetics of selected volatile organic compounds from field contaminated soils", Environ. Sci. Technol. 2, pp. 532-538.

RIFAI, H.S., R.C. BORDEN, J.T. WILSON and C.H. WARD (1995), "Intrinsic bioattenuation for subsurface restoration", in R.E. Hinchee, J.T. Wilson, and D.C. Downey (eds.), *Intrinsic bioremediation*, Battelle Press, Columbus, Richland.

RIJNAARTS, H.H.M., A. BACHMANN, J.C. JUMELET and A.J.B. ZEHNDER (1990), "Effect of desorption and intraparticle mass transfer on the aerobic biomineralization of α-hexachlorocyclohexane in a contaminated calcareous soil", Environ. Sci. Technol. 24, pp. 1349-1354.

SALKINOJA-SALONEN, M.S., P.J.M. MIDDELDORP, M. BRIGLIA, R.J. VALO, M.M. HÄGGBLOM, A. McBAIN and J.H.A. APAJALAHTI (1989), "Clean-up of old industrial sites", in D. Kamely, A. Chakrabarty, and G. Omenn (eds.), *Advances in Applied Biotechnology*, p. 347-365,.Gulf Publishing Company, Houston, TX.

SCHWARZENBACH, R.P. and J. WESTALL (1981), "Transport of non-polar organic compounds from surface water to ground water: laboratory sorption studies", Environ. Sci. Technol. 15, pp. 1360-1367.

STEINBERG, S.M., J.J. PIGNATELLO and B.L. SAWHNEY (1987), "Persistence of 1,2-dibromoethane in soils: entrapment in intraparticle micropores", Environ. Sci. Technol. 21, pp. 1201-1208.

VAN DER MEER, J.R., W.M. DE VOS, S. HARAYAMA and A.J.B. ZEHNDER (1992), "Molecular mechanisms of genetic adaptation to xenobiotic compounds", Microbiol. Rev. 56, pp. 677-694.

VOLKERING, F., A.M. BREURE and J.G. VAN ANDEL (1993), "Effect of micro-organisms on the bioavailability and biodegradation of crystalline naphthalene", Appl. Microbiol Biotechnol. 40, pp. 535-540.

WEBER, W.J. Jr, P.M. McGINLEY and L.E. KATZ (1991), "Sorption phenomena in subsurface systems: concepts, models and effects on contaminant fate and transport", Wat. Res. 25, pp. 499-528.

WU, S. and P.M. GSCHWEND (1986), "Sorption kinetics of hydrophobic organic compounds to natural sediments and soils", Environ. Sci. Technol. 20, pp. 717-725.

BIOREMEDIATION, STANDARDISATION AND BEST PRACTICE: A REGULATOR'S POINT OF VIEW

by

Hans E.N. Bergmans
Netherlands Committee on Genetic Modification, Utrecht, The Netherlands

Introduction

Bioremediation is an environmental technology that makes use of the emerging knowledge of microbial ecology. Its aim is to exploit the beneficial impact of micro-organisms on polluted environments.

Environmental regulations for the introduction of foreign micro-organisms into the environment often require detailed predictive data, from which it can be concluded that the introduction will have no significant negative impact on the environment. However, the knowledge of microbial ecology needed for such predictions still is fragmentary, and the processes going on in an actual introduction into the environment can often only be understood from monitoring such projects. Prediction of environmental impact is therefore difficult, if not impossible. Regulation may clearly end up here in a vicious circle.

What is needed here is an approach to risk analysis of a situation where clear predictions can be made, e.g. risk analysis of projects where an impact on the microbial environment is an actual aim, not something to be avoided. Bioremediation probably offers such opportunities, and, from a regulator's point of view, the risk evaluation of bioremediation projects might be helpful in gaining familiarity with risk assessment of the environmental applications of micro-organisms in general.

Risk/safety analysis in environmental applications of micro-organisms

Risk/safety analysis of biotechnological activities consists of two phases: hazard identification and subsequent risk assessment if hazards are identified and, when necessary, the application of adequate risk management measures.

In the case of environmental applications of genetically engineered micro-organisms (GEMs), the regulations in many countries require a strictly formalised risk analysis. The hazard identification is based on:

- properties of the GEM, deduced from the properties of the host organism, the properties of the genetic material introduced into the host organism, and the expected interaction between these two;

- properties of the receiving environment;

- interaction between the GEM and the receiving environment, e.g biogeochemical consequences, possible selective advantage of the GEM, gene flow, non-target effects and in general the impact of the release on biodiversity.

The subsequent risk assessment is often hampered because of difficulties in the estimation of the frequency at which the identified hazardous situations will occur. If no reasonable estimation can be given, the main factor determining the need for risk management practices will be the severity of the hazardous situation should it occur.

This view of risk analysis is highly biased from the point of view of the bioengineer constructing, supposedly, highly efficient strains for environmental applications. In such cases, data on the strains and on their behaviour under laboratory conditions will be available, but usually data on the interaction of the organism with the environment are available only to a limited extent, and the predictive value of data can be disputed, e.g. in the case of data from microcosm studies.

A main impediment for environmental risk analysis is the fact that it is generally not possible to provide a description of "the receiving environment" in microbial terms. It is mainly the lack of familiarity with the microbial environment that makes it questionable whether this type of reasoning can lead to a useful environmental risk analysis.

Bioremediation and risk/safety analysis

In bioremediation risk assessment is approached from a description of the process: the environment is described in terms of availability of substrates (pollutants, and other substrates for general metabolism) to micro-organisms with a suitable metabolism to perform the desired environmental chemical processes. It is usually supposed that micro-organisms capable of the processes are present, as such or co-operating in consortia; alternatively it is supposed that the genetic information coding for the enzymes involved in the metabolic processes is available in the population as exchangeable information, so that a suitable organism can emerge from an environmental genetical process.

Bioavailability of the substrate, i.e. the environmental pollutant to be remediated, is a limiting factor in the process; the process may be rendered more efficient by the addition of substrates for growth and metabolism of the micro-organisms. Depending on bioavailability of the substrates, selective pressure will favour the emergence of well-adapted micro-organisms or consortia for the desired remediation process. However, this situation will be transient, as the bioavailability of substrates will decrease because of the activity of the well-adapted micro-organisms.

The process of bioremediation as described above is meant to have an environmental impact, and as such may be evaluated as potentially risky. There are hazards in the bioconversion of xenobiotic substances, e.g. non-mobile insoluable substances may be converted to mobile soluable metabolites. Also the process of bioremediation may have a clear, though probably transient, impact on microbial biodiversity at the remediation site or even in the wider environment.

Although the risk analysis of bioremediation processes will be hampered by the lack of familiarity with the microbial environment, it will ususally be possible to make a number of clear predictions, e.g. on the role of selective pressure, on the movement of genetic information, on the impact on biodiversity, and on the transient character of eventual disturbances. These predictions can be tested during the remediation process.

Best practice

The formulation of "best practice" for bioremediation is still at a very early stage, because we are dealing here with a newly emerging technology. Succesful approaches in particular situations require careful analysis before attempts can be made to extrapolate best practice protocols from them. In this respect it is ominous that succesful protocols can only be transferred to other situations to a limited extent.

Still, careful analysis of succesful applications of bioremediation must eventually produce the necessary know-how of the process necessary to formulate general rules. Careful monitoring of the process is of great importance to create this know-how. Monitoring should not be limited to the chemical description of the process. "Bioremediation" implies a biological component that must be understood in order to have full knowledge and experience, or "familiarity", with the process.

Therefore, careful monitoring, also of the biology of the process, is of great importance. Microbiological monitoring should concentrate on identifying the active partners in natural processes, the flow of genetic material, and the effect of selective pressure. It should be remembered that micro-organisms active in the process in the field may turn out not to be culturable in the laboratory.

When micro-organisms are added in the process, monitoring should focus on the survival and dissemination of these organisms, and on the particular properties of the substrains that do survive.

The formulation of best practice, starting with careful analysis of the actual process, should clearly facilitate scientifically sound risk analysis. In fact, the analysis of the actual process is the validation of risk analyses done beforehand.

From a regulator's point of view, it is clear that we need results from environmental microbiological processes that are predicted to have measurable impacts on, and therefore inherent risks for, the environment in order to get the familiarity needed for the performance of risk analysis. The risk/benefit ratio in bioremediation, a process offering large potential advantages at relatively limited risks, would appear to be favourable on the side of the benefits. Regulators should therefore take great interest in the recent developments in the field of bioremediation.

Normalisation

The development of best practice is a first step towards normalisation. Normalisation in biotechnology is deemed necessary by regulatory authorities as an aid to the development of harmonised legislation. Industry is also keen to arrive at normalisation wherever possible. On the other hand, normalisation efforts in fields where no clear standard practices have been developed will only be counterproductive. In the field of biotechnology, development of new or improved techniques is going on at an extremely fast and ever increasing rate. Attempts at normalisation in such a situation will be hampered by the fact that norms will be outdated before they have even been established.

This is not to say that no normalisation efforts should be made. The European normalisation office CEN, for instance, has started a large programme for normalisation of biotechnology in its technical working committee TC 233. Working Group 3 in this programme concentrates on modified organisms for application in the environment. Items for normalisation taken on board by Working Group 3 are: characterisation of GEMs in terms of description of the genetic modification, expression and stability; sampling and monitoring strategies; and quality control of microbiological products in terms of purity, biological activity and stability.

Technically these fields are not all at the same stage of development, and the norms that can be formulated for the different subjects will be very disparate.

It is the author's view that normalisation of environmental biotechnology should be approached with utmost care. Again, bioremediation is probably a field where developments will first reach the stage where meaningful normalisation can be attempted. Once this stage is reached, it should be clear that expeditious development of norms is of great importance to all partners in the field, science and industry, potential customers and regulatory authorities.

Questions, comments, and answers

Q: Monitoring is an important activity but in environmental biotechnology applications are by definition in open systems and the micro-organisms multiply - where and what do you monitor? Is it only at the application site?

A: The naïve idea is that as long as you can trace organisms and discriminate from others, you start at the point of application and work out. This is not practical and you will miss much. You need to take into account knowledge of the movement of the organisms.

Q: There seems to be a shift in emphasis - in the past it was testing on microcosms and certain model systems to predict the behaviour in the environment. Is there now more emphasis on the organism itself, geno and phenotype, and its probable behaviour?

A: In fact, this is the only thing you can do before release. Extrapolation from microcosms has only limited value. Eventually you have to take the next step with a certain amount of confidence based on all the data you can collect. There is no way of answering all questions beforehand. The risk should be limited as much as the state of the art allows.

Q: We should discuss gene flow as it is a normal process in nature. We should define the quality of introduced genes; similar ones may already be in the environment and they may be spreading. What is different between the spreading of natural genes and ones introduced via GMOs?

A: At this moment a committee is advising the Dutch government that if genes could have gone in naturally, there is no reason to call them GMOs in the regulatory sense (the same applies to mutants). It may be some time before this is politically acceptable.

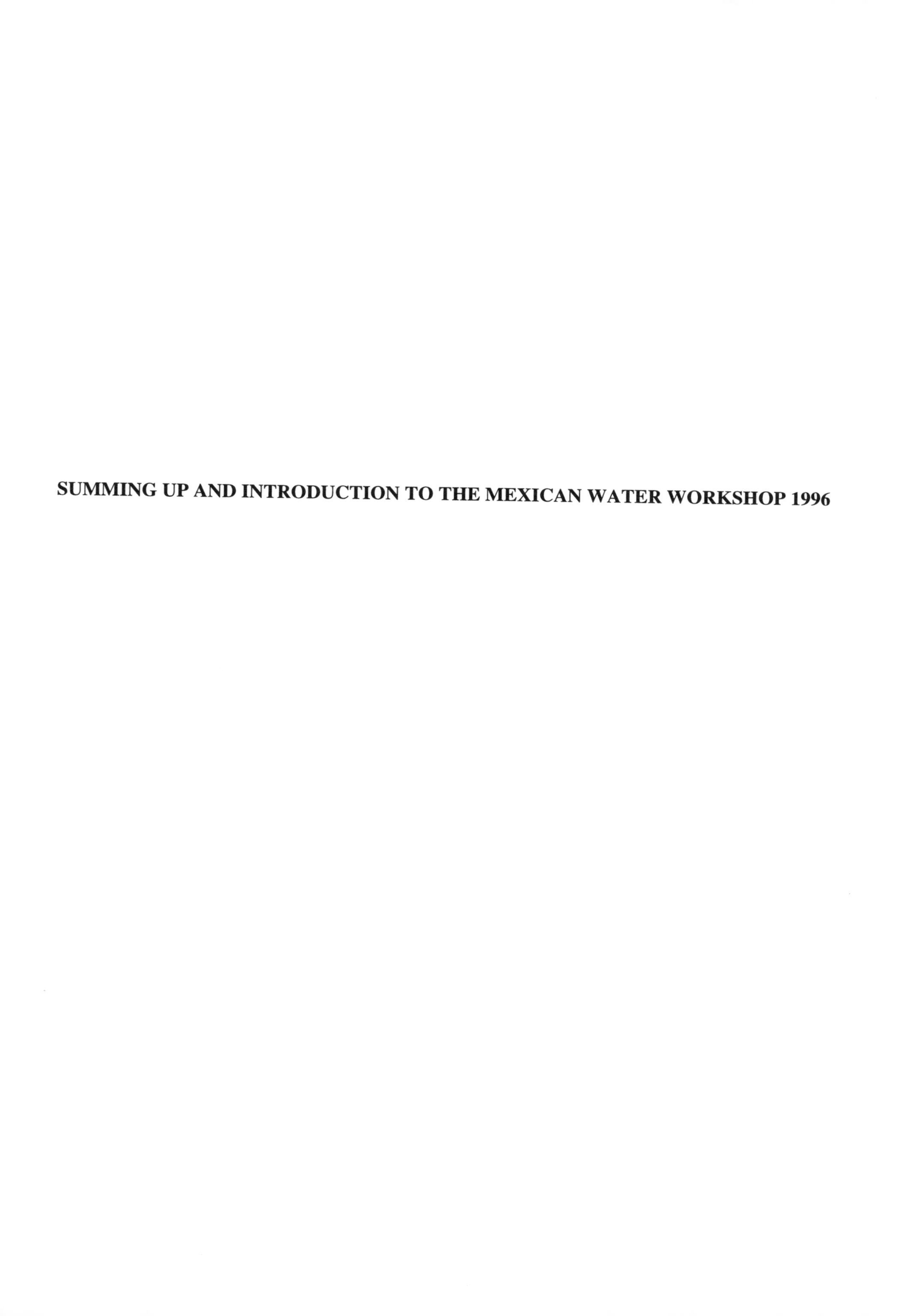

SUMMING UP AND INTRODUCTION TO THE MEXICAN WATER WORKSHOP 1996

THE ROLE OF BIOTECHNOLOGY IN ENVIRONMENTAL REMEDIATION

by

Álvaro A. Aldama-Rodríguez
Instituto Mexicano de Tecnología del Agua, Morelos, México

Introduction

Bioremediation, the use of living organisms to degrade wastes or to prevent pollution through treatment, is emerging as one of several alternative technologies for removing undesirable compounds from the environment, restoring contaminated sites, and preventing further deterioration.

The OECD report issued in 1994, "Biotechnology for a clean environment", responded to a need for a scientific review of the potential of biotechnology to contribute to the prevention, detection and remediation of environmental pollution, providing a broad spectrum review of the state of art in this emerging field. The report estimates that the worldwide market potential for all environmental biotechnologies would rise from $40 billion in 1990 to $75 billion by the year 2000.

The Tokyo OECD workshop, held in November 1994, covered not only bioremediation but also long-term applications of biotechnology to protect environmental quality. Participants recognised that bioremediation can have global, regional, and local applications and that both naturally-occurring and genetically-modified micro-organisms may play an important role for this purpose.

The Amsterdam workshop has been held with the purpose of promoting a wider application and diffusion of bioremediation technologies, and emphasizing industrial aspects of bioremediation in air and soil. The presentations have given us insight into biotechnologies applied to clean-up problems in different countries. Toxic organic and inorganic chemicals from point and non-point sources have been neutralised or removed by bioremediation or by combined technologies. The efficiencies of these techniques, as well as their threat to the environment, are currently being investigated by groups from several countries as we have learned from the speakers at this workshop (Figure 1).

Although biological remediation appears to be a relatively clean and cost-effective procedure, for the long-term goal of reducing and eliminating the negative impacts of contaminants on the environment, prevention represents the most cost-effective procedure (see article by Nakajima in this volume).

Removal of contaminants from polluted soil

Almost one third of the papers were related to the remediation of contaminated soil through biotechnology (Figure 2), using a great variety of techniques . Discussions covered topics from risk-assessment of bioaugmentation and stimulation procedures to the use of indigenous micro-organisms to mobilise toxic heavy metals from soil particles and as prototypes for the degradation of organic pollutants.

Figure 1. Percentage of papers by nationalities presented during the 1995 OECD workshop on Bioremediation Technologies

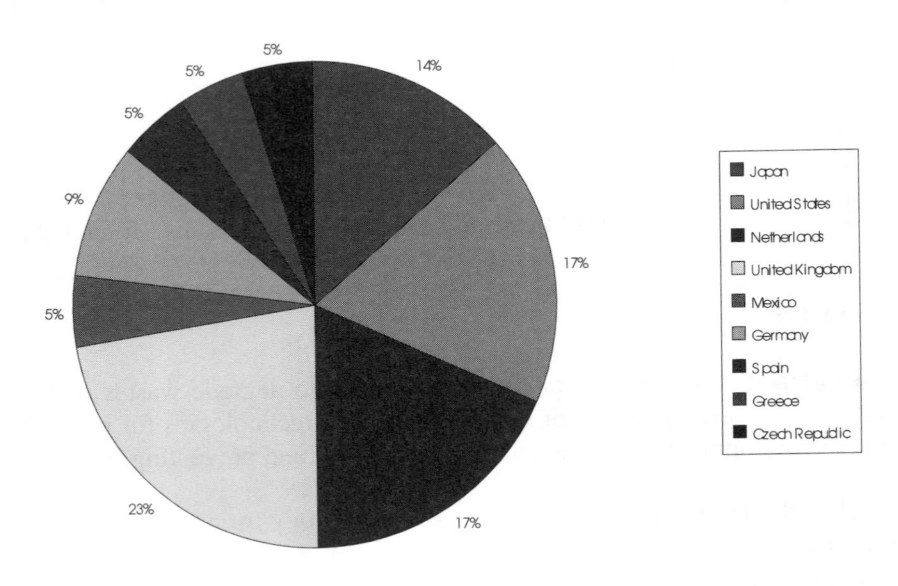

Source: Author.

The next difficulty to overcome is the selection of the most suitable means of treatment considering site geology, hydrogeology and contaminant distribution. Each site has its own conditions, redox environment, and microbial characteristics that must be accounted for in the selection of the most appropriate remediation program.

Monitoring

During and after the treatment process, a monitoring program is required to verify that target concentrations have been attained in heterogeneous media (soil or air) . For monitoring strategies to be useful during the bioremediation processes, they must be highly sensitive and robust, as well as offering reproducible, specific and rapid on-line analysis with little or no sampling processing bias. New molecular monitoring technologies, including polymerase chain reaction and enzyme activity, open the way. Microtox and Ames tests, with the ability to reveal a wider range of compounds, can be used to detect accumulations of toxic intermediates during the treatment process.

Removal of gases during industrial processes

The removal of gases from industrial processes was another major topic in this workshop (Figure 2). Researchers are confident that over 95 per cent of the real world waste gas problems can be solved using available biotechnology. Treatments are in the mill that are commercially attractive and well received in North America and Europe.

Figure 2. Percentage of papers by subject presented during the 1995 OECD workshop on Bioremediation Technologies.

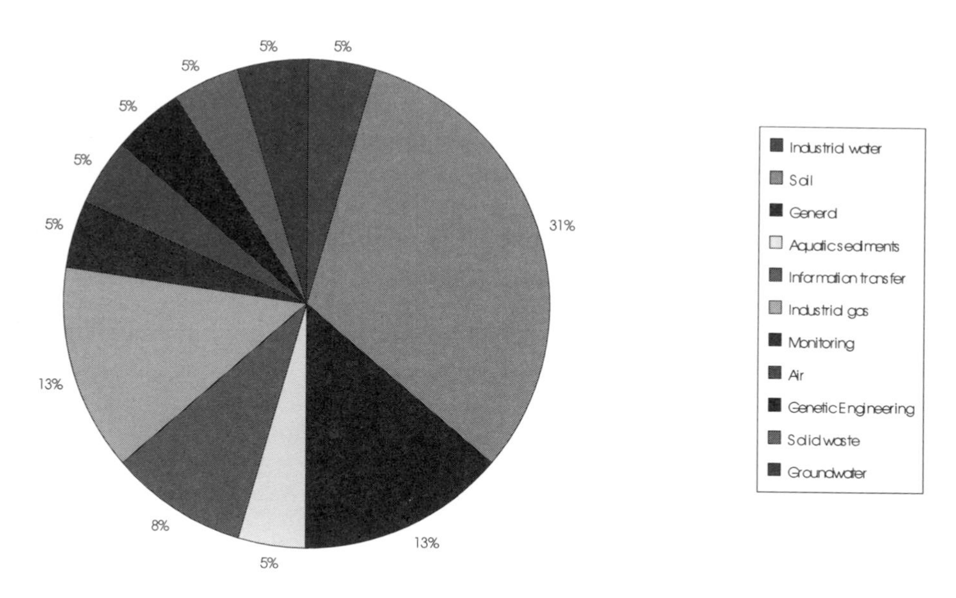

Source: Author.

They deal with air borne pollutants from sources as varied as agriculture, paintshops or industry, as well as to removing sulfur and nitrogen from petroleum before they get aloft.

Modelling and optimisation

Models can be used to optimise biotechnological processes. After calibration using real systems, they can predict rates, interactions, secondary reactions and costs. The range of reactions that may be tested is unlimited. The models are useful in optimisation considerations. Calibrated models can help with technology transference. Site-specific characteristics must be evaluated each time that *in situ* bioremediation is contemplated. The more we know about the technology, the better we can use it and develop it.

Information transfer

The general public and legislators must be informed of the risks and benefits of using genetically engineered micro-organisms. Precise and understandable facts should be offered so that these methods can be accepted and used.

Bioremediation in Mexico

Three different aspects of bioremediation research and development are emerging worldwide. In Europe, efforts are made to broaden traditional waste treatment systems to cope with specific chemical pollutants. In the United States of America, site-specific clean-up of soil and water contaminated with petroleum and xenobiotic pollutants is the focal point. In Japan, bioremediation is considered a means of addressing global environmental concerns.

Bioremediation in Mexico is an emerging area and its development depends upon the creation of specific standards and human resources. The few research groups in the area of bioremediation and environmental biotechnology are concentrated at the National Autonomous University of Mexico (UNAM), the Metropolitan Autonomous University (UAM), the Technological Institute of Sonora (ITSON), the National Polytechnic Institute (IPN) and the Mexican Institute of Water Technology (IMTA).

At the moment, the studies have been directed mainly to bioremediation in three sectors (Quintero, 1995):

- soils contaminated with oil and their by-products (xenobiotics);
- agricultural soils contaminated with pesticides;
- air contaminated by odors and toxic compounds.

Studies have been done to identify hazardous pollutants in the country. Research by the Technological Institute of Sonora in the ecosystem of the Yaqui Valley (Sonora), the most modern irrigation district in the country, indicated that high agricultural productions are obtained, to a large extent, through the use of massive doses of pesticides. Muñoz and Remi (1986) reported that in southern Sonora, 163 different types of pesticides were applied in that year. In 1989, about 2 000 tons of pesticides were applied.

Moeller and Fernandez, (1986*a*, 1986*b*), at the Environmental Engineering Department of the Graduate School of Engineering, UNAM, studied the possibility of acclimating a strain of *Pseudomonas* to use the surfactants in detergents as a carbon and energy source. A mixed acclimated population was also tested with promising results.

Torres *et al.* (1995) at the Institute of Biotechnology, UNAM, have studied the possibility of biodegrading polynuclear aromatic hydrocarbons using biocatalysts. At the Insitute of Engineering, UNAM, two research teams (Rosas *et al.*, 1995; González and Buitron, 1995) have examined specific micro-organisms for the ability to degrade toxic compounds.

Work at the National Polytechnic Institute (IPN-CINVESTAV) focussed on using microalgae to remove heavy metals, and fungi for aromatic chlorohydrocarbons.

The Metropolitan Autonomous University (UAM) is investigating the degradation of amides and nitriles by *Bacillus* sp, bacterial dynamics in different environments, plastic biodegradation, anaerobic biodegradation of polyaromatic hydrocarbons and biofiltration for the removal of toxic compounds from air.

Bioremediation of the Silva Dam sediment bed

The Silva Dam in the state of Guanajuato, Mexico, has been receiving industrial and municipal wastewater, as well as rainwater and agricultural runoff, for several decades. IMTA participated in the evaluation of contaminants in these sediments and in the development of recommendations for their final disposal. Sediment samples were extracted and analysed using physico-chemical tests (pH, Redox potential), isotopic dating, trace contaminant determinations and bioassay techniques (Microtox). Even when metals (chrome), phthalates and chlorinated pesticides were found in relatively high concentrations in the upper layers of the deposited sediments, biological tests indicated that toxicity remained high at depths up to one meter into the sediment column.

Considering these results, IMTA suggested *in situ* remediation technologies to reduce the toxicity of the sediments. Biological materials were obtained from natural substrates and from external sources. These will be evaluated for their ability to reduce the levels of toxic compounds. The project includes scaling up the technology from the laboratory to a limited area in the Silva Dam, as well as the evaluation of the risks associated with the application of the remediation technology.

Wastewater treatment

IMTA has developed an aerobic bioxidation technology using powdered activated carbon to reduce the load of toxic substances and organic compounds in wastewater from the chemical-pharmaceutical industry. Micro-organisms of the genera *Pseudomonas* and *Alcaligenes* are being isolated, identified and acclimated for treatment of this mixed product wastewater. Solvents under study as food sources are phenol, toluene and benzene (Ramírez *et al.*, 1994). Efforts have also been made to enhance the anaerobic digestion of sludge by adding growth stimulants and improving mixing and reactor design (Moeller, 1994).

Mexican Water Workshop 1996

The upcoming workshop to be hosted by Mexico in 1996, "Biotechnology R & D for Water Conservation and Quality", will touch upon issues such as remediation and prevention technologies, with the long-term goal of reducing and eliminating the negative impacts of human activities on the environment and the global climate. Researchers will discuss the growing importance of new approaches to prevention and review the contribution of biotechnological research and development toward addressing the problems of water conservation and quality in Mexico and other countries. The goal is improved environmental conditions for sustainable development. The relationships between detection, prevention and bioremediation will be explored.

Conservation is a key word in the administration of any resource. New approaches must be explored as the old ones become obsolete or insufficient. Water use must become more efficient, as must the processes to remove wastes from industrial and agricultural effluents. The recycling of effluents for agricultural and industrial uses and even human consumption must be optimised through biotechnology. This will help slow down the desertification and climatic changes discussed by experts today.

Water quality, an equally important consideration at the 1996 Mexico workshop, will cover the main remediation applications for aquifer and groundwater pollution and illustrate classic cases. The removal of contaminating biological and chemical agents with new *in situ* and *ex situ* technologies will be closely examined. Bioremediation of water systems will cover topics on microbial ecology, ground water and aquifers, river sediments and marine oil spills. *Ex situ* water treatment will cover new technologies for water treatment and remediation, biological disinfection processes, advanced water treatment, water treatment for developing countries and biological water quality.

Bioremediation has many potential applications in cleaning up contaminated environments and treating wastes. However, to effectively use these applications, both further R & D and clarification of government policies are needed. Research must be done to isolate or genetically engineer micro-organisms that can attack the most resistant compounds in our environment, such as polynuclear aromatic hydrocarbons and chlorinated compounds, develop methods of efficiently delivering micro-organisms to the site and establish environmental conditions under which they can function effectively.

One of the tools for a better and cleaner tomorrow is in our hands. We must work with that tool, towards understanding its limitations and its horizons. Our environment, the combination of soil, air and water, gave birth to life as we know it. We have taxed its ability to absorb the wastes we create. We must work now to clean it with the same organisms from which life is supposed to have come.

REFERENCES

GONZÁLEZ, A. and G. BUITRON, (1995), "Study of population dynamics in activated sludge in a biological process used for the treatment of phenol-contaminated effluents", presentation at the VI Congreso Nacional de Biotecnología y Bioingeniería, Ixtapa, Guerrero, Mexico, September 10-14, 1995.

MOELLER, G and G. FERNÁNDEZ, (1986*a*), "Acclimation of micro-organismos for biodegradability of ABS detergent type", Proceedings of the 4th International Symposium on Microbial Ecology. Ljubljana, Yugoslavia.

MOELLER, G and G. FERNÁNDEZ, (1986*b*), "Adaptation of micro-organisms for the biodegradation of ABS detergents", Proceedings of the 5th International Congress of the Sociedad Mexicana de Ingeniería Sanitaria y Ambiental, Puebla, Puebla, Mexico, October 20-25.

MUÑOZ, A. and C. REMI, (1986), "The Agrochemical Market: Sonora", (sp), Shell Co., Cd. Obregon, Mexico.

QUINTERO, R., R. CASTAÑON, and J. SOLLEIRO (1995), "Soil bioremediation: Development opportunities in Mexico?", Proceedings of the National Course on Bioremediation of Soil and Aquifers, Programa Universitario del Medio Ambiente, UNAM.

RAMÍREZ, E., P. MIJAYLOVA, L. CARDOSO and S. LÓPEZ, (1994) "Final project report UI9413: Reduction of toxicity of organic synthesis effluents. IMTA/CTCA. Subcoordination of Wastewater Treatment.", December 1994.

ROSAS, D., A. SANCHEZ DE LA VEGA, L. TORRES and B. JIMÉNEZ (1995) "Immobilization of *Pseudomonas* in four matrices for the removal of chlorophenols from water", Proceedings of the VI Congreso Nacional de Biotecnología y Bioingeniería, Sociedad Mexicana de Biotecnología y Bioingeniería, Ixtapa, Guerrero, Mexico, September 10-14, 1995.

TORRES, E., J. SANDOVAL, F. ROSELL, A. MAUK and R. VAZQUEZ-DUHALT (1995), "Site-directed mutagenesis improves the biocatalytic activity of iso-1-cytochrome c in polycyclic hydrocarbon oxidation", *Enzyme and Microb.Technol.*, (in print).

LIST OF PARTICIPANTS

OECD WORKSHOP AMSTERDAM '95 ON WIDER APPLICATION AND DIFFUSION OF BIOREMEDIATION TECHNOLOGIES:

LIST OF PARTICIPANTS

CHAIRMAN

Mr. Wim HARDER
TNO Institute for Environmental Sciences
P.O. Box 6011
DELFT 2600 JA
THE NETHERLANDS

Tel: (31 55)54 93500
Fax: (31 55)54 21458

CO-CHAIRS

Mr. Alan BULL
University of Kent
Department of Biosciences
CANTERBURY
KENT CT2 7NJ
UNITED KINGDOM

Tel: (44 1227) 764000
Fax: (44 1227) 475498

Mr. Thomas EGLI
EAWAG, Microbiology
8600 DÜBENDORF
SWITZERLAND

Tel: (41) 1823 5158
Fax: (41) 1823 5547
E-mail: egli@eawag.ch

Mr. D. Jay GRIMES
US Department of Energy
Office of Energy Research, er-74
19901 Germantown Road
GERMANTOWN MD 20874-1290
UNITED STATES

Tel: (1 301) 903 4183
Fax: (1 301) 903 8519
E-mail:
darrell.grimes@oer.doe.gov

Mr. Dick JANSSEN
University of Groningen
Nijenborgh 4
9747 AG GRONINGEN
THE NETHERLANDS

Tel: (31 50) 363 4209
Fax: (31 50) 363 4165
E-mail: d.b.janssen@chem.rug.nl

RAPPORTEURS

Mr. Michael GRIFFITHS
Mike Griffiths Associates
The Pantiles, Ivy Lane
WOKING, SURREY, GU22 JB4
UNITED KINGDOM

Tel: (44 1483) 767 818
Fax: (44 1483) 756 136

Ms. Debbie POOLE
BBSRC
Polaris House
North Star Avenue
SWINDON, SN1 1ET
UNITED KINGDOM

Tel: (44 1793) 414 657
Fax: (44 1793) 414 674

SPEAKERS AND OTHER PARTICIPANTS

BELGIUM

Mr. Willy VERSTRAETE
University of Gent
Lab. Microbial Ecology
Coupure L 653
9000 GENT

Tel: (32 92) 645 976
Fax: (32 92) 646 248

CANADA

Mr. Desmond MAHON
Environment Canada
Pollution Prevention Directorate
14th floor, Place Vincent Massey
351 St. Joseph Boulevard
HULL, QUEBEC K1A OH3

Tel: (1 819) 997 4336
Fax: (1 819) 953 7155

DENMARK

Mr. Holger PEDERSEN
Danish Environmental Protection Agency
Strandgade 29
1401 COPENHAGEN

Tel: (45) 326 60350
Fax: (45) 326 60479

GERMANY

Mr. Wolfgang BABEL
Umweltforschungszentrum Leipzig-Halle GmbH
Permoserstrasse 15
04318 LEIPZIG

Tel: (49) 3412 352225
Fax: (49) 3412 352247

GERMANY (continued)

Mr. Wulf CRUEGER
Bayer AG
WV-UWS/UP, Geb. 9115/1
51368 LEVERKUSEN

Tel: (49 21) 430 62156
Fax: (49 21) 430 62826

Ms. Beatrix DAEI
Lobbe Xenex GmbH &Co
Stenglingser Weg 4 - 12
58642 ISERLOHN

Tel: (49 23) 745 04155
Fax: (49 23) 745 04171

Mr. Hans EGGERS
Bundesministerium für Bildung
Wissenschaft, Forschung und Technologie
BMBF, Ref 514
53170 BONN

Tel: (49 228) 573647
Fax: (49 228) 573605

Mr. Karl-Heinich ENGESSER
University of Stuttgart
ISWA
Biol. Abluftreinigung/Biotransformation
70569 STUTTGART

Tel: (49) 711 685 3734
Fax: (49) 711 685 3734
E-mail:
karl-h.engesser@iswa.uni-stuttgart.de

Mr. Wolfgang FRITSCHE
Institut für Mikrobiologie
Universität Jena
Philosophenweg 12
07743 JENA

Tel: (49) 364 163 1237
Fax: (49) 364 163 0956

Mr. Uwe GAUGLITZ
BASF AG
DUR/EG - K357
67056 LUDWIGSHAFEN

Tel: (49) 621 607 9215
Fax: (49) 621 607 9641

Mr. Anton HARTMANN
GSF-Research Center for Environment
Health & Neuherberg
Ingolstädter, Landstrasse 1
85764 OBERSCHLEISSHEIM

Tel: (49) 893 187 4064
Fax: (49) 893 187 3376

Ms. Claudia JUNGE
Forschungzentrum Julich GmbH
Projekttrages BEO
Breite Strasse 3
10178 BERLIN

Tel: (49) 302 319 9466
Fax: (49) 302 319 9470

GERMANY (continued)

Mr. Bernd MAHRO
Bremen Polytechnic
Inst. of Environmental Technology
Neustadtswall 30
28199 BREMEN
Tel: (49) 421 590 5305
Fax: (49) 421 590 5292

Ms. Marion POEL
Federal Ministry for Environment
BMU, Ref. NII4
Postfach 120629
53048 BONN
Tel: (49 228) 305 2646
Fax: (49 228) 305 2697

Mr. Hans-Peter RATZKE
Umweltschutz Nord GbmH & Co
Industriepark 6
27767 GANDERKESEE
Tel: (49) 422 247106
Fax: (49) 422 247203

Mr. Dieter SELL
Dechema
Technology-Transfer Envir. Biotechnology
Theodor Heuss Allee 25
60486 FRANKFURT / M
Tel: (49) 697 56 4370
Fax: (49) 697 56 4388

IRELAND

Mr. Barry KIELY
National Food Biotechnology Centre
University College
CORK
Tel: (35 3) 2127 3803
Fax: (35 3) 2127 6318

Mr. Tim PAUL
JB Land Rehabilitation
c/o John Barnett & Associates
Parkview House
Beech Hill, Cionskeagh
DUBLIN 4
Tel: (35 3) 1269 4077
Fax: (35 3) 1269 4424
E-mail: csa@iol.ie

ITALY

Mr. Marco P. NUTI
Universita di Padova
Dip. Biotecnologie agrarie
via Gradenigo 6
35131 PADOVA
Tel: (39 49) 807 1442
Fax: (39 49) 807 0517
E-mail:
biotmpn@idpunidx.unipd.it

ITALY (continued)

Ms. Andrea ROBERTIELLO
Enicerche S.p.A.
c/o Mr. Silvia Piacentini
Via Maritano 26
20097 SAN DONATO MILANESE

Tel: (39) 252 022 614
Fax: (39) 252 024 611

JAPAN

Mr. Ryuichiro KURANE
Nat. Inst. of Bioscience & Human-Tech.
Applied Microbiology Department
1-1, Higashi, Tsukuba, Ibaraki
305 TSUKUBA

Tel: (81 298) 546030
Fax: (81 298) 546005

Mr. Yukinori KUSAKA
Fukui Medical School
Department of Environmental Health
Matsuoka-cho
Fukui Prefecture
910-11 FUKUI

Tel: (81 776) 613111
Fax: (81 776) 618107

Mr. Kiyotaka MIYASHITA
National Inst. of Agro-Environmental Sciences
3-1-1 kan-nondai
305 TSUKUBA

Tel: (81 298) 388256
Fax: (81 298) 388199
E-mail:
kmiya@cc.niaes.affrc.go.jp

Mr. Kunio NAKAJIMA
Ministry of International Trade & Industry
Basic Industries Bureau
1-3-1 Kasumigaseki, Chiyoda-ku
100 TOKYO

Tel: (81 3) 3501 3848
Fax: (81 3) 3501 6588

Mr. Takeshi OGAWA
Environment Agency, Japan
Nat. Institute for Environmental Studies
1-2-2, Kasumigaseki, Chiyoda-ku
100 TOKYO

Tel: (81 3) 3593 0042
Fax: (81 3) 3580 3542

Mr. Osami YAGI
Environmental Agency, Japan
Nat. Institute for Environmental Studies
16-2 Onogawa Tsukuba, Ibaraki
305 TSUKUBA

Tel: (81 298) 502542
Fax: (81 298) 514732

JAPAN (continued)

Mr. Tomoho YAMADA
Ministry of International Trade and Industry
Biochemical Industry Division
1-3-1 Kasumigaseki Chiyoda-ku
100 TOKYO

Tel: (81 3) 3501 8625
Fax: (81 3) 3501 0197

MEXICO

Mr. Alvaro ALDAMA-RODRIQUEZ
Mexican Institute of Water Technology
Paseo Cuauhnahuac 8532
62550 JUITEPEC, MORELUS

Tel: (52) 1942 41
Fax: (52) 7319 3422
E-mail: aaldama@tlaloc.imta.mx

Mr. Enrique MARROQUIN
Grupo Cydsa, Sa de CV
Ricardo Margain Z 325
66220 GARZA GARCIA N.L.

Tel: (52) 833 58302
Fax: (52) 833 55841
E-mail: emarroqu@infosel.net.mx

Mr. Elia RAXO-FLORES
Institute Mexicano del Petroleo
WAV-Environmental Technology
Bomenweg 2,
6700 EV WAGENINGEN
THE NETHERLANDS

Tel: (52) 317 482431
Fax: (52) 317 482100
E-mail: razomt@rcl.wav.nl

Mr. Sergio REVAH
Department of Chemical Engineering
Universidad Autonoma Metropolitana Iztapalapa
Apdo. Postal 55-534
09340 MEXICO CITY

Tel: (52) 572 44648
Fax: (52) 572 44900
E-mail: srevah@xanum.uam.mx

NORWAY

Mr. Tormod BRISEID
SINTEF
P.O. Box 124, Blindern
0314 OSLO

Tel: (47) 220 67627
Fax: (47) 220 67350

SPAIN

Mr. Eloy GARCIA CALVO
CICYT
MADRID

Tel: (34 1) 885 4939

Mr. Juan L. RAMOS
C.S.I.C.
P.O. Box 419
18008 GRANADA

Tel: (34 58) 121 011
Fax: (34 58) 129 600
E-mail: jlramos.eez.csic.es

SWITZERLAND

Mr. Tom N.P. BOSMA
EAWAG
Seestrasse
6047 KASTANIENBAUM

Tel: (41) 41 3492115
Fax: (41) 41 3492168
E-mail: bosma@eawag.ch

THE NETHERLANDS

Mr. J.A.M. VAN BALKEN
DSM Research
Environmental and Safety Research
P.O. Box 18
6160 MD GELEEN

Tel: (31) 46 476 1434
Fax: (31) 46 476 0700

Mr. Hans BERGMANS
Committee on Genetic Modification
P.O. Box 80022
3508 TA UTRECHT

Tel: (31 30) 253 2432
Fax: (31 30) 253 2721

Mr. Ger BEUMING
Shell Int. Oil Products bv
P.O. Box 162
2501 AN THE HAGUE

Tel: (31 70) 377 6047
Fax: (31 70) 377 2833

Mr. J. BOVENDEUR
Heidemij Realisatie B.V.
Research and Development
P.O. Box 660
5140 AR WAALWIJK

Tel: (31) 41 63 44 044
Fax (31) 41 63 44 080

Mr. Cees BUISMAN
Paques B.V.
P.O. Box 52
8560 AB BALK

Tel: (31) 51 408 500
Fax: (31) 51 408 666

Mr. Hans DODDEMA
TNO
Environmental Biotechnology
P.O. Box 6011
2600 JA DELFT

Tel: (31 15) 269 6022
Fax: (31 15) 256 9670

Mr. Ton HEDDEMA
Heidemij Realisatie BV
P.O. Box 139
6800 AC ARNHEM

Tel: (31) 26 377 8413
Fax: (31) 26 443 7386

THE NETHERLANDS (continued)

Mr. Jef HEIJNEN
Delft University of Technology
Vakgroep Bioprocestechnologie
Julianalaan 67
2628 BC DELFT
Tel: (31 15) 278 2341
Fax: (31 15) 278 2355

Mr. Paul G.M. HESSELINK
SGS Redwood NL
P.O. Box 200
3200 AE SPIJKENISSE
Tel: (31) 18 807 6460
Fax: (31) 79 331 6554

Mr. L.P. JANSSEN
University of Groningen
Technische Scheikunde
P.O. Box 72
9700 AB GRONINGEN
Netherlands

Mr. Sytze KEUNING
Bioclear Environmental Biotechnology
P.O. Box 2262
9704 CG GRONINGEN
Tel: (31) 50 571 8455
Fax: (31) 50 571 7920

Mr. B. KOERS
Europe CVT Bioway
P.O. Box 166
6740 AD LUNTEREN
Tel: (31) 83 882 513
Fax: (31) 83 886 088

Mr. Aldert VAN DER KOOIJ
DHV Milieu en Infrastructure
P.O. Box 1076
3800 BB AMERSFOORT
Tel: (31) 334 682 933
Fax: (31) 334 682 804

Mr. Iman W. KOSTER
Ecotechniek bv
P.O. Box 1330
3600 BH MAARSSEN
Tel: (31) 346 557 700
Fax: (31) 346 554 452

Mr. J. Gijs KUENEN
Delft University of Technology
Kluyver Lab. for Biotechnology
Julianalaan 67
2628 BC DELFT
Tel: (31 15) 278 5308
Fax: (31 15) 278 2355

Mr. Chris VAN LITH
ClairTech bv
P.O. Box 65
3931 EB WOUDENBERG
Tel: (31) 332 867376
Fax: (31) 322 865736

THE NETHERLANDS (continued)

Mr. John MARKS
Ministry Education, Culture and Science
Environment, Life Sciences and R&D Str.
P.O. Box 25000
2700 LX ZOETERMEER

Tel: (31 79) 532294
Fax: (31 79) 531953

Mr. Robert VAN DER MEER
NIABA
P.O. Box 443
2260 AK LEIDSCHENDAM

Tel: (31 70) 327 0464
Fax: (31 70) 320 5765

Mr. G.C. MOLENKAMP
KPMG
Environmental Consulting
P.O. Box 96919
2509 JH THE HAGUE

Tel: (31 70) 33 82 309
Fax: (31 70) 33 82 141

Mr. Eduard W.H. MOSMULLER
Senter
P.O. Box 30732
2500 GS DEN HAAG

Tel: (31 70) 361 0393
Fax: (31 70) 361 0417

Mr. J.J. OLIE
Grondmechanica Delft
P.O. Box 69
2600 AB DELFT

Tel: (31 15) 269 3662
Fax: (31 15) 261 0821

Mr. J.H.J. PAQUES
Paques B.V.
P.O. Box 52
8560 AB BALK

Tel: (31) 31 51 40 8600
Fax: (31) 31 51 40 3342

Mr. Pieter G. PAUL
Stork Comprimo bv
P.O. Box 58026
1040 HA AMSTERDAM

Tel: (31 20) 580 7379
Fax: (31 20) 580 7050

Mr. Huub RIJNAARTS
TNO-Inst. of Environmental Sciences
P.O. Box 6011
2600 JA DELFT

Tel: (31 15) 269 7368
Fax: (31 15) 256 9670
E-mail:
rijnaarts@tno.mw.nl

Mr. Bert SATIJN
Iwaco B.V.
P.O. Box 8520
3009 AM ROTTERDAM

Tel: (31) 10 286 5526
Fax: (31) 10 220 0025

THE NETHERLANDS (continued)

Mr. Maarten SMITS
University of Amsterdam
Chemical Engineering Department
Nieuwe Achtergrachts 166
1018 WV AMSTERDAM
Tel: (31 20) 525 5265
Fax: (31 20) 525 5604

Ms. Esther SOCZO
RIVM
P.O. Box 1
3720 BA BILTHOVEN
Tel: (31 30) 743065
Fax: (31 30) 293651

Mr. Pieter VER LOREN VAN THEMAAT
Ministry of Economic Affairs
Directoraat-Generaal voor Industrie en Diensten
P.O. Box 20101
2500 EC The Hague
Tel: (31 70) 379 7526
Fax: (31 70) 379 7746

Mr. Harry VERMEULEN
NOBIS
P.O. Box 420
2800 AK GOUDA
Tel: (31) 18 257 3925
Fax: (31) 18 253 0046

Mr. P.G. WINTERS
Ministry of Economic Affairs
Plv. dir.-gen. Industrie en Diensten
P.O. Box 20101
2500 EC THE HAGUE
Tel: (31 70) 379 6177
Fax: (31 70) 379 6529

Mr. Han DE WIT
Tauw Milieu bv
P.O. Box 133
3931 EB WOUDENBERG
Tel: (31) 57 069 9653
Fax: (31) 57 069 9666

UNITED KINGDOM

Mr. Richard BEWLEY
Dames & Moore
Blackfriars House, St. Mary's Parsonage
MANCHESTER, M3 2JA
Tel: (44 1618) 320166
Fax: (44 1618) 321493

Mr. William DAY
Silsoe Research Institute
Process Engineering Division
Wrest Park, Silsoe
BEDFORD MK45 4HS
Tel: (44 1525) 860 000
Fax: (44 1525) 860 156
E-mail: bill.day@bbsrc.ac.uk

UNITED KINGDOM (continued)

Mr. Harry ECCLES
British Nuclear Fuels Plc
Company Research Laboratory
Springfields
PRESTON, PR4 OXJ, LANCS

Tel: (44 1772) 763 911
Fax: (44 1772) 762 470

Mr. Alwyn HART
British Gas
Gas Research Centre
Ashby Road
LOUGHBOROUGH LE11 3QU

Tel: (44 1509) 282 208
Fax: (44 1509) 283 115
E-mail: alwyn.hart@bggrc.co.uk

Ms. Daphne THOMPSON
Chemicals & Biotechnology Division
Department of Trade & Industry
151 Buckingham Palace Road
LONDON, SW1W 9SS

Tel: (44 171) 215 4110
Fax: (44 171) 215 1379

UNITED STATES

Mr. Daniel ABRAMOWICZ
GE Corporate Research and Development
P.O. Box 8, K1 3C7
SCHENECTADY NY 12301-0008

Tel: (1 518) 387 7072
Fax: (1 518) 387 7611
E-mail: abramda@crd.ge.com

Mr. Ronald ATLAS
University of Louisville
Dept. of Biology
LOUISVILLE KY 40292

Tel: (1 502) 852 8962
Fax: (1 502) 852 0725

Mr. Perry McCARTY
Stanford University
Department of Civil Engineering
STANFORD, CA 94305-4020

Tel: (1 415) 723 4131
Fax: (1 415) 725 9474

Mr. Gary SAYLER
University of Tennessee
Center for Environmental Biotechnology
10515 Research Drive, Suite 100
KNOXVILLE TN 37932

Tel: (1 423) 914 8080
Fax: (1 423) 974 8086
E-mail: sayler@utkvx.utk.edu

Mr. Ronald UNTERMAN
Envirogen, Inc.
4100 Quakerbridge Road
Princeton Research Center
LAWRENCEVILLE NJ 08648

Tel: (1 609) 936 9300
Fax: (1 609) 936 9221

CEC

Mr. Ioannis ECONOMIDIS
European Commission
200 Rue de la Loi
1049 BRUSSELS

Tel: (32 2) 295 1574
Fax: (32 2) 299 1860

Mr. Victor DE LORENZO
CIB-CSIC
Velazquez 144
28006 MADRID
SPAIN

Tel: (34 1) 561 1800
Fax: (34 1) 562 7518
E-mail: ciblp70@cc.csic.es

Mr. Bernard DIXON
Bio Technology
130, Cornwall Road
RUISLIP MANOR, MIDDLESEX, H14 6AW
UNITED KINGDOM

Tel: (44 1895) 632 390
Fax: (44 1895) 678 645

Mr. Gerasimos LYBERATOS
University of Patras
Dept. of Chemical Engineering
26110 PATRAS, GREECE

Tel: (30) 619 97573
Fax: (30) 619 97848

CZECH REPUBLIC

Mr. Jaroslav DROBNIK
Charles University
Inst. of Biotechnology
Vinicna 5
128 44 PRAHA 2

Tel: (42 2) 2491 5520
Fax: (42 2) 2936 43
E-mail: drobnik@earn.cvut.cz

Mr. Jan PACA
University of Chemical Technology
Dept. Ferment. Chemistry & Bioengineering
Technicka 5
16628 PRAGUE 6

Tel: (42 2) 243 53785
Fax: (42 2) 243 11082

Ms. Helena STEPANKOVA
Charles University
Institute of Biotechnology
Vinicna 5
12844 PRAGUE 2

Tel: (42 2) 2491 5520
Fax: (42 2) 2936 43
E-mail: hstepank@earn.cvut.cz

HUNGARY

Mr. Kornel L. KOVACS
Hungarian Academy of Science
Biological Research Center
P.O.Box 521
6710 SZEGED

Tel: (36) 624 32232
Fax: (36) 624 33133
E-mail: kornel@szbk.u-szeged.hu

KOREA

Mr. Sang-Jon KIM
Department of Microbiology
Seoul National University
SEOUL 151-742

Tel: (82) 288 06704
Fax: (82) 288 99474

OECD SECRETARIAT

Mr. Risaburo NEZU
Director
Directorate for Science, Technology and Industry
OECD
2, rue André Pascal
75775 PARIS CEDEX 16
FRANCE

Tel: (33 1) 45 24 94 20
Fax: (33 1) 45 24 93 99

Mr. Salomon WALD
Deputy-Head Biotechnology Unit
Directorate for Science, Technology and Industry
OECD
2, rue André Pascal
75775 PARIS CEDEX 16
FRANCE

Tel: (33 1) 45 24 92 32
Fax: (33 1) 45 24 18 25

Mr. Tadashi HIRAKAWA
Biotechnology Unit
Directorate for Science, Technology and Industry
OECD
2, rue André Pascal
75775 PARIS CEDEX 16
FRANCE

Tel: (33 1) 45 24 93 42
Fax: (33 1) 45 24 18 25

Mr. Yoshinobu MIYAMURA
Biotechnology Unit
Directorate for Science, Technology and Industry
OECD
2, rue André Pascal
75775 PARIS CEDEX 16
FRANCE

Tel: (33 1) 45 24 93 62
Fax: (33 1) 45 24 18 25

OECD SECRETARIAT (continued)

Ms. Sonia GUIRAUD
Biotechnology Unit
Directorate for Science, Technology and Industry
OECD
2, rue André Pascal
75775 PARIS CEDEX 16
FRANCE

Tel (33 1) 45 24 16 54
Fax: (33 1) 45 24 18 25

Mr. Peter KEARNS
Environment Directorate
OECD
15, Boulevard Amiral Bruix
75016 PARIS
FRANCE

Tel: (33 1) 45 24 16 77
Fax: (33 1) 45 24 16 50

MAIN SALES OUTLETS OF OECD PUBLICATIONS
PRINCIPAUX POINTS DE VENTE DES PUBLICATIONS DE L'OCDE

AUSTRALIA – AUSTRALIE
D.A. Information Services
648 Whitehorse Road, P.O.B 163
Mitcham, Victoria 3132 Tel. (03) 9210.7777
Fax: (03) 9210.7788

AUSTRIA – AUTRICHE
Gerold & Co.
Graben 31
Wien I Tel. (0222) 533.50.14
Fax: (0222) 512.47.31.29

BELGIUM – BELGIQUE
Jean De Lannoy
Avenue du Roi, Koningslaan 202
B-1060 Bruxelles Tel. (02) 538.51.69/538.08.41
Fax: (02) 538.08.41

CANADA
Renouf Publishing Company Ltd.
1294 Algoma Road
Ottawa, ON K1B 3W8 Tel. (613) 741.4333
Fax: (613) 741.5439

Stores:
61 Sparks Street
Ottawa, ON K1P 5R1 Tel. (613) 238.8985
12 Adelaide Street West
Toronto, ON M5H 1L6 Tel. (416) 363.3171
Fax: (416)363.59.63

Les Éditions La Liberté Inc.
3020 Chemin Sainte-Foy
Sainte-Foy, PQ G1X 3V6 Tel. (418) 658.3763
Fax: (418) 658.3763

Federal Publications Inc.
165 University Avenue, Suite 701
Toronto, ON M5H 3B8 Tel. (416) 860.1611
Fax: (416) 860.1608

Les Publications Fédérales
1185 Université
Montréal, QC H3B 3A7 Tel. (514) 954.1633
Fax: (514) 954.1635

CHINA – CHINE
China National Publications Import
Export Corporation (CNPIEC)
16 Gongti E. Road, Chaoyang District
P.O. Box 88 or 50
Beijing 100704 PR Tel. (01) 506.6688
Fax: (01) 506.3101

CHINESE TAIPEI – TAIPEI CHINOIS
Good Faith Worldwide Int'l. Co. Ltd.
9th Floor, No. 118, Sec. 2
Chung Hsiao E. Road
Taipei Tel. (02) 391.7396/391.7397
Fax: (02) 394.9176

DENMARK – DANEMARK
Munksgaard Book and Subscription Service
35, Nørre Søgade, P.O. Box 2148
DK-1016 København K Tel. (33) 12.85.70
Fax: (33) 12.93.87

J. H. Schultz Information A/S,
Herstedvang 12,
DK – 2620 Albertslung Tel. 43 63 23 00
Fax: 43 63 19 69

Internet: s-info@inet.uni-c.dk

EGYPT – ÉGYPTE
Middle East Observer
41 Sherif Street
Cairo Tel. 392.6919
Fax: 360-6804

FINLAND – FINLANDE
Akateeminen Kirjakauppa
Keskuskatu 1, P.O. Box 128
00100 Helsinki

Subscription Services/Agence d'abonnements :
P.O. Box 23
00371 Helsinki Tel. (358 0) 121 4416
Fax: (358 0) 121.4450

FRANCE
OECD/OCDE
Mail Orders/Commandes par correspondance :
2, rue André-Pascal
75775 Paris Cedex 16 Tel. (33-1) 45.24.82.00
Fax: (33-1) 49.10.42.76
Telex: 640048 OCDE

Internet: Compte.PUBSINQ@oecd.org

Orders via Minitel, France only/
Commandes par Minitel, France exclusivement :
36 15 OCDE

OECD Bookshop/Librairie de l'OCDE :
33, rue Octave-Feuillet
75016 Paris Tél. (33-1) 45.24.81.81
(33-1) 45.24.81.67

Dawson
B.P. 40
91121 Palaiseau Cedex Tel. 69.10.47.00
Fax: 64.54.83.26

Documentation Française
29, quai Voltaire
75007 Paris Tel. 40.15.70.00

Economica
49, rue Héricart
75015 Paris Tel. 45.75.05.67
Fax: 40.58.15.70

Gibert Jeune (Droit-Économie)
6, place Saint-Michel
75006 Paris Tel. 43.25.91.19

Librairie du Commerce International
10, avenue d'Iéna
75016 Paris Tel. 40.73.34.60

Librairie Dunod
Université Paris-Dauphine
Place du Maréchal-de-Lattre-de-Tassigny
75016 Paris Tel. 44.05.40.13

Librairie Lavoisier
11, rue Lavoisier
75008 Paris Tel. 42.65.39.95

Librairie des Sciences Politiques
30, rue Saint-Guillaume
75007 Paris Tel. 45.48.36.02

P.U.F.
49, boulevard Saint-Michel
75005 Paris Tel. 43.25.83.40

Librairie de l'Université
12*a*, rue Nazareth
13100 Aix-en-Provence Tel. (16) 42.26.18.08

Documentation Française
165, rue Garibaldi
69003 Lyon Tel. (16) 78.63.32.23

Librairie Decitre
29, place Bellecour
69002 Lyon Tel. (16) 72.40.54.54

Librairie Sauramps
Le Triangle
34967 Montpellier Cedex 2 Tel. (16) 67.58.85.15
Fax: (16) 67.58.27.36

A la Sorbonne Actual
23, rue de l'Hôtel-des-Postes
06000 Nice Tel. (16) 93.13.77.75
Fax: (16) 93.80.75.69

GERMANY – ALLEMAGNE
OECD Bonn Centre
August-Bebel-Allee 6
D-53175 Bonn Tel. (0228) 959.120
Fax: (0228) 959.12.17

GREECE – GRÈCE
Librairie Kauffmann
Stadiou 28
10564 Athens Tel. (01) 32.55.321
Fax: (01) 32.30.320

HONG-KONG
Swindon Book Co. Ltd.
Astoria Bldg. 3F
34 Ashley Road, Tsimshatsui
Kowloon, Hong Kong Tel. 2376.2062
Fax: 2376.0685

HUNGARY – HONGRIE
Euro Info Service
Margitsziget, Európa Ház
1138 Budapest Tel. (1) 111.62.16
Fax: (1) 111.60.61

ICELAND – ISLANDE
Mál Mog Menning
Laugavegi 18, Pósthólf 392
121 Reykjavik Tel. (1) 552.4240
Fax: (1) 562.3523

INDIA – INDE
Oxford Book and Stationery Co.
Scindia House
New Delhi 110001 Tel. (11) 331.5896/5308
Fax: (11) 332.5993

17 Park Street
Calcutta 700016 Tel. 240832

INDONESIA – INDONÉSIE
Pdii-Lipi
P.O. Box 4298
Jakarta 12042 Tel. (21) 573.34.67
Fax: (21) 573.34.67

IRELAND – IRLANDE
Government Supplies Agency
Publications Section
4/5 Harcourt Road
Dublin 2 Tel. 661.31.11
Fax: 475.27.60

ISRAEL – ISRAËL
Praedicta
5 Shatner Street
P.O. Box 34030
Jerusalem 91430 Tel. (2) 52.84.90/1/2
Fax: (2) 52.84.93

R.O.Y. International
P.O. Box 13056
Tel Aviv 61130 Tel. (3) 546 1423
Fax: (3) 546 1442

Palestinian Authority/Middle East:
INDEX Information Services
P.O.B. 19502
Jerusalem Tel. (2) 27.12.19
Fax: (2) 27.16.34

ITALY – ITALIE
Libreria Commissionaria Sansoni
Via Duca di Calabria 1/1
50125 Firenze Tel. (055) 64.54.15
Fax: (055) 64.12.57

Via Bartolini 29
20155 Milano Tel. (02) 36.50.83

Editrice e Libreria Herder
Piazza Montecitorio 120
00186 Roma Tel. 679.46.28
Fax: 678.47.51

Libreria Hoepli
Via Hoepli 5
20121 Milano Tel. (02) 86.54.46
Fax: (02) 805.28.86

Libreria Scientifica
Dott. Lucio de Biasio 'Aeiou'
Via Coronelli, 6
20146 Milano Tel. (02) 48.95.45.52
Fax: (02) 48.95.45.48

JAPAN – JAPON
OECD Tokyo Centre
Landic Akasaka Building
2-3-4 Akasaka, Minato-ku
Tokyo 107 Tel. (81.3) 3586.2016
Fax: (81.3) 3584.7929

KOREA – CORÉE
Kyobo Book Centre Co. Ltd.
P.O. Box 1658, Kwang Hwa Moon
Seoul Tel. 730.78.91
Fax: 735.00.30

MALAYSIA – MALAISIE
University of Malaya Bookshop
University of Malaya
P.O. Box 1127, Jalan Pantai Baru
59700 Kuala Lumpur
Malaysia Tel. 756.5000/756.5425
Fax: 756.3246

MEXICO – MEXIQUE
OECD Mexico Centre
Edificio INFOTEC
Av. San Fernando no. 37
Col. Toriello Guerra
Tlalpan C.P. 14050
Mexico D.F. Tel. (525) 665 47 99
Fax: (525) 606 13 07

Revistas y Periodicos Internacionales S.A. de C.V.
Florencia 57 - 1004
Mexico, D.F. 06600 Tel. 207.81.00
Fax: 208.39.79

NETHERLANDS – PAYS-BAS
SDU Uitgeverij Plantijnstraat
Externe Fondsen
Postbus 20014
2500 EA's-Gravenhage Tel. (070) 37.89.880
Voor bestellingen: Fax: (070) 34.75.778

NEW ZEALAND – NOUVELLE-ZÉLANDE
GPLegislation Services
P.O. Box 12418
Thorndon, Wellington Tel. (04) 496.5655
Fax: (04) 496.5698

NORWAY – NORVÈGE
NIC INFO A/S
Bertrand Narvesens vei 2
P.O. Box 6512 Etterstad
0606 Oslo 6 Tel. (022) 57.33.00
Fax: (022) 68.19.01

PAKISTAN
Mirza Book Agency
65 Shahrah Quaid-E-Azam
Lahore 54000 Tel. (42) 735.36.01
Fax: (42) 576.37.14

PHILIPPINE – PHILIPPINES
International Booksource Center Inc.
Rm 179/920 Cityland 10 Condo Tower 2
HV dela Costa Ext cor Valero St.
Makati Metro Manila Tel. (632) 817 9676
Fax: (632) 817 1741

POLAND – POLOGNE
Ars Polona
00-950 Warszawa
Krakowskie Przedmieácie 7 Tel. (22) 264760
Fax: (22) 268673

PORTUGAL
Livraria Portugal
Rua do Carmo 70-74
Apart. 2681
1200 Lisboa Tel. (01) 347.49.82/5
Fax: (01) 347.02.64

SINGAPORE – SINGAPOUR
Gower Asia Pacific Pte Ltd.
Golden Wheel Building
41, Kallang Pudding Road, No. 04-03
Singapore 1334 Tel. 741.5166
Fax: 742.9356

SPAIN – ESPAGNE
Mundi-Prensa Libros S.A.
Castelló 37, Apartado 1223
Madrid 28001 Tel. (91) 431.33.99
Fax: (91) 575.39.98

Mundi-Prensa Barcelona
Consell de Cent No. 391
08009 – Barcelona Tel. (93) 488.34.92
Fax: (93) 487.76.59

Llibreria de la Generalitat
Palau Moja
Rambla dels Estudis, 118
08002 – Barcelona
(Subscripcions) Tel. (93) 318.80.12
(Publicacions) Tel. (93) 302.67.23
Fax: (93) 412.18.54

SRI LANKA
Centre for Policy Research
c/o Colombo Agencies Ltd.
No. 300-304, Galle Road
Colombo 3 Tel. (1) 574240, 573551-2
Fax: (1) 575394, 510711

SWEDEN – SUÈDE
CE Fritzes AB
S–106 47 Stockholm Tel. (08) 690.90.90
Fax: (08) 20.50.21

Subscription Agency/Agence d'abonnements :
Wennergren-Williams Info AB
P.O. Box 1305
171 25 Solna Tel. (08) 705.97.50
Fax: (08) 27.00.71

SWITZERLAND – SUISSE
Maditec S.A. (Books and Periodicals - Livres et périodiques)
Chemin des Palettes 4
Case postale 266
1020 Renens VD 1 Tel. (021) 635.08.65
Fax: (021) 635.07.80

Librairie Payot S.A.
4, place Pépinet
CP 3212
1002 Lausanne Tel. (021) 320.25.11
Fax: (021) 320.25.14

Librairie Unilivres
6, rue de Candolle
1205 Genève Tel. (022) 320.26.23
Fax: (022) 329.73.18

Subscription Agency/Agence d'abonnements :
Dynapresse Marketing S.A.
38, avenue Vibert
1227 Carouge Tel. (022) 308.07.89
Fax: (022) 308.07.99

See also – Voir aussi :
OECD Bonn Centre
August-Bebel-Allee 6
D-53175 Bonn (Germany) Tel. (0228) 959.120
Fax: (0228) 959.12.17

THAILAND – THAÏLANDE
Suksit Siam Co. Ltd.
113, 115 Fuang Nakhon Rd.
Opp. Wat Rajbopith
Bangkok 10200 Tel. (662) 225.9531/2
Fax: (662) 222.5188

TRINIDAD & TOBAGO
SSL Systematics Studies Limited
9 Watts Street
Curepe
Trinadad & Tobago, W.I. Tel. (1809) 645.3475
Fax: (1809) 662.5654

TUNISIA – TUNISIE
Grande Librairie Spécialisée
Fendri Ali
Avenue Haffouz Imm El-Intilaka
Bloc B 1 Sfax 3000 Tel. (216-4) 296 855
Fax: (216-4) 298.270

TURKEY – TURQUIE
Kültür Yayinlari Is-Türk Ltd. Sti.
Atatürk Bulvari No. 191/Kat 13
Kavaklidere/Ankara
Tel. (312) 428.11.40 Ext. 2458
Fax: (312) 417 24 90

Dolmabahce Cad. No. 29
Besiktas/Istanbul Tel. (212) 260 7188

UNITED KINGDOM – ROYAUME-UNI
HMSO
Gen. enquiries Tel. (0171) 873 0011
Postal orders only:
P.O. Box 276, London SW8 5DT
Personal Callers HMSO Bookshop
49 High Holborn, London WC1V 6HB
Fax: (0171) 873 8463
Branches at: Belfast, Birmingham, Bristol, Edinburgh, Manchester

UNITED STATES – ÉTATS-UNIS
OECD Washington Center
2001 L Street N.W., Suite 650
Washington, D.C. 20036-4922 Tel. (202) 785.6323
Fax: (202) 785.0350
Internet: washcont@oecd.org

Subscriptions to OECD periodicals may also be placed through main subscription agencies.

Les abonnements aux publications périodiques de l'OCDE peuvent être souscrits auprès des principales agences d'abonnement.

Orders and inquiries from countries where Distributors have not yet been appointed should be sent to: OECD Publications, 2, rue André-Pascal, 75775 Paris Cedex 16, France.

Les commandes provenant de pays où l'OCDE n'a pas encore désigné de distributeur peuvent être adressées aux Éditions de l'OCDE, 2, rue André-Pascal, 75775 Paris Cedex 16, France.

5-1996

OECD PUBLICATIONS, 2, rue André-Pascal, 75775 PARIS CEDEX 16
PRINTED IN FRANCE
(93 96 03 1 ISBN 92-64-14869-8 – No. 48730 1996